Power-Flow Modelling of HVDC Transmission Systems

This book deals exclusively with the power-flow modelling of HVDC transmission systems. Different types of HVDC transmission systems, their configurations/connections and control techniques are covered in detail. Power-Flow modelling of both LCC- and VSC-based HVDC systems is covered in this book. Both the unified and the sequential power-flow methods are addressed. DC grid power-flow controllers and renewable energy resources like offshore wind farms (OWFs) are also incorporated into the power-flow models of VSC-HVDC systems. The effects of the different power-flow methods and HVDC control strategies on the power-flow convergence are detailed along with their implementation.

Features:

- Introduces the power-flow concept and develops the power-flow models of integrated AC/DC systems.
- Different types of converter control are modelled into the integrated AC/DC power-flow models developed.
- Both unified and the sequential power-flow methods are addressed.
- DC grid power-flow controllers like the IDCPFC and renewable energy resources like offshore wind farms (OWFs) are introduced and subsequently modelled into the power-flow algorithms.
- Integrated AC/DC power-flow models developed are validated by implementation in the IEEE 300-bus and European 1354-bus test networks incorporating different HVDC grids.

This book aims at researchers and graduate students in Electrical Engineering, Power Systems, and HVDC Transmission.

Power-Flow Modelling of HVDC Transmission Systems

Shagufta Khan
Suman Bhowmick

CRC Press
Taylor & Francis Group
Boca Raton London New York

CRC Press is an imprint of the
Taylor & Francis Group, an **informa** business

MATLAB® is a trademark of The MathWorks, Inc. and is used with permission. The MathWorks does not warrant the accuracy of the text or exercises in this book. This book's use or discussion of MATLAB® software or related products does not constitute endorsement or sponsorship by The MathWorks of a particular pedagogical approach or particular use of the MATLAB® software.

First edition published 2023
by CRC Press
6000 Broken Sound Parkway NW, Suite 300, Boca Raton, FL 33487-2742

and by CRC Press
2 Park Square, Milton Park, Abingdon, Oxon, OX14 4RN

CRC Press is an imprint of Taylor & Francis Group, LLC

ISBN: 9781032171661 (hbk)
ISBN: 9781032171678 (pbk)
ISBN: 9781003252078 (ebk)

DOI: 10.1201/9781003252078

Typeset in Times
by codeMantra

Contents

Preface

Over the past few decades, the construction of generation facilities and new transmission lines has been delayed in light of rising energy costs, environmental concerns, rights-of-way (RoW) restrictions and other legislative and cost problems. In addition, system stability issues may render long-distance AC transmission infeasible. In this respect, high-voltage DC (HVDC) transmission requires a smaller RoW, simpler, lighter and cheaper transmission towers, reduced conductor and insulator costs, reduced losses and is not limited by stability considerations. A HVDC link can augment system reliability by interconnecting two asynchronous AC grids and can be used to integrate offshore wind farms with onshore AC grids.

The first commercial application of HVDC transmission took place between the Swedish mainland and the island of Gotland in 1954, using mercury arc valves. Subsequently, the first 320 MW thyristor-based HVDC system was commissioned in 1972 between the Canadian provinces of New Brunswick and Quebec. Continuous development in conversion equipment led to reduced size and costs which resulted in more widespread use of HVDC transmission. The thyristor-based line commutated converter (LCC)–high-voltage DC (LCC-HVDC) technology now constitutes the bulk of the installed HVDC transmission corridors over the world.

With LCC-HVDC, for controlling the active power, both the rectification and the inversion processes consume reactive power. This necessitates the use of reactive power sources to match the reactive power demand at both ends. To reduce the effects of harmonic voltages and currents generated by the converters, harmonic filters are used on both the AC and DC sides. Also, a minimum short circuit level is required to avoid voltage instability. However, despite its limitations, LCC-HVDC possesses high reliability, good overload capability and lower converter losses. It requires low maintenance and capital costs, and is robust to DC fault currents due to its current-regulating nature.

Subsequently, the development of the Insulated Gate Bipolar Transistor (IGBT) paved the way for the Voltage-Sourced Converter (VSC)-based HVDC (VSC-HVDC) technology, which offered significant advantages over the LCC-HVDC. VSC-HVDC facilitates independent active and reactive power control, along with reduction in filter size. VSC-HVDC also enables the integration of offshore wind farms with AC grids. Compact, modular designs of the VSCs enable rapid installation, commissioning and relocation. Unlike LCC-HVDC, fixed DC voltage polarity in the VSC-HVDC enables the use of stronger and lighter XLPE cables, suitable for under-sea environments and attractive for offshore transmission. In addition, VSC-HVDC systems can be integrated with AC systems having low short circuit ratios.

The first 3-MW, VSC-HVDC link was commissioned at Hellsjon in Sweden in 1997. Subsequently, rapid development in the VSC technology has now resulted in the availability of higher rated (up to 2000 MW) VSC-HVDC links. This has resulted in the installation and commissioning of a large number of VSC-HVDC systems worldwide. Present VSC-HVDC solutions use the modular multilevel converter (MMC) technology, which is more advantageous than two- or three-level VSCs in terms of reduced converter losses, increased modularity and scalability along with elimination of filter requirements.

Now, in both LCC-HVDC and VSC-HVDC systems, the converter stations can be connected in two ways—back-to-back (BTB) and point-to-point (PTP). Most of the MTDC systems installed worldwide are in PTP configurations, their DC sides being interconnected through DC links or cables.

Unlike a 2-terminal HVDC interconnection, a multi-terminal HVDC (MTDC) system is more versatile and better capable of utilising the economic and technical advantages of HVDC technology. Moreover, sources of renewable energy can be easily integrated with a MTDC system, as and when the need arises.

For proper MTDC operation, DC voltage control is an essential requirement. In this respect, several control techniques have been envisaged. These include DC slack-bus control (also known as DC master-slave control), distributed DC voltage droop control, power synchronization control, hierarchical power control and transient management control.

However, among all the DC voltage control techniques, the DC slack-bus control and distributed DC voltage droop control have been the more popular and widely employed ones.

In DC slack-bus control, the voltage of one DC terminal, known as the DC slack bus, is maintained constant by the master converter. The main disadvantage of this control scheme is the DC grid instability following a failure of the master converter.

The above problem can be tackled by ensuring that individual converters contribute to the DC voltage regulation scheme by adjusting their active power flow in response to changes in the DC voltage with the operating point, known as DC voltage droop control. For MTDC control, both linear and nonlinear types of DC voltage droop characteristics have been envisaged to ensure proper sharing based on the converter ratings. Voltage-Power (V-P), Voltage-Current (V-I), Voltage Margin (VM), V-P droop with power Dead-Band (DB) and V-P droop with voltage limits are some of the more widely used characteristics.

To manage power-flows within the DC grids, DC power-flow control devices have been conceptualized and developed. They include the use of DC transformers, variable resistors, current flow controllers (CFCs), thyristor power-flow controllers (TPFCs), DC series voltage sources and Interline DC Power-Flow Controllers (IDCPFCs) for power-flow control in meshed DC grids. The IDCPFC is a DC power-flow controller without an external AC or DC source and is used for power-flow management of DC grids, similar to its AC counterpart—the flexible AC transmission systems (FACTS)-based Interline Power-Flow Controller (IPFC).

Now, for proper planning, design and operation of AC power systems integrated with multi-terminal DC grids, the development of suitable power-flow models of both LCC- and VSC-based integrated AC–DC systems is a fundamental requirement.

The requirement of suitable power-flow models of both LCC and VSC-based hybrid AC–DC systems along with the adoption of the Newton-Raphson algorithm as the de-facto standard for industrial power-flow solutions has resulted in a lot of interest in the development of Newton-Raphson power-flow models of such hybrid AC–DC systems.

Now, the development of Newton-Raphson power-flow models of both LCC- and VSC-based integrated AC–DC systems has resulted in two distinctly different approaches known as the unified and the sequential Newton algorithms, respectively. In the former, the AC and DC quantities are solved simultaneously, while in the latter, the AC and DC systems are solved separately in each iteration. Unlike the unified method, the sequential method is easier to implement and poses lesser computational burden due to the smaller size of the Jacobian matrix.

Although a number of books on modelling, analysis, control and applications of HVDC systems do exist, very few books dwell on their power-flow modelling. This book intends to deal exclusively with the power-flow modelling of HVDC systems. The book starts by detailing the different types of HVDC systems, their configuration and connections and the control techniques adopted. Next, the power-flow modelling concept is gradually built up. At first, the power-flow modelling of LCC-based HVDC systems is covered, followed by that of VSC-based HVDC systems. Subsequently, the introduction and incorporation of DC grid power-flow controllers like IDCPFC into the power flow modelling of VSC-HVDC systems is addressed. Finally, the power-flow modelling of HVDC systems integrated with renewable energy sources is covered. For each of these technologies (LCC / VSC), both the unified and the sequential power-flow models are developed. The different types of HVDC control strategies employed are also incorporated into the LCC and the VSC-based HVDC power-flow models. For realistic analysis, the losses in the converter transformer and the VSCs have been incorporated in all the power-flow models. The effect of the power-flow methods (unified or sequential) as well as the HVDC control strategies adopted on the convergence of the power-flow algorithm, is detailed clearly, along with their implementation in the IEEE 300-bus and the European 1354-bus test networks.

This book is intended for senior undergraduate and graduate students in electrical power systems, design engineers and researchers in the area of integrated AC–DC systems. The reader is expected to have an undergraduate-level background in electric circuits, electric power systems, engineering mathematics and power electronics.

The proposed book is organized into six chapters. Chapter 1 provides a brief introduction to the HVDC transmission systems, the power-flow problem and the Newton-Raphson method for solving power-flow problems.

Chapter 2 deals with the development of power-flow models of LCC-based integrated AC–DC systems, in light of different per-unit AC–DC system models and diverse DC link control strategies employed.

Chapter 3 addresses the development of power-flow models of VSC-based integrated AC–DC systems for both the back-to-back (BTB) and the point-to-point

(PTP) VSC-HVDC configurations, employing DC slack-bus (master-slave) control for the MTDC grid.

Chapter 4 details the development of power-flow models of VSC-based integrated AC–DC systems employing DC voltage droop control. The DC voltage droop control comprises both linear {voltage-power (V-P) and voltage-current (V-I)} as well as nonlinear {power dead-band and voltage limits} droop characteristics. Based on the terminal end line active and reactive power specifications of the VSCs, two different droop control models are considered.

Chapter 5 addresses the development of power-flow models of VSC-based integrated AC–DC systems incorporating IDCPFC(s) for the power-flow management of the DC grid. The IDCPFC(s) employs both DC link current and DC link power controls.

Chapter 6 details the development of Newton power-flow models of VSC-based integrated AC–DC systems interfaced with offshore wind farms (OWFs). The VSCs employ both linear and nonlinear DC voltage droop controls. The effects of OWFs on the DC grid voltage profile and the power-flow convergence, vis-à-vis varying wind farm powers, are explained in detail.

Converter transformer and VSC losses are included in all the power-flow models. All the models developed in this book have been implemented in the IEEE 300-bus and European 1354-bus test systems.

The appendix at the end of this book presents the derivations of all the difficult formulae used in the different chapters.

The authors would like to express sincere thanks to all the reviewers for their critical review and suggestions in the proposal of this book. We would also like to thank the publisher and our families for their efforts in pursuing us to take up the project of writing this book.

Shagufta Khan
Department of Electrical, Electronics and Communication Engineering
Galgotias University, Gr. Noida UP, India

Suman Bhowmick
Department of Electrical Engineering
Delhi Technological University, Delhi, India

MATLAB® is a registered trademark of The MathWorks, Inc. For product information,
    please contact:
    The MathWorks, Inc.
    3 Apple Hill Drive
    Natick, MA 01760-2098 USA
    Tel: 508-647-7000
    Fax: 508-647-7001
    E-mail: info@mathworks.com
    Web: www.mathworks.com

Authors

Dr. Shagufta Khan received her Ph.D. in Electrical Engineering from Delhi Technological University, Delhi, India. She is currently an Assistant Professor with the School of Electrical, Electronics and Communication Engineering, Galgotias University, Greater Noida, India. Her research interests include power systems and renewable energy. She has several publications in national and international journals and conferences including *IEEE Transactions on Sustainable Energy, Electrical Power System Research* (Elsevier), *International Journal of Electrical Power and Energy Systems* (Elsevier), *Electrical Energy Journal* (Springer), *AIN Shams Engineering Journal* (Elsevier) and *Arabian Journal for Science and Engineering* (Springer) to her credit.

Prof. Suman Bhowmick received his Ph.D. in Electrical Engineering in 2010. He has been working as a Professor in the Department of Electrical Engineering, Delhi Technological University since 2012. His areas of interest are power systems in general, and FACTS and HVDC systems in particular. He has several publications in national and international journals and conferences to his credit. He has also authored a book on FACTS- which was published by the CRC Press, USA, in 2016.

List of Abbreviations

BTB	Back-to-back
CT	Computational time in seconds taken by the algorithm to converge to a specified tolerance
FACTS	Flexible AC transmission systems
HVDC	High-voltage direct current
IDCPFC	Interline direct current power-flow controller
IGBT	Insulated gate bipolar transistor
IPFC	Interline power-flow controller
LCC	Line commutated converter
MLDC	Multi-terminal LCC-based HVDC
MTDC	Multi-terminal Direct Current
MVDC	Multi-terminal VSC-based HVDC
NI	Number of iterations taken by the algorithm to converge to a specified tolerance
NR	Newton-Raphson
OWF	Offshore wind farm
PTP	Point-to-point
PWM	Pulse width modulation
RoW	Right of way
VSC	Voltage source converter
XLPE	Cross-linked poly ethylene

List of Symbols

CAPITALS

$\mathbf{E}$	Vector of mismatch error
$\mathbf{E_{AC}}$	Vector of mismatch error in AC network
$\mathbf{E_{DC}}$	Vector of mismatch error in DC network
$I_{(AC\,base)}$	AC base current
$I_{(DC\,base)}$	DC base current
I_{DC}	DC current
$\mathbf{I_{sha}}$	Current through the converter transformer of the ath VSC
I^*_{DCa}	DC current reference for linear V-I droop line of the ath VSC
$\mathbf{J}$	Jacobian matrix
$\mathbf{J_{old}}$	Conventional power-flow Jacobian sub-block
P_{Di}	Active power demand at bus 'i'
P_i	Net active power injection at bus 'i'
$\mathbf{P}$	Bus active power injection vector
P_i^{sp}	Specified active power injection at bus 'i'
P_{DCR}	Active power associated with the rectifier
P_{DCI}	Active power associated with the inverter
P_{sha}	Active power flow in the line connecting the ath VSC to its AC terminal bus
P_{sha}^{sp}	Specified active power flow in the line connecting the ath VSC to its AC terminal bus
P_{sha}^{cal}	Calculated active power flow in the line connecting the ath VSC to its AC terminal bus
P_{lossa}	Losses of the ath VSC
P^*_{DCa}	DC power reference for linear V-P droop line of the ath VSC
P_{IDCPFC}	Power delivered by the IDCPFC
P_{DCWF}	Rectifying power of wind farm injected into the DC grid
$\mathbf{P_{DCWF}}$	Vector of rectifying powers of wind farms
$\mathbf{Q}$	Bus reactive power injection vector
Q_{DCR}	Reactive power associated with the rectifier
Q_{DCI}	Reactive power associated with the inverter
Q_{Di}	Reactive power demand at bus 'i'
Q_i	Net reactive power injection at bus 'i'
Q_i^{sp}	Specified reactive power injection at bus 'i'
Q_{sha}	Reactive power flow in the line connecting the ath VSC to its AC terminal bus
Q_{sha}^{sp}	Specified reactive power flow in the line connecting the ath VSC to its AC terminal bus
Q_{sha}^{cal}	Calculated reactive power flow in the line connecting the ath VSC to its AC terminal bus

R_{DC}	Resistance of DC link
R_a	Droop control gain of the ath VSC
R_{sha}	Resistance of the ath VSC transformer
$R_{(DC\ base)}$	Base value of DC resistance
$\mathbf{R}$	Mismatch vector of control specifications
R_{max}	Maximum droop control gain
S_{base}	Base MVA
S_{sha}	Complex line power flow at the terminal end of the line connecting the ath VSC to its AC bus
$\mathbf{V}$	AC bus voltage vector
$V_{(AC\ base)}$	Base value of AC voltage
V_{DC}	DC bus voltage
$V_{(DC\ base)}$	Base value of DC voltage
V_{DCI}	DC bus voltage at the inverter side
V_{DCI1}	DC bus voltage at the inverter-1 side
V_{DCI2}	DC bus voltage at the inverter-2 side
V_{DCR}	DC bus voltage at the rectifier side
V_i	AC bus voltage magnitude (rms) at i^{th} bus
$\mathbf{V_{sha}}$	Voltage phasor representing the output (fundamental) of the ath VSC
V_{doR}	No load direct voltage at the rectifier side
V_{DCa}^*	DC voltage reference for the droop line of the ath VSC
V_{doI}	No load DC voltage at the inverter side
V_{DCav}^*	Average value of DC voltage references in a DC grid
V_{DCav}	Average value of the DC voltages in a DC grid
V_{Bus}	Specified AC bus voltage
V_{DCs}	DC voltage source of IDCPFC
V_{DChigh}^*	Upper DC voltage threshold of nonlinear DC voltage droop characteristics
V_{DClow}^*	Lower DC voltage threshold of nonlinear DC voltage droop characteristics
$V_{(DC\ max)}$	Maximum DC voltage threshold of nonlinear DC voltage droop characteristics
$V_{(DC\ min)}$	Minimum DC voltage threshold of nonlinear DC voltage droop characteristics
X_c	Commutating reactance
$X_{(c\ base)}$	Base value of the commutating reactance
X_{sha}	Leakage reactance of the ath converter transformer
Y_{ik}	Magnitude of the element in the ith row and k^{th} column of the bus admittance matrix
$\mathbf{Y_{dc}}$	Admittance matrix of DC grid
$Z_{(AC\ base)}$	Base value of AC impedance
$\mathbf{Z_{sha}}$	Leakage impedance of the ath converter transformer
$Z_{(DC\ base)}$	Base value of DC side impedance

LOWERCASE

a_1	Constant representing no load VSC losses
a_I	Converter transformer tap ratio at the inverter side
a_R	Converter transformer tap ratio at the rectifier side
b_1	Constant representative of the linear dependency of the VSC losses on the converter current magnitude
c	Constant representative of the VSC architecture
c_1	Constant representative of the quadratic dependency of the VSC losses on the converter current magnitude
f	Vector of control functions
g	Number of generators in the AC system
k	Constant which depends on the type of converter in the LCC-HVDC system
m	Modulation index of the VSC
n	Total number of buses in the AC system
n_b	Number of bridges in the LCC-HVDC system
p.u.	Per unit
p	Total number of DC terminals
q	Total number of VSCs
$\mathbf{y_{sha}}$	Admittance of the converter transformer of the ath VSC
z	Total number of DC voltage sources in IDCPFC

UPPERCASE GREEK

Σ	Summation symbol
Δ	Mismatch in electrical quantity of interest; mismatch vector

LOWERCASE GREEK

α_R	Firing angle of the rectifier in the LCC-HVDC system
γ_I	Extinction angle of the inverter in the LCC-HVDC system
θ_i	Phase angle of voltage at AC bus 'i'
$\boldsymbol{\theta}$	Vector comprising phase angles of AC bus voltages
$\boldsymbol{\theta_{sh}}$	Vector of phase angles of output voltage phasors of the VSC
θ_{sha}	Phase angle of the output voltage phasor of the ath VSC
ϕ_R	Power factor angle at the rectifier end of the LCC-HVDC system
ϕ_I	Power factor angle at the inverter end of the LCC-HVDC system
ϕ_{sha}	Phase angle of $\mathbf{y_{sha}}$

SUBSCRIPTS

i	Bus 'i' quantity
AC	AC side quantity

DC	DC side quantity
DCR	DC quantity at rectifier end
DCI	DC quantity at inverter end
AC_{base}	AC base values
DC_{base}	DC base values
b	Number of bridges
R	Rectifier
I	Inverter
sha	Shunt connected quantity of ath VSC
DCs	DC voltage source of IDCPFC
IDCPFC	IDCPFC quantities
Loss	VSC loss quantity
DC_{min}	Minimum DC quantity
DC_{max}	Maximum DC quantity
DCavg	Average value of DC voltage

SUPERSCRIPTS

$()^T$	Transpose of a matrix
$()^{cal}$	Calculated or unknown quantity
$()^*$	Conjugate of a complex quantity
$()^{sp}$	Specified or known quantity
$()^{old}$	Quantity in the original network without any HVDC link

1 HVDC Transmission Systems

1.1 INTRODUCTION

In recent years, the global demand of electric power has increased exponentially. Therefore, the generation and transmission facilities have to be upgraded from time to time to match the peak demand. In this respect, HVDC transmission systems provide additional transmission capacity along with power-flow controllability. Unlike AC transmission, for the same power, HVDC transmission requires less right-of-way (RoW), cheaper towers, smaller number of conductors and insulator costs along with reduced losses. In addition, the length of the HVDC transmission line is not limited by stability considerations. For transmission line lengths exceeding about 500 km, HVDC transmission is more economical as compared to AC [1–16]. In recent times, rapid, large-scale integration of renewable energy sources with the existing power network has been taking place globally to fulfil the requirement of increased electricity demand. In this respect, the integration of offshore wind farms (OWFs) with onshore AC grids is possible using HVDC links [4,12,15,16].

The first 10 MW HVDC transmission system using mercury arc valve was commissioned between the Swedish mainland and the island of Gotland in 1954. In due course, significant technical advancement with solid-state valves (thyristors) paved the way for the first 320 MW thyristor-based HVDC system commissioned between the Canadian provinces of New Brunswick and Quebec in 1972 [2–4]. Subsequently, there has been rapid development of this HVDC technology with further reduced size and costs and popularly known as line commutated converter (LCC)-based HVDC (LCC-HVDC) technology. Based on this technology, several LCC-HVDC links were installed worldwide, and some of these are listed in Table 1.1.

In a LCC-based HVDC system, the commutation process is achieved using the source voltage and the leakage reactance of the converter transformer. Thus, for controlling the active power, both the rectification and inversion processes consume reactive power. Also, the reactive power consumption varies with load. This necessitates the use of reactive power sources to match the reactive power demand at both ends [1,6,7]. If nearby generators are not capable of accounting for the reactive power, additional shunt capacitors or other reactive power sources are needed to match the requirement of reactive power. Also, in LCC-HVDC systems, a minimum short circuit level is required to avoid voltage instability. Due to the switching operation of the thyristor, harmonics are introduced in the power system voltages and currents. This influences the use of filters at both AC and DC

DOI: 10.1201/9781003252078-1

TABLE 1.1

LCC-Based HVDC Systems Throughout the World

S.N.	HVDC Link	Transmission Line Length (km)			Rated Voltage (kV)	Nominal Capacity (MW)	Commissioning Date	Remark
		OHL	Cable	Total				
1	New Brunswick-Eel River (Canada)	-	-	-	80×2	320	1972	BTB
2	Skagerrak (Denmark-Norway)	113	127	240	±250	500	1977	
3	David A. Hamil (United States of America)	-	-	-	50	100	1977	BTB
4	Square Butte (Centre, North Dakota-Arrowhead, Minnesota), US	749	0	749	±250	500	1977	
5	Shin-Shinano (Japan)	-	-	-	125×2	300	1977	BTB (50/60 Hz)
6	Nelson River Bipole 2 (Sundane-Rosser), Canada	930	0	930	±250	900	1978	
7	Cabora Bassa – Apollo (Songo, Mozambique-Apollo, South Africa)	1414	0	1414	±533	1920	1977/79	
8	Vancouver Pole 2 (Delta-North Cowichan), British Columbia	41	33	74	−280	370	1977/79	
9	Cu (Underwood Minneapolis) (Coal Creek, North Dakota-Dickinson, Minesota), US	710	0	910	±400	1000	1979	

(Continued)

TABLE 1.1 (*Continued*)
LCC-Based HVDC Systems Throughout the World

S.N.	HVDC Link	Transmission Line Length (km) OHL	Cable	Total	Rated Voltage (kV)	Nominal Capacity (MW)	Commissioning Date	Remark
10	Hokkaido-Honshu (Japan)	124	44	158	250	300	1979/80	
11	Acaray (Paraguay-Brazil)	-	-	-	26	50	1981	BTB (50/60 Hz)
12	EPRI Compact Station (USA)	-	0.6	0.6	100/400	100	1981	
13	Vyborg (USSR-Finland)	-	-	-	±85×3	170	1982	BTB
14	Inga Shaba (Kolwezi-Inga), Zaire	1700	0	1700	±500	560	1982	
15	Dumrohr (Lower Austria)	-	-	-	±145	550	1983	BTB
16	Gotland 2 Swedish Mainland (Vastervik-Yigne), Sweden	7	91	98	150	130	1983	
17	Eddy County (USA)	-	-	-	82	200	1983	BTB
18	Itaipu (Brazil)	783/806	0	783/806	±300	1575	1984	
19	Chateauguay (Canada)	-	-	-	140	1000	1984	BTB
20	Oklaunion (US)	-	-	-	82	200	1984	BTB
21	Pacific intertie (US)	-	-	-	±500	400	1985	
22	Madawaska (Canada)	-	-	-	144	350	1985	BTB
23	Miles City (US)	-	-	-	82	200	1985	BTB
24	Walker Co. (US)	246	0	246	±400	500–1500	1985	
25	Black water (US)	-	-	-	56	200	1985	BTB
26	Highgate (US)	-	-	-	56	200	1985	BTB
27	Cross-Channel 2 (Les Mandarins, France-Sellindge, UK)	0	72	72	±270×2	2000	1985/86	

(Continued)

TABLE 1.1 (*Continued*)
LCC-Based HVDC Systems Throughout the World

S.N.	HVDC Link	Transmission Line Length (km) OHL	Cable	Total	Rated Voltage (kV)	Nominal Capacity (MW)	Commissioning Date	Remark
28	Corsica Tap (France)	-	-	-	200	50	1986	
29	Des Cantons Camerford (Canada-USA)	175		175	±450	690	1986	
30	Sidney (US)	-	-	-		200	1986	BTB
31	Wien Sud Ost (Austria)	-	-	-	145	550	1987	BTB
32	Intermountain (intermountain, Utah-Adelanto, California), US	794	0	794	±500	1600	1987	
33	Gotland 3-Swedish Mainland	-	98	98	150	130	1987	
34	Itaipu (Foz do Iguacu, Parana-Sao Roque, Sao Paulo), Brazil	783/806	0	783/806	±600×2	6300	1985–87	
35	Uruguiana (Brazil-Argentina)					50	1986/87	BTB
36	Ekibastus Centre (USSR)	2400	0	2400	±250	6000	1985–88	
37	Greece (Bulgaria)	-	-	-	NA	300		BTB
38	Virginia Smith (US-Sidney Nebraska)				150	200	1988	BTB
39	Vindhyachal (India)	-	-	-	70	250×2	1988	BTB
40	Kanti-Skan 2 (Sweden-Denmark)	95	85	160	250	270	1988/89	
41	Skagerrak 2 (Denmark-Norway)	113	127	240	300	320	1988–89	

(*Continued*)

TABLE 1.1 (*Continued*)
LCC-Based HVDC Systems Throughout the World

S.N.	HVDC Link	Transmission Line Length (km)			Rated Voltage (kV)	Nominal Capacity (MW)	Commissioning Date	Remark
		OHL	Cable	Total				
42	Fenno-Skan (Dannebo, Sweden-Rauma, Finland)	33	200	233	400	500	1989	
43	HVDC Sileru-Barsoor (India)	196	-		±200	400	1989	
44	Store Baelt (Denmark)	35	30	55	280	350	1989–90	
45	Liberty Mead (US)	400	0	400	±364/±500	1600/2200	1989–90	
46	Chicoasen (Mexico)	720	0	720	±500	900/1800	1985/90	
47	Quebec New England	175/375		175/375	±450	690/2070	1986/92	
48	SACOI 2 (Suvereto, Italia-ucciana, France; Codrongianos)	304	118		200	300	1992	MT
49	HVDC Inter-Island 2 (Benmore Dam-Haywards), New Zealand	570	40		350	640	1992	
50	Pacific Intertie II (US)				±500	1100		
51	South Finland Fast Sweden	35	185	220	350	420	1989/90	
52	Cameford-Sandy Pond	200				1400	1990	
53	Rihand-Dadri (India)	814	-	814	±500	1500	1990	
54	Gezhouba-Nan Qiao (China)	1080	-	1080	±500	1200	1987–91	

(*Continued*)

TABLE 1.1 (*Continued*)
LCC-Based HVDC Systems Throughout the World

S.N.	HVDC Link	Transmission Line Length (km)			Rated Voltage (kV)	Nominal Capacity (MW)	Commissioning Date	Remark
		OHL	Cable	Total				
55	Cross-Skagerrak 3 Tjele, Denmark-Kristiansand, Norway	100	130		350	500	1993	
56	Etzenricht (Germany)				160	600	1993	BTB
57	Nelson River Bipole 3 (Canada)	930	0	930	±500	2000	1992/97	
58	Chandrapur-Pdghe (India)	900			±500	1500	1997	
59	Minami Fukumitsu (Japan)				125	300	1999	BTB
60	SwePol (Starno, Sweden-Slupsk, Poland)		245		450	600	2000	
61	(Galatina, Italy-Arachthos Greece)	110	200		400	500	2001	
62	East South 2 Talcher (Orissa)-Kolar (Karnataka), India	1450			±500	2000	2002	
63	HVDC Three Gorges-Changzhou Longquan-Zhengping (China)	890	-		±500	3000	2003	
64	HVDC Three Gorges-Guangdong Station 1: Jingzhou, Iluizhou (China)	940	-	-	±500	3000	2003	

(Continued)

TABLE 1.1 (*Continued*)
LCC-Based HVDC Systems Throughout the World

S.N.	HVDC Link	Transmission Line Length (km)			Rated Voltage (kV)	Nominal Capacity (MW)	Commissioning Date	Remark
		OHL	Cable	Total				
65	Basslink (Loy Yang-George Town), Australia	71.8	298.3		400	600	2005	
66	Vizag-2 Eastern Grid and Southern grid, India				176	500	2005	BTB
67	Norned (Feda, Norway-Eemshaven Netherlands)	-	580		±450	700	2007	
68	Sharyland (Texas, USA)				21	150	2007	BTB
69	Al Fdhili (Saudi Arabia)					1800	2008	BTB
70	SAPEI (Latina, Italy-Fiume Santo, Sardinia)	-	435		±500	1000	2008/9	
71	Xianjiba, Shanghai (China)	2071	-		800	6400	2010	
72	Yannan– Guangdong (China)	1400	-		±800	5000	2010	
73	Ballia (UP)-Bhiwadi (Rajasthan) (India)	780			±500	2500	2010	
74	North East Agra (Biswanath Chariali) (Assam), Agra (UP), Alipurduar (West Bengal) (India)	1775			±800	6000	2016	MT
75	Champa Kurukshetra (India)			1365	800	2×3000	2016	

BTB, back-to-back; MT, multi-terminal; OHL, overhead line.

ends. In LCC-HVDC transmission, the reversal of power is carried out by reversing the polarity of the DC voltage. Hence, stronger and lighter cross-linked polyethylene (XLPE) cables cannot be used, which are suited for harsh environmental conditions as encountered in the ocean beds [4,12]. However, despite its limitations, LCC-HVDC possesses high reliability, good overload capability and lower converter losses. It requires low maintenance and capital costs, and is robust to DC fault currents due to its current regulating nature [1,6,7].

Subsequently, the development of the Insulated Gate Bipolar Transistor (IGBT) paved the way for the voltage-sourced converter (VSC)-based HVDC (VSC-HVDC) technology, which offered significant advantages over the LCC-HVDC. VSC-HVDC facilitates independent active and reactive power controls, along with reduction in filter size [8–18]. VSC-HVDC also enables the integration of offshore wind farms with AC grids. Compact, modular designs of the VSCs enable rapid installation, commissioning and relocation. Unlike LCC-HVDC, fixed DC voltage polarity in the VSC-HVDC enables the use of stronger and lighter XLPE cables, suitable for under-sea environment and attractive for offshore transmission. In addition, VSC-HVDC systems can be integrated with AC systems having low short circuit ratios [4,12,16,19–21].

The first 3-MW, VSC-HVDC link was commissioned at Hellsjon in Sweden in 1997. Subsequently, rapid development in the VSC technology resulted in the availability of higher rated VSC-HVDC links (up to 2000 MW). This has resulted in the installation and commissioning of a large number of VSC-HVDC systems worldwide. Present VSC-HVDC solutions use the modular multi-level converter (MMC) technology, which is more advantageous than two- or three-level VSCs in terms of reduced converter losses, increased modularity and scalability along with elimination of filter requirements [4,12]. Table 1.2 shows the list of some VSC-HVDC systems installed worldwide.

1.1.1 LINE COMMUTATED CONVERTER (LCC)-BASED HVDC TRANSMISSION

In LCC-HVDC transmission, thyristors are used in Current Source Converter (CSC) topology. The schematic diagram of a typical LCC-based HVDC system (monopolar type) interconnecting two AC systems is shown in Figure 1.1. As shown in Figure 1.1, the LCC-HVDC system consists of converters, converter transformers, filters, reactive power sources, smoothing reactors, electrodes and other equipment. The converter is an important component of the HVDC system. It performs the conversion from AC to DC or from DC to AC. In a six-pulse Graetz bridge converter, six thyristor valves are connected and fired sequentially. For large systems, such types of bridges are connected in series in 12 or 24 pulse configurations. The converter transformers are special type of transformers which operate with high harmonic currents and deals with AC and DC voltage stresses. Tap changers are available to manage the level of voltage and to reduce losses. In comparison with the typical transformers, the converter transformers are more expensive for the same rating. They transform the AC voltage to be supplied to the DC system.

TABLE 1.2

VSC-Based HVDC Systems Installed Throughout the World

S.N.	HVDC Link	Transmission Line Length (km)	Rated Voltage (kV)	Nominal Capacity (MW)	Year	Remark/ Converter/ Supplier
1	HVDC Tjaereborg Denmark	4.3 Cable	±9	7.2	2000	Interconnection of wind park/ ABB
2	HVDC Eagle Pass, Texas United States of America	-	±15.9	36	2000	BTB ABB
3	Directlink (Mullumbimby– Bungalora), Australia	59 Land cable	±80	180	2000	ABB
4	Murray Link (Berri, South Australia-Red Cliff, Victoria) Australia	180 Land cable	±150	220	2002	Three-level converters ABB
5	Cross sound cable (New Haven, Connecticut- Shoreham, Long Island) United States of America	40 Buried under water cable	±150	330	2002	ABB
6	HVDC Troll (Kollnes– Offshore Platform Troll A), Norway	70 Cable	±60	84	2005	To supply power for offshore gas compressor ABB
7	Estlink (Espoo, Finland– Harku, Estonia)	105 Cable	±150	350	2006	Two level ABB
8	Transbay (East Bay Oakland–San Francisco), United States of America	88 Cable	200	400	2010	Multi-level Siemens
9	Caprivi Link (Gerus– Zambari), Namibia	970 Overhead line	500	300	2010	Two level ABB
10	HVDC Valhall (Lista– Valhall, Offshore platform), Norway	292 Cable	150	78	2011	ABB, Nexans
11	BorWin1 (Diele–BorWin Alpha Platform), Germany	200 Cable	±150	400	2012	ABB Two level
12	BorWin2 (Diele–BorWin Beta), Germany	200 Cable	±300	800	2015	Multi-level Siemens
13	HelWin1 (Buttel–HelWin Beta), Germany	130 Cable	±250	575	2015	Multi-level Siemens
14	DolWin1 (Heede–DolWin Alpha), Germany	165 Cable	±320	800	2015	Multi-level ABB
15	INELFE Baixas, France–Santa Llogaia Spain	64 Cable	±320	2000	2015	Multi-level Siemens
16	SylWin1 (Buttel–SylWin Alpha), Germany	205 Cable	±320	864	2015	Multi-level Siemens

(Continued)

TABLE 1.2 (*Continued*)
VSC-Based HVDC Systems Installed Throughout the World

S.N.	HVDC Link	Transmission Line Length (km)	Rated Voltage (kV)	Nominal Capacity (MW)	Year	Remark/ Converter/ Supplier
17	HelWin1 (Buttel–HelWin Beta), Germany	130 Cable	±320	690	2015	Multi-level Siemens
18	DolWin2 (Heede–DolWin Beta), Germany	135 Cable	±320	900	2016	Multi-level ABB
19	DolWin3 (Heede–DolWin Gamma Platform), Germany	160 Cable	±320	900	2017	Alstom
20	New HVDC Hokkaido-Honshu (Imabetsu–Hokuto), Japan	24 Cable 98 Overhead line	250	300	2019	Toshiba
21	IFA-2 Tourbe Normandie, France–Chilling Hampshire, UK	204 Cable	320	1000	2021	ABB

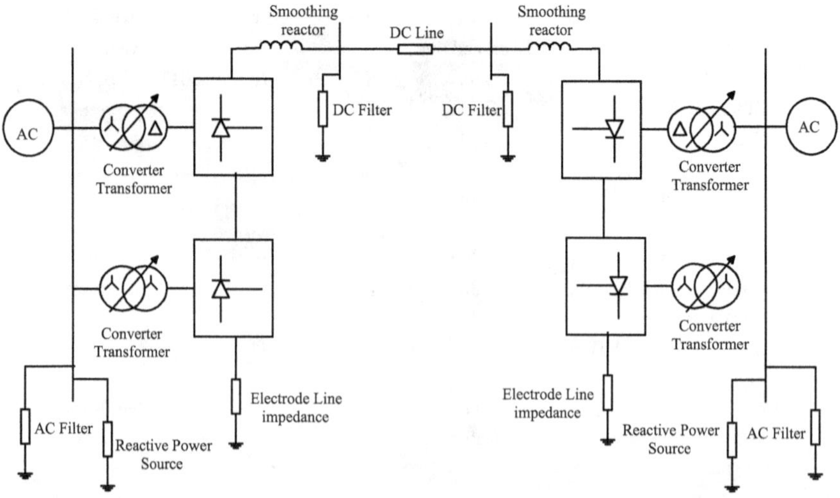

FIGURE 1.1   Schematic diagram of LCC-HVDC system (monopolar type).

In LCC-HVDC systems, several filters are required to mitigate the voltage and current harmonics produced by converter operation at both the ends of DC transmission lines. The reactive power sources are also required for power conversion at both the line ends. Smoothing reactors are used to minimize the DC ripple current. They also limit the rapid rise of current passing into the converter in case of commutation failure. In case of pole failure, the bipolar HVDC links operate in monopolar configuration which requires ground return path at half power for short duration. Therefore, the electrodes are designed to flow full current. However, in case of normal operation, the electrodes carry no current [1,6,7].

1.1.2 VOLTAGE SOURCE CONVERTER (VSC)-BASED HVDC TRANSMISSION

In a VSC-HVDC system, self-commutating devices like IGBTs and GTOs are used which provide the fully controlled AC voltage. This technique allows independent control of active and reactive powers along with reduction in filter size. The schematic diagram of a VSC-HVDC link is shown in Figure 1.2. The power flow can be reversed without changing the DC voltage polarity. Therefore, simple extruded cross-linked polyethylene (XLPE) DC cables can be used to integrate multiple DC buses. The VSC-HVDC system has the capability of black start. This feature allows VSC-HVDC systems to integrate with offshore wind farms [4,12].

1.2 INTERCONNECTION OF HVDC SYSTEMS

Based on the location of the converters, two types of interconnections are possible: back-to-back (BTB) and point-to-point (PTP) [4,10,12].

1.2.1 BACK-TO-BACK (BTB) HVDC

All the converters are located at the same location, and as a consequence, the DC cables are very short. The objective of the BTB HVDC configuration is to provide controllable power flow between two AC systems which are operating at different frequencies. This type of configuration is designed at low voltages and high

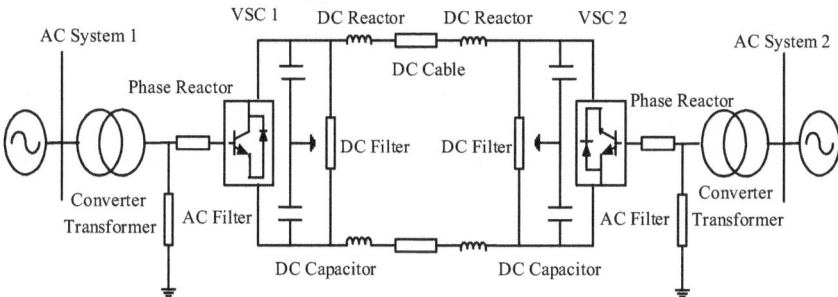

FIGURE 1.2   Schematic diagram of VSC-HVDC system (two level).

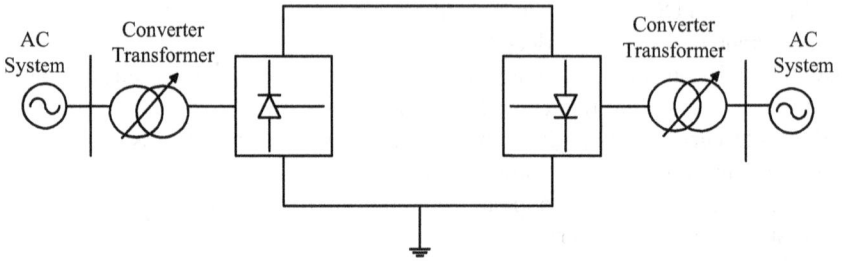

FIGURE 1.3   BTB HVDC.

currents to reduce the insulation costs. Also, the transmission losses are low due to the very short length of the cable. Back-to-back HVDC configuration operates with variable DC voltages and limited reactive power control capability. The BTB interconnection is depicted in Figure 1.3.

1.2.2   POINT-TO-POINT (PTP) HVDC

In PTP HVDC configuration, the converters are located at different locations. This type of configuration is mostly used for long-distance DC power transmission and consists of either cables or overhead lines or a combination of both. In the PTP HVDC system, several configurations are possible. These are monopole with ground return, monopole with metallic return, bipole with ground return, bipole with metallic return and homopole. Monopole configuration has one conductor at high DC voltage. The return path is provided as ground/sea or metallic return wire. The monopolar link has conductor of negative polarity to reduce corona losses. It is economical as it requires only one conductor. However, this connection is always not preferable due to the effect of corrosion on metallic return path and negative effect on the environment. The monopole with ground return and metallic return are depicted in Figures 1.4 and 1.5, respectively. The bipolar link is the most preferable configuration for high power ratings. In bipolar configuration, each terminal has two converters of the same ratings and grounded at mid-point. If the grounding is provided at both ends, each pole can operate independently. In case of the failure of one line, the bipolar link can operate as a monopolar one and can supply half the load. After the disturbance is cleared, it can act as a bipolar link. Therefore, the bipolar link is more reliable than the monopolar link. The bipole configuration with ground return and metallic return are depicted in Figures 1.6 and 1.7, respectively. The homopolar link has two or more conductors with the same polarity, mostly negative polarity with ground return. In case of a fault in one conductor, the healthy conductors can supply the power [22].

1.2.3   MULTI-TERMINAL HVDC

A MTDC system is an extension of two terminal HVDC systems. A MTDC system includes more than two converters: some converters act as rectifiers, while

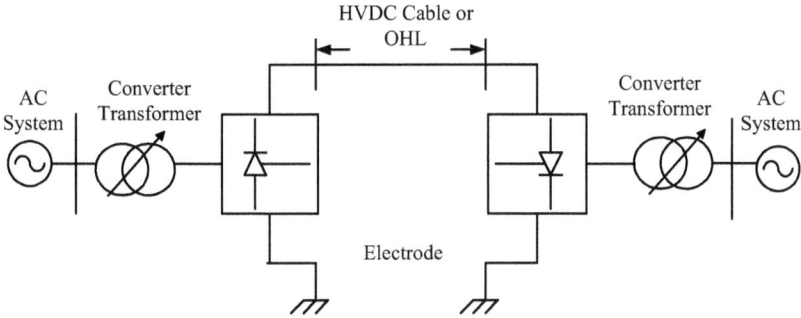

FIGURE 1.4 Monopole with ground return (PTP).

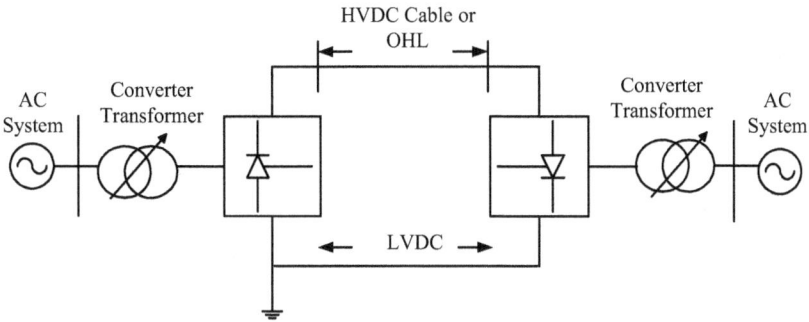

FIGURE 1.5 Monopole with metallic return.

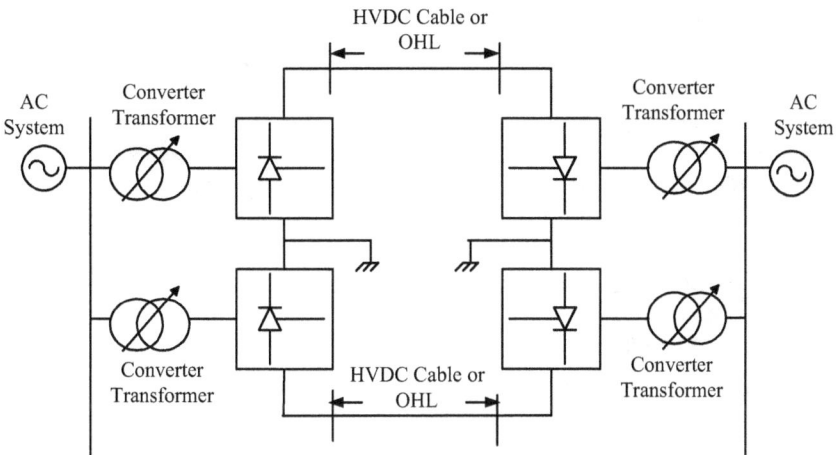

FIGURE 1.6 Bipole with ground return.

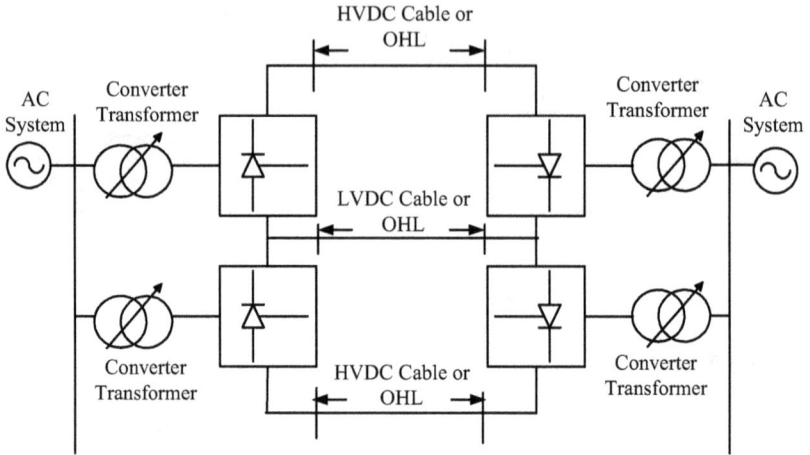

FIGURE 1.7   Bipole with metallic return.

others act as inverters. MTDC system is used to interconnect several generating stations at remote locations to several load centres to transfer the bulk power. The MTDC system is more economic and flexible than several two terminal HVDC links to transfer the same power. The MTDC system is more advantageous to inject powers at several points so that the AC network is not overloaded. In this respect, the MTDC system can include one rectifier station and several inverter stations [19–22].

The MTDC system is classified as

1. Series MTDC system
2. Parallel MTDC system

Series MTDC system: In this system, the current is controlled by one converter and it is common for all other converters. Rest of the converter stations can control the angle (firing angle or extinction angle) or voltage at their respective terminals. The system is grounded at only one suitable point. The grounding capacitors can also be used to improve system performance during transients and insulation coordination. The direction of power flow can change without providing the mechanical switches. The power rating is related to the voltage rating of the valve. The insulation coordination is a problem due to the different voltage ratings along the line. In case of line faults, the system will be completely shut down. The series connected MTDC system is shown in Figure 1.8.

Parallel MTDC: The voltage remains constant at all converter stations. The current at the converter stations can be adjusted based on the power requirement. Mostly, bipolar arrangement is used for the connection. In case of increased power demands, addition of parallel converters to the existing parallel MTDC system is easy. To change the direction of power flow, mechanical switching is required as the voltage cannot be reversed. Failure of a bridge in a converter

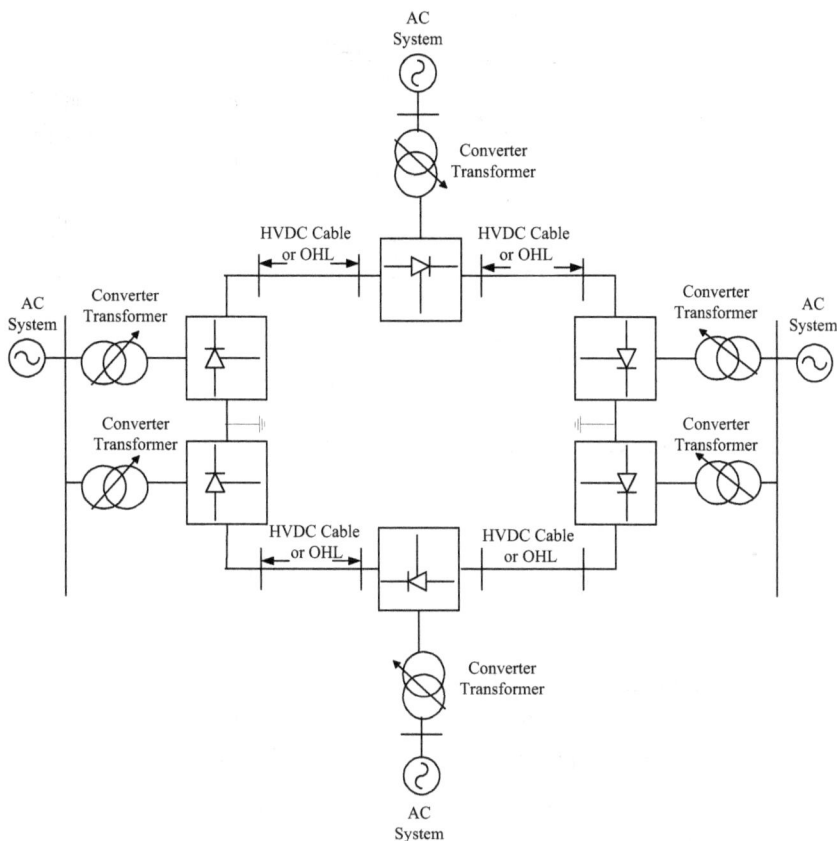

FIGURE 1.8   Series connected MTDC system.

would require either disconnection of a bridge in all converters or disconnection of the affected station. The parallel MTDC systems are of two types: radial and mesh. In radial system, the disconnection of line will lead to interruption of power from converter. The disconnection of any link in mesh MTDC system would not lead to interruption of power if the remaining links are capable of carrying the required power. Hence, a mesh-connected MTDC system is more reliable than the radial system. The radial and mesh-connected MTDC systems are depicted in Figures 1.9 and 1.10, respectively.

1.3   CONTROL OF HVDC SYSTEMS

Multiple converters can be interconnected to create a multi-terminal DC (MTDC) grid. For proper MTDC operation in DC grids, DC voltage control is an essential requirement. In this respect, several control techniques have been envisaged. An extensive review of these control techniques has been presented in [23].

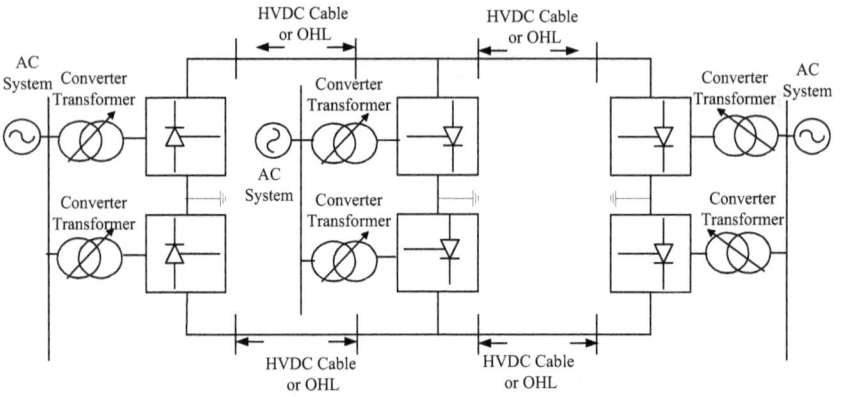

FIGURE 1.9   Parallel connected radial MTDC system.

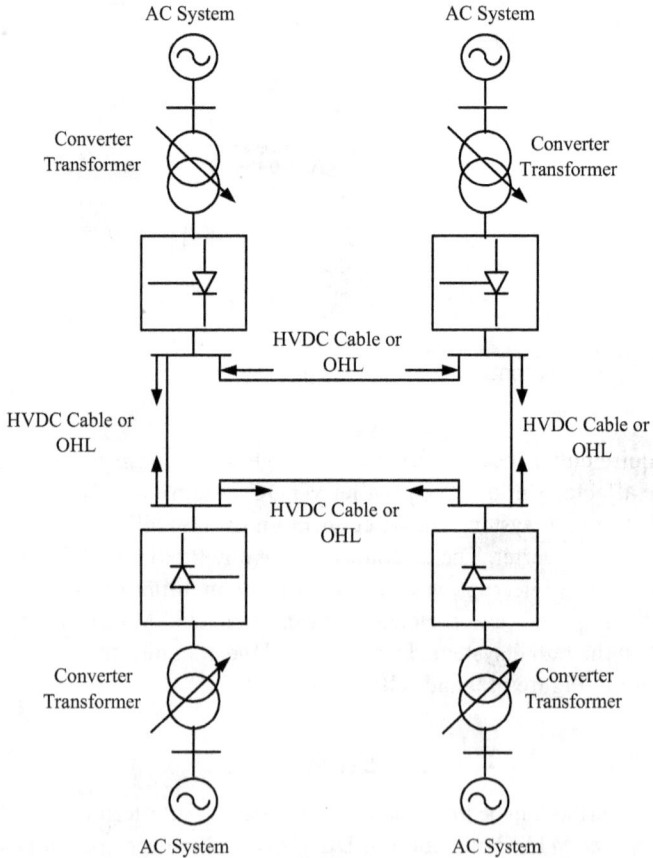

FIGURE 1.10   Parallel connected mesh MTDC system.

References [24–37] detail some of the comprehensive research works carried out in the area of DC voltage control. Commonly used DC voltage control modes are classified as follows.

1.3.1   DC Master-Slave Control

In this type of control mode, one converter operates as a master converter, while the rest operate as slave converters. The master converter is used to control the voltage of its DC terminal bus (acting as the DC slack bus in the MTDC grid) and sustain the power balance of the DC grid. The slave converters control the active power flows at the terminal ends of the lines connecting the VSCs to their respective AC system buses. In addition, all the converters can control either the voltage magnitudes at their AC terminal buses or the reactive power flows at the terminal ends of the lines connecting the VSCs to their respective AC system buses [10,12,24]. Reference [24] has employed DC slack-bus control in two- and six-terminal DC systems. The main disadvantage of this control scheme is the DC grid instability following a failure of the master converter. In DC master-slave control mode, the converters can be operated in different control modes as given below.

PQ control: In this control mode, the converter is used to control both the active and reactive power flows at the terminal end of the line connecting the VSC to its respective AC system bus.

PV control: In this control mode, the converter is used to control the voltage magnitude at its AC terminal bus and the active power flow at the terminal end of the line connecting the VSC to its respective AC system bus.

V_{DC} **Q control:** The converter acts as a master converter and controls the voltage of its DC terminal bus and the reactive power flow at the terminal end of the line connecting the VSC to its AC system bus.

V_{DC} **V control:** The converter acts as a master converter and controls the voltage of its DC terminal bus as well as the voltage magnitude at its AC terminal bus.

1.3.2   DC Voltage Droop Control

In a MTDC grid, with master-slave control, the DC grid becomes unstable in case of a failure or outage of the master converter. To maintain the stability of the DC grid under such a situation, DC voltage droop control is a better option. In DC voltage droop control, all converters participate in the DC voltage regulation scheme by adjusting their active power flow in response to changes in the DC voltage with the operating point [12,25–32,38]. For MTDC systems, different types of DC voltage droop control have been envisaged to ensure proper sharing based on the converter ratings. These include both linear and nonlinear voltage droop characteristics. Among the linear ones, Voltage-Power (V-P) and Voltage-Current (V-I) droops have been the two most popular strategies for DC voltage droop control. Nonlinear voltage droop control characteristics include dead-bands and limits. Among the nonlinear ones, Voltage Margin (VM), V-P droop with power Dead-Band (DB) and V-P droop with voltage limits are some of the more widely used characteristics [12,25–32,38].

1.4 INTRODUCTION TO DC POWER-FLOW CONTROLLERS

Management of power flow within a complex DC grid has been one of the main challenges in VSC-based integrated AC–DC systems. Although the VSCs control the power injections into a DC grid, the power flows within the DC grid depend upon the resistances of the DC links/cables [38,39]. In this respect, DC power-flow control devices [40–45] have been conceptualised and developed, similar to FACTS controllers developed for AC grids. These DC power-flow controllers can be classified into resistance and voltage types. A variable resistance is inserted in series with the DC line to control the DC power flow, but it leads to higher line losses [38,46]. The voltage type of DC power-flow controllers includes DC transformers, series adjustable voltage sources, multi-port DC power-flow control (MDCPFC) and the interline DC power-flow controller (IDCPFC) [47–52]. In multi-phase AC coupling transformer and MMC structure, the MDCPFC is utilized to control the two lines' active power flow for a three-terminal DC grid [19]. The interline DC power-flow controller (IDCPFC) reported in Refs. [51,52] is a DC power-flow controller without an external AC or DC source which can be used for power-flow management of MTDC grids. It is similar to the interline power-flow controller (IPFC), which is a VSC-based flexible AC transmission systems (FACTS) device developed for AC grids [53–55]. The IDCPFC comprises variable DC voltage sources which are incorporated in series with the DC links and regulates their power flow by controlling the power exchanged with these links. The three-terminal MTDC incorporating IDCPFC is shown in Figure 1.11. V_{DCS1} and V_{DCS2} shown in Figure 1.11 are series connected variable DC voltage sources with the DC links.

1.5 INTEGRATION OF RENEWABLE ENERGY SOURCES (RES) TO HVDC GRID

In recent years, the HVDC transmission is associated with the integration of large-scale renewable energy sources. The HVDC technology can also be utilized to integrate wind farms which are located far away from the consumer. In a MTDC grid, all the converters should contribute to automatic power balancing for the integration of intermittent power from wind farms. In MTDC grid, the control scheme of droop control should be able to balance the DC power without communicating between converter station terminals [3,4,7,8,11,15,16,19].

1.6 INTRODUCTION TO THE POWER-FLOW PROBLEM AND THE NEWTON-RAPHSON METHOD

In a power system network, the power is transferred from generators (mostly synchronous machines) to the loads through transmission lines. Under normal conditions, the power systems operate in steady state and the basic calculation needed to observe the performance of this state is known as power-flow analysis or load flow. The information regarding the voltage magnitudes and phase angles, active and reactive power flows in transmission lines, losses and the reactive

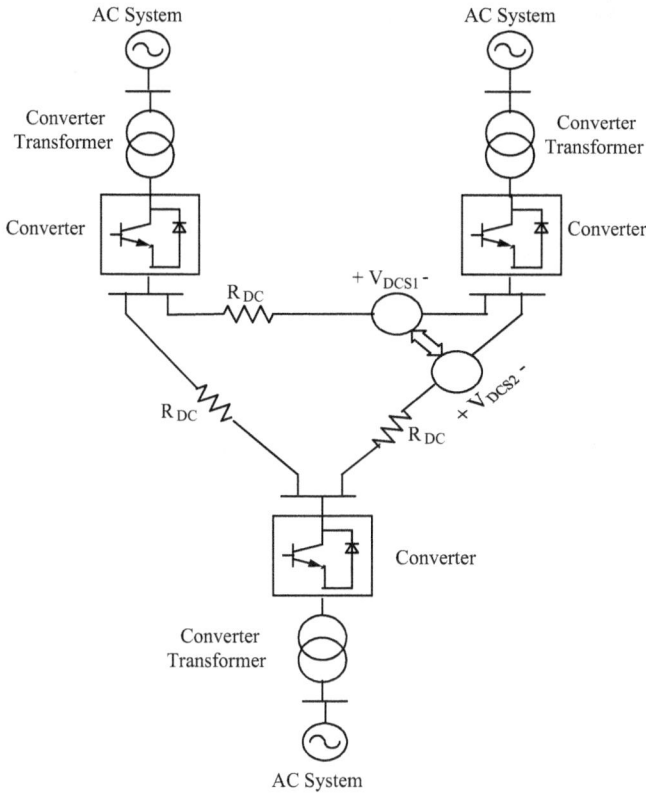

FIGURE 1.11 Three-terminal AC-MTDC system incorporating IDCPFC.

power generated and absorbed at buses is expected to be provided from power-flow analysis [56].

To understand the power-flow problem, let us consider an n-bus power system network consisting of 'm' generators (m ≤ n). Without any loss of generality, it is assumed that there are 'g' generators connected at the first 'g' buses of this system with bus 1 being the slack bus.

For further analysis, consider the ith bus in a power system network. The net complex power (per phase) at the ith bus injected into the transmission system is denoted by

$$S_i = S_{Gi} - S_{Di}$$

where S_{Gi} is the total complex power (per phase) produced by the generator connected at bus 'i' and denoted by $S_{Gi} = P_{Gi} + jQ_{Gi}$. S_{Di} is the total complex power (per phase) absorbed by the loads connected at bus 'i' and denoted by $S_{Di} = P_{Di} + jQ_{Di}$.

Also, the net current injection at the bus 'i' can be written as

$$I_i = \sum_{k=1}^{n} Y_{ik}\ V_k$$

where Y_{ik} is the element in the ith row and k^{th} column of the bus admittance matrix Y. V_k is the bus voltage phasor at bus 'k'.

The complex power injected at bus 'i' is given below:

$$S_i = P_i + j\,Q_i = V_i\ I_i^* = \sum_{k=1}^{n} V_i\ V_k^* Y_{ik}^*$$

The real part of the net complex power injected bus 'i' is

$$P_i = \sum_{k=1}^{n} V_i\ V_k\ Y_{ik}\ \cos\left(\theta_i - \theta_k - \phi_{ik}\right)$$

$$= \sum_{k=1,\,k\neq i}^{n} V_i\ V_k\ Y_{ik}\ \cos\left(\theta_i - \theta_k - \phi_{ik}\right) + V_i^2\ Y_{ii}\mathrm{Cos}\phi_{ii}$$

Similarly, the net reactive power injected at bus 'i' is

$$Q_i = \sum_{k=1}^{n} V_i\ V_k\ Y_{ik}\ \sin\left(\theta_i - \theta_k - \phi_{ik}\right)$$

$$= \sum_{k=1,\,k\neq i}^{n} V_i\ V_k\ Y_{ik}\ \sin\left(\theta_i - \theta_k - \phi_{ik}\right) - V_i^2 Y_{ii}\mathrm{Sin}\phi_{ii}$$

The objective of power flow is to calculate the four variables $\left(P_i,\ Q_i,\ V_i,\ \theta_i\right)$ at each bus 'i' in the power system network. Only two variables are known prior to solving the power flow, and the remaining two variables are solved using the power flow solution.

Based on the steady-state assumptions of constant frequency and constant voltages, the power system buses are classified as follows:

Voltage-controlled or generator bus: the total injected active power P_i and the voltage magnitude V_i of bus 'i' are specified. This type of the bus corresponds to the generators where active power is fixed by the turbine governor setting and voltage is fixed by the excitation system.

Load or PQ bus: the total injected active and reactive powers (P_i, Q_i) are specified at this bus. This type of bus corresponds to load centres.

Slack or swing bus: prior to the power-flow solution, the system losses are not known. Therefore, the total injected power cannot be specified at every individual bus. It is needed to select a voltage-controlled bus as a slack bus, and the bus voltage magnitude and phase angle at the slack bus (V_i, θ_i) are specified. It is also considered as a reference bus.

So, power-flow problem becomes

Solve: θ, **V**

Given: **P, Q**

where $\theta = [\theta_2 \cdots \theta_n]^T$, $V = [V_{g+1} \cdots V_n]^T$

$$P = [P_2 \cdots P_n]^T, Q = [Q_{g+1} \cdots Q_n]^T$$

The active and reactive power injections ($P_2 \cdots P_n$ and $Q_{g+1} \cdots Q_n$) are functions of $\theta_2 \cdots \theta_n$ and $V_{g+1} \cdots V_n$.

The specified active and reactive powers of any bus 'i' can be rewritten as

$$P_i = P_i^{sp}$$

$$Q_i = Q_i^{sp}$$

It is known that the injected active and reactive powers at any bus 'i' are functions of bus voltage magnitudes and phase angles of the bus voltages. Therefore, the above equations can be written in the form of unknown variables:

$$P_i \left(\theta_2 \cdots \theta_n \quad V_{g+1} \cdots V_n \right) = P_i^{sp}$$

$$Q_i \left(\theta_2 \cdots \theta_n \quad V_{g+1} \cdots V_n \right) = Q_i^{sp}$$

The initial guesses of voltage magnitudes and phase angles of the bus voltages are selected as 1 p.u. and 0^0, respectively. It is assumed that the actual solution lies at an infinitesimal distance away from the initial guess. Therefore, it is assumed that the actual solution is at

$$P_i \left(\theta_2^0 + \Delta\theta_2, \theta_3^0 + \Delta\theta_3, \ldots, \theta_n^0 + \Delta\theta_n, V_{g+1}^0 + \Delta V_{g+1}, V_{g+2}^0 + \Delta V_{g+2}, \ldots, V_n^0 + \Delta V_{n+1} \right)$$

where $\theta_2^0, \theta_3^0 \cdots \theta_n^0 \ V_{g+1}^0, V_{g+2}^0 \cdots V_n^0$ are the initial guesses of bus voltage angle and magnitudes.

By using the Taylor series expansion of the nonlinear system of equations mentioned above and with higher order terms neglected, we get

$$P_i\left(\theta_2^0, \theta_3^0,\ldots,\theta_n^0, V_{g+1}^0, V_{g+2}^0,\ldots V_n^0\right) + \frac{\partial P_i}{\partial \theta_2}\,\Delta\theta_2 + \frac{\partial P_i}{\partial \theta_3}\,\Delta\theta_3$$

$$+\cdots\frac{\partial P_i}{\partial \theta_n}\,\Delta\theta_n + \frac{\partial P_i}{\partial V_{g+1}}\,\Delta V_{g+1} + \frac{\partial P_i}{\partial V_{g+2}}\,\Delta V_{g+2} + \cdots\frac{\partial P_i}{\partial V_n}\,\Delta V_n = P_i^{sp}$$

or

$$\frac{\partial P_i}{\partial \theta_2}\,\Delta\theta_2 + \frac{\partial P_i}{\partial \theta_3}\,\Delta\theta_3 + \cdots\frac{\partial P_i}{\partial \theta_n}\,\Delta\theta_n + \frac{\partial P_i}{\partial V_{g+1}}\,\Delta V_{g+1} + \frac{\partial P_i}{\partial V_{g+2}}\,\Delta V_{g+2} + \cdots\frac{\partial P_i}{\partial V_n}\,\Delta V_n$$

$$= P_i^{sp} - P_i\left(\theta_2^0, \theta_3^0,\ldots,\theta_n^0, V_{g+1}^0, V_{g+2}^0,\ldots V_n^0\right)$$

$$= P_i^{sp} - P_i^{cal}$$

Similarly, the equation of reactive power injection at bus 'i' can also be expanded. In matrix form, the system equations can now be written as

$$
\begin{bmatrix}
\Delta\theta_2 \\
\Delta\theta_3 \\
\\
\Delta\theta_n \\
\Delta V_{g+1} \\
\Delta V_{g+2} \\
\\
\Delta V_{g+n}
\end{bmatrix}
$$

$$
=
\begin{bmatrix}
\frac{\partial P_2}{\partial \theta_2} & \frac{\partial P_2}{\partial \theta_3} & \cdots & \frac{\partial P_2}{\partial \theta_2} & \frac{\partial P_2}{\partial \theta_2} & \frac{\partial P_2}{\partial \theta_2} & \cdots & \frac{\partial P_2}{\partial V_n} \\
\frac{\partial P_3}{\partial \theta_2} & \frac{\partial P_2}{\partial \theta_3} & \cdots & \frac{\partial P_2}{\partial \theta_2} & \frac{\partial P_2}{\partial \theta_2} & \frac{\partial P_2}{\partial \theta_2} & \cdots & \frac{\partial P_3}{\partial V_n} \\
\frac{\partial P_n}{\partial \theta_2} & \frac{\partial P_n}{\partial \theta_3} & \cdots & \frac{\partial P_n}{\partial V_{g+1}} & \frac{\partial P_n}{\partial V_{g+2}} & \frac{\partial P_n}{\partial \theta_2} & \cdots & \frac{\partial P_n}{\partial V_n} \\
\frac{\partial Q_2}{\partial \theta_2} & \frac{\partial Q_2}{\partial \theta_3} & \cdots & \frac{\partial Q_2}{\partial \theta_n} & \frac{\partial Q_2}{\partial V_{g+1}} & \frac{\partial Q_2}{\partial V_{g+2}} & \cdots & \frac{\partial Q_2}{\partial V_n} \\
\frac{\partial Q_3}{\partial \theta_2} & \frac{\partial Q_3}{\partial \theta_3} & \cdots & \frac{\partial Q_3}{\partial \theta_n} & \frac{\partial Q_3}{\partial V_{g+1}} & \frac{\partial Q_3}{\partial V_{g+2}} & \cdots & \frac{\partial Q_3}{\partial V_n} \\
\frac{\partial Q_n}{\partial \theta_2} & \frac{\partial Q_n}{\partial \theta_3} & \cdots & \frac{\partial Q_n}{\partial \theta_n} & \frac{\partial Q_n}{\partial V_{g+1}} & \frac{\partial Q_n}{\partial V_{g+2}} & \cdots & \frac{\partial Q_n}{\partial V_n}
\end{bmatrix}^{-1}
\begin{bmatrix}
P_2^{sp} - P_2^{cal} \\
P_2^{sp} - P_3^{cal} \\
P_n^{sp} - P_n^{cal} \\
Q_{g+1}^{sp} - Q_{g+1}^{cal} \\
Q_{g+2}^{sp} - Q_{g+2}^{cal} \\
Q_n^{sp} - Q_n^{cal}
\end{bmatrix}
$$

Hence, the solution lies at

$$
\begin{bmatrix}
\theta_2^1 \\
\theta_3^1 \\
\vdots \\
\theta_n^1 \\
V_{g+1}^1 \\
V_{g+2}^1 \\
\vdots \\
V_n^1
\end{bmatrix}
=
\begin{bmatrix}
\theta_2^0 + \Delta\theta_2 \\
\theta_3^0 + \Delta\theta_3 \\
\vdots \\
\theta_n^0 + \Delta\theta_n \\
V_{g+1}^0 + \Delta V_{g+1} \\
V_{g+2}^0 + \Delta V_{g+2} \\
\vdots \\
V_n^0 + \Delta V_n
\end{bmatrix}
=
\begin{bmatrix}
\theta_2^0 \\
\theta_3^0 \\
\vdots \\
\theta_n^0 \\
V_{g+1}^0 \\
V_{g+2}^0 \\
\vdots \\
V_n^0
\end{bmatrix}
+
$$

$$
\begin{bmatrix}
\dfrac{\partial P_2}{\partial \theta_2} & \dfrac{\partial P_2}{\partial \theta_3} & \cdots & \dfrac{\partial P_2}{\partial \theta_2} & \dfrac{\partial P_2}{\partial \theta_2} & \dfrac{\partial P_2}{\partial \theta_2} & \cdots & \dfrac{\partial P_2}{\partial V_n} \\[2mm]
\dfrac{\partial P_3}{\partial \theta_2} & \dfrac{\partial P_2}{\partial \theta_3} & \cdots & \dfrac{\partial P_2}{\partial \theta_2} & \dfrac{\partial P_2}{\partial \theta_2} & \dfrac{\partial P_2}{\partial \theta_2} & \cdots & \dfrac{\partial P_3}{\partial V_n} \\[2mm]
\dfrac{\partial P_n}{\partial \theta_2} & \dfrac{\partial P_n}{\partial \theta_3} & \cdots & \dfrac{\partial P_n}{\partial V_{g+1}} & \dfrac{\partial P_n}{\partial V_{g+2}} & \dfrac{\partial P_n}{\partial \theta_2} & \cdots & \dfrac{\partial P_n}{\partial V_n} \\[2mm]
\dfrac{\partial Q_2}{\partial \theta_2} & \dfrac{\partial Q_2}{\partial \theta_3} & \cdots & \dfrac{\partial Q_2}{\partial \theta_n} & \dfrac{\partial Q_2}{\partial V_{g+1}} & \dfrac{\partial Q_2}{\partial V_{g+2}} & \cdots & \dfrac{\partial Q_2}{\partial V_n} \\[2mm]
\dfrac{\partial Q_3}{\partial \theta_2} & \dfrac{\partial Q_3}{\partial \theta_3} & \cdots & \dfrac{\partial Q_3}{\partial \theta_n} & \dfrac{\partial Q_3}{\partial V_{g+1}} & \dfrac{\partial Q_3}{\partial V_{g+2}} & \cdots & \dfrac{\partial Q_3}{\partial V_n} \\[2mm]
\dfrac{\partial Q_n}{\partial \theta_2} & \dfrac{\partial Q_n}{\partial \theta_3} & \cdots & \dfrac{\partial Q_n}{\partial \theta_n} & \dfrac{\partial Q_n}{\partial V_{g+1}} & \dfrac{\partial Q_n}{\partial V_{g+2}} & \cdots & \dfrac{\partial Q_n}{\partial V_n}
\end{bmatrix}^{-1}
\begin{bmatrix}
P_2^{sp} - P_2^{cal} \\[2mm]
P_2^{sp} - P_3^{cal} \\[4mm]
P_n^{sp} - P_n^{cal} \\[2mm]
Q_{g+1}^{sp} - Q_{g+1}^{cal} \\[2mm]
Q_{g+2}^{sp} - Q_{g+2}^{cal} \\[4mm]
Q_n^{sp} - Q_n^{cal}
\end{bmatrix}
$$

The above values correspond to the phase angles and the voltage magnitudes of the bus voltages at the end of the first power-flow iteration. It may be noted that the values of the partial derivatives in the above equation also correspond to the first iteration of the power-flow algorithm. The obtained values of the phase angles and voltage magnitudes of the system buses after the first iteration are treated as the starting values (initial guess) for the second iteration. The partial derivatives are also evaluated corresponding to these new starting values (guess), and we get the updated values of the unknown variables (phase angles and voltage magnitudes of the system buses) after completion of the second iteration. These values are further taken as the starting values for the third iteration and so on. The algorithm is made to terminate when the error in the specified and calculated values of the active and reactive powers falls below a given tolerance, say 10^{-4} pu.

The square matrix is called as the Jacobian matrix. It consists of four sub-matrices:

$$
J = \begin{bmatrix}
\dfrac{\partial \mathbf{P}}{\partial \theta} & \vdots & \dfrac{\partial \mathbf{P}}{\partial \mathbf{V}} \\
\cdots \quad \cdots & \vdots & \cdots \quad \cdots \\
\dfrac{\partial \mathbf{Q}}{\partial \theta} & \vdots & \dfrac{\partial \mathbf{Q}}{\partial \mathbf{V}}
\end{bmatrix}
$$

The diagonal and off-diagonal elements of submatrix $\dfrac{\partial \mathbf{P}}{\partial \theta}$ can be calculated by following expressions:

$$
\frac{\partial P_i}{\partial \theta_k} = -Q_i - V_i^2 Y_{ii} \; \phi_{ii} \quad \text{if } i = k
$$

$$
= V_i V_k Y_{ik} \left(\theta_i - \theta_k - \phi_{ik} \right) \quad \text{if } i \neq k
$$

Similarly, the diagonal and off-diagonal elements of submatrix $\dfrac{\partial \mathbf{P}}{\partial \mathbf{V}}$ can be calculated by following expressions:

$$
\frac{\partial P_i}{\partial V_k} = V_i Y_{ik} \; \cos\left(\theta_i - \theta_k - \phi_{ik}\right), \quad \text{if } i \neq k
$$

$$
= \frac{P_i}{V_i} + V_i Y_{ii} \cos f_{ii}, \quad \text{if } i = k
$$

$$
\frac{\partial Q_i}{\partial \theta_k} = - V_i \; V_k \; Y_{ik} \; \cos\left(\theta_i - \theta_k - \phi_{ik}\right), \quad \text{if } i \neq k
$$

$$
= P_i - V_i^2 \; Y_{ii} \; \cos f_{ii}, \quad i = k
$$

$$
\frac{\partial Q_i}{\partial V_k} = V_i \; Y_{ik} \; \sin\left(\theta_i - \theta_k - \phi_{ik}\right), \quad \text{if } i \neq k
$$

$$
= \frac{Q_i}{V_i} - V_i \; Y_{ii} \; \sin\phi_{ii}, \quad \text{if } i = k
$$

1.7 INTRODUCTION TO THE POWER-FLOW MODELLING OF LCC-BASED INTEGRATED AC–DC SYSTEMS

For proper planning, design and operation of AC power systems integrated with multi-terminal LCC-based DC grids, the power-flow solution of such integrated

AC–DC systems are required. Therefore, the development of suitable power-flow models of such systems is a fundamental requirement.

Usually, in LCC-based integrated AC–DC systems, as detailed in subsequent chapters, the power-flow unknowns may comprise AC bus voltage magnitudes and their phase angles, DC bus voltages, firing angles of rectifiers, extinction angle of inverters, converter transformer tap settings and power factors at rectifier and inverter ends. Due to the increased number of variables involved, the power-flow solution becomes more complex than that of a normal AC power system. Now, two different algorithms have generally been reported in the literature for the power-flow solution of LCC-based integrated AC–DC systems using the Newton-Raphson method. These are detailed below.

1.7.1 THE UNIFIED METHOD

In this method, the AC and DC quantities are solved simultaneously, which increases the size of the Jacobian matrix.

1.7.2 THE SEQUENTIAL METHOD

In this method, the AC and DC systems are solved separately in each iteration and are coupled by injecting an equivalent amount of real and reactive power at the terminal AC buses [1,6,7]. Unlike the unified method, the sequential method is easier to implement and poses lesser computational burden due to the smaller size of the Jacobian matrix.

Some comprehensive power-flow and optimal power-flow (OPF) models of LCC-based integrated AC–DC systems have been presented in Refs. [60–85].

1.8 INTRODUCTION TO THE POWER-FLOW MODELLING OF VSC-BASED INTEGRATED AC–DC SYSTEMS

Similar to LCC-based integrated AC–DC systems, for VSC-based integrated AC–DC systems, both AC and DC quantities are involved. Both the power-flow unknowns and the specified quantities depend upon a number of factors, e.g., the DC grid voltage control strategy adopted (DC master-slave control, linear and/or nonlinear DC voltage droop control), integration of offshore wind farms (OWFs) with the DC grid and inclusion of interline DC power-flow controllers (IDCPFCs). So, for power-flow solution of VSC-based integrated AC–DC systems too, both unified and sequential power-flow methods can be adopted.

References [86–106] present some comprehensive research works carried out in the area of power-flow and optimal power-flow (OPF) modelling of VSC-based integrated AC–DC systems.

Typical flow charts for unified and sequential power-flow algorithms of integrated AC-HVDC systems are shown in Figures 1.12 and 1.13, respectively.

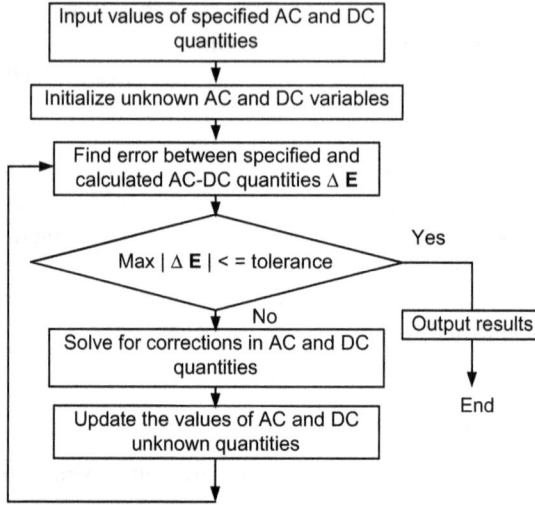

FIGURE 1.12 Flow chart for unified method of integrated AC-HVDC power flow.

1.9 ORGANIZATION OF THE BOOK

In this book, the power-flow modelling of integrated AC–DC systems is investigated systematically. Both unified and sequential power-flow models of LCC- and VSC-based HVDC systems are addressed.

The outline of the remaining chapters of the book is as follows:

Chapter 2 presents the development of power-flow models of LCC-based integrated AC–DC systems [107–109]. Both unified and sequential power-flow models of such systems are developed in this chapter. The effects of the different per-unit AC–DC system models on the convergence of the unified and sequential AC–DC power-flow algorithms are investigated in light of diverse DC link control strategies employed. The power-flow models are implemented in the IEEE 300-bus and the European 1354-bus test systems integrated with DC grids. The convergence characteristics validate the model.

Chapter 3 addresses the development of unified and sequential power-flow models of VSC-based integrated AC–DC systems. DC slack-bus (master-slave) control is employed for the MTDC grid. The models are applicable for both the back-to-back (BTB) and point-to-point (PTP) VSC-HVDC configurations [110,111]. Both the MTDC grid topology and the number of VSCs can be arbitrarily chosen in the proposed model. Several case studies were carried out in the IEEE 300-bus and the European 1354-bus test systems with diverse topologies of MTDC networks, employing different DC grid control strategies.

Chapter 4 addresses the development of unified and sequential power-flow models [112,113] of VSC-based integrated AC–DC systems employing DC voltage droop control. The DC voltage droop control comprises both linear {voltage-power (V-P) and voltage-current (V-I)} and nonlinear {power dead-band and

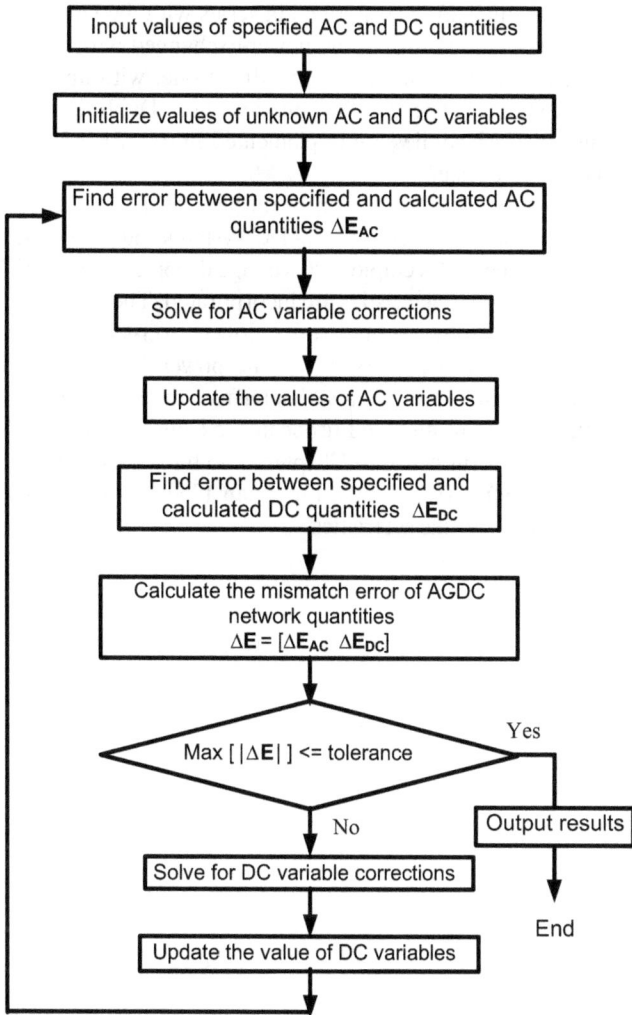

FIGURE 1.13 Flow chart of sequential method of integrated AC-HVDC power flow.

voltage limits} droop characteristics. Voltage margin control is also employed in the proposed models, as a specific case of the voltage droop with a power dead-band. Based on the terminal end line active and reactive power specifications of the VSCs, two different droop control models have been developed. Multiple case studies have been carried out in the IEEE 300-bus and the European 1354-bus test systems with diverse topologies of MTDC networks, employing different types (linear and nonlinear) of DC voltage droop controls.

Chapter 5 addresses the development of unified [114] and sequential power-flow models of VSC-based integrated AC–DC systems incorporating an IDCPFC for the power-flow management of the DC grid. The IDCPFC comprises variable

DC voltage sources which are incorporated in series with the DC links and regulates their power flow by controlling the power exchanged with these links. The IDCPFC considered in the model is a generalized one, with an arbitrary number of DC voltage sources. The IDCPFC employs both DC link current and DC link power controls. Case studies are implemented in the IEEE 300-bus and the European 1354-bus test systems.

Chapter 6 addresses the development of unified [113] and sequential power-flow models of VSC-based integrated AC–DC systems integrated with offshore wind farms (OWFs). The VSCs employ DC voltage droop control. Both linear and nonlinear voltage droop characteristics are incorporated. The effects of the OWFs on the DC grid voltage profile and the power-flow convergence are investigated, vis-à-vis varying wind farm powers. In all the power-flow models developed, VSC losses have been included. In this chapter too, the case studies are implemented in the IEEE 300-bus and the European 1354-bus test systems. Feasibility studies of all the models proposed in Chapters 2–6 have been carried out on the IEEE 300 bus test system [115] and the European 1354 bus test-network [116] to validate their convergence characteristics.

2 Power-Flow Modelling of AC Power Systems Integrated with LCC-Based Multi-Terminal DC (AC-MLDC) Grids

2.1 INTRODUCTION

As already mentioned in Chapter 1, the LCC-based HVDC technology constitutes the bulk of the DC transmission capacity installed globally. A multi-terminal LCC-based HVDC (MLDC) interconnection is more versatile and better capable of utilising the economic and technical advantages of the LCC-HVDC technology than a two-terminal one.

For planning, operation and control of AC power systems incorporating MLDC networks, the power-flow solution of integrated AC-MLDC systems is required. This necessitates suitable power-flow models of such systems. Now, for developing such models, the base values of the various DC quantities can be defined in several ways, each comprising separate sets of system equations in per-unit. It is observed that different per-unit system models affect the convergence of the AC-DC power-flow algorithm in different ways. Although several choices are feasible, only two different per-unit system models are considered in this book.

Now, to solve the power-flow equations in integrated AC-MLDC systems, two different algorithms are available in the literature. These are known as the unified and the sequential power-flow methods, respectively [1,6]. In the unified method, the AC and DC quantities are solved simultaneously, which increases the size of the Jacobian matrix. References [62,71–74] present some comprehensive research works on the unified method.

In the sequential AC-DC Newton power-flow algorithm, the AC and DC systems are solved separately in each iteration and are coupled by injecting an equivalent amount of real and reactive power at the AC terminal buses [1,6]. Unlike the unified method, the sequential method is easier to implement and poses lesser computational burden due to the smaller size of the Jacobian matrix. References [64,75–85] present some comprehensive research works on the sequential method.

DOI: 10.1201/9781003252078-2

For power-flow solution of integrated AC-MLDC systems, five quantities are required to be solved per converter [1,6]. These include the DC voltage, the DC current, the control angle, the converter transformer tap ratio and the converter power factor. On the other hand, only three independent equations comprising two basic converter equations and one DC network equation exist per converter. Thus, for solution, two additional equations are usually required. These two equations are derived from the control specifications adopted for the DC links. Thus, mathematically, the control specifications are used to bridge the gap between the number of independent equations and the number of unknown quantities. Control specifications usually include specified values of converter transformer tap ratio, converter control angle, DC voltage, DC current or DC power. Depending on the application, several combinations of valid control specifications are possible. Each combination of a set of valid control specifications is known as a control strategy [1,6]. The number of possible control strategies increases drastically with increase in the number of the DC terminals or converters. Of a myriad of combinations, only some control strategies are practically adopted in practice. In this book, six control strategies have been considered for a three-terminal DC network.

It is to be noted that in this chapter as well as in the subsequent chapters, bold quantities represent complex variables and equivalent-pi models are used to represent the transmission lines.

2.2 MODELLING OF INTEGRATED AC-MLDC SYSTEMS

Figure 2.1 depicts a typical AC power system incorporating a three-terminal DC network. The DC network contains two HVDC links. The first link is connected in the branch 'i–j' between any two system buses 'i' and 'j' of the AC network, while the second one is connected in the branch 'i–k' between system buses 'i' and 'k'. The three converters representing one rectifier and two inverters are connected to

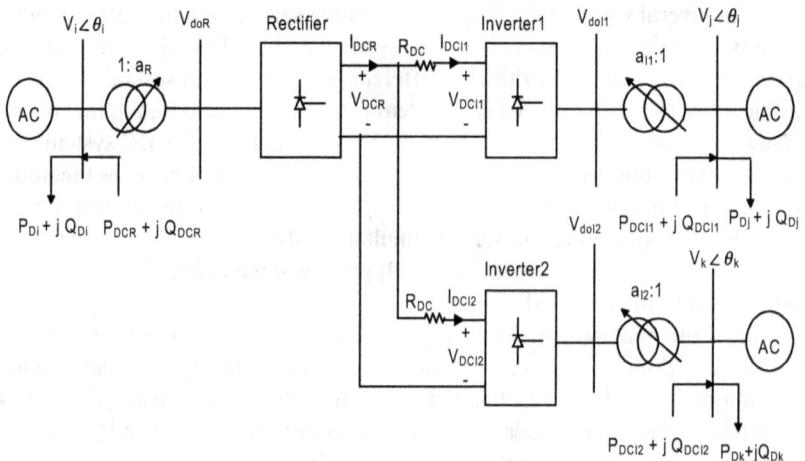

FIGURE 2.1 Three-terminal HVDC link between buses 'i', 'j' and 'k' of an existing AC power system network.

the AC system at buses 'i', 'j' and 'k,' respectively, through their respective converter transformers. The complex load powers at the AC buses 'i', 'j' and 'k' are represented as $S_{Di} = P_{Di} + j\,Q_{Di}$, $S_{Dj} = P_{Dj} + j\,Q_{Dj}$ and $S_{Dk} = P_{Dk} + j\,Q_{Dk}$, respectively. The effects of the DC links are accounted for as equivalent amount of real and reactive power injections P_{DCR}, Q_{DCR}, P_{DCI1}, Q_{DCI1}, P_{DCI2} and Q_{DCI2} at the converters' AC terminal buses 'i', 'j' and 'k', respectively. These power injections are included in the analysis by appropriate modifications of the power-flow equations, as detailed later.

Prior to the selection of variables and formulation of the equations, several basic assumptions are made for the analysis of steady-state DC converter operation [1–3,6]. These are

1. The three AC voltages at the terminal bus bars are balanced and sinusoidal.
2. The operation of the converters is perfectly balanced.
3. The direct currents and voltages are smooth.
4. The converter transformers are lossless and their magnetizing admittances are ignored.

Subsequently, for analysis of the integrated AC-DC system, the DC and AC equations are combined together which requires the translation of the converter equations into the per-unit system in order to use them with AC system per-unit equations. The base values for the DC system are defined in Table 2.1 which culminate in the per-unit AC-DC system equations as given in Table 2.2.

From Table 2.2, it can be observed that 12 independent equations exist against a total of 18 unknowns. Thus, for a complete solution, six quantities (two per converter) are needed to be specified. These are derived from the control specifications adopted for the DC links. Each combination of a set of valid control specifications is known as a control strategy. Theoretically, the number of possible control strategies increases rapidly with increase in the number of the DC terminals or converters. However, only a few of the possible control strategies are practically adopted in practice. This is explained in the next section.

TABLE 2.1
Different Base Values for DC System

Convention 1 (Model 1)	Convention 2 (Model 2)
$V_{DC\,base} = k\,V_{AC\,base}$; where $k = \dfrac{3\,\sqrt{2}}{\pi}\,n_b$	$V_{DC\,base} = V_{AC\,base}$
$I_{DC\,base} = \dfrac{\sqrt{3}}{k}\,I_{AC\,base}$	$I_{DC\,base} = \sqrt{3}\,I_{AC\,base}$
$Z_{DC\,base} = k^2 Z_{AC\,base}$	$Z_{DC\,base} = Z_{AC\,base}$
$R_{DC\,base} = \dfrac{3}{\pi}\,n_b X_{c\,base}$	$R_{DC\,base} = X_{c\,base}$

TABLE 2.2

Basic Converter Equations for a Three-Terminal DC Network in Per-Unit System

Per-Unit System 1 (Model 1)	Per-Unit System 2 (Model 2)	Eq. No.
$V_{DCR} = a_R\ V_i\ \cos\alpha_R - X_c\ I_{DCR}$	$V_{DCR} = \dfrac{3\sqrt{2}}{\pi}a_R\,n_b V_i\,\cos\alpha_R - \dfrac{3X_c}{\pi}\,n_b I_{DCR}$	(2.1)
$V_{DCI1} = a_{I1}\ V_j\ \cos\gamma_{I1} - X_c\ I_{DCI1}$	$V_{DCI1} = \dfrac{3\sqrt{2}}{\pi}a_{I1}\,n_b V_j\,\cos\gamma_{I1} - \dfrac{3X_c}{\pi}\,n_b I_{DCI1}$	(2.2)
$V_{DCI2} = a_{I2}\ V_k\ \cos\gamma_{I2} - X_c\ I_{DCI2}$	$V_{DCI2} = \dfrac{3\sqrt{2}}{\pi}a_{I2}\,n_b V_k\,\cos\gamma_{I2} - \dfrac{3X_c}{\pi}\,n_b I_{DCI2}$	(2.3)
$V_{DCR} = a_R\ V_i\ \cos\phi_R$	$V_{DCR} = \dfrac{3\sqrt{2}}{\pi}a_R\,n_b V_i\,\cos\phi_R$	(2.4)
$V_{DCI1} = a_{I1}\ V_j\ \cos\phi_{I1}$	$V_{DCI1} = \dfrac{3\sqrt{2}}{\pi}a_{I1}\,n_b V_j\,\cos\phi_{I1}$	(2.5)
$V_{DCI2} = a_{I2}\ V_k\ \cos\phi_{I2}$	$V_{DCI2} = \dfrac{3\sqrt{2}}{\pi}a_{I2}\,n_b V_k\,\cos\phi_{I2}$	(2.6)
$I_{DCI1} = \dfrac{V_{DCR} - V_{DCI1}}{R_{DC}}$		(2.7)
$I_{DCI2} = \dfrac{V_{DCR} - V_{DCI2}}{R_{DC}}$		(2.8)
$I_{DCR} = I_{DCI1} + I_{DCI2}$		(2.9)
$P_{DCR} = V_{DCR}\ I_{DCR}$		(2.10)
$P_{DCI1} = V_{DCI1} * \left[\dfrac{V_{DCR} - V_{DCI1}}{R_{DC}}\right]$		(2.11)
$P_{DCI2} = V_{DCI2} * \left[\dfrac{V_{DCR} - V_{DCI2}}{R_{DC}}\right]$		(2.12)

2.3   CONTROL STRATEGIES FOR MLDC GRIDS

For a three-terminal DC network, several control strategies are possible [1]. However, due to a lack of space, only six typical control strategies are considered in this chapter. These are detailed in Table 2.3.

A. Control Strategy 1

In this control strategy, the DC voltage and the firing angle are specified for the rectifier. On the other hand, the active powers for both the inverters are specified along with their extinction angles.

TABLE 2.3

Some Control Strategies for a Three-Terminal DC Link

Control Strategies	Specified Quantities	Unknown Quantities
1	$P_{DCI1}, P_{DCI2}, V_{DCR}$ $\alpha_R, \gamma_{I1}, \gamma_{I2}$	$V_{DCI1}, V_{DCI2}, P_{DCR}, I_{DCI1}, I_{DCI2}, I_{DCR}, a_R, a_{I1}, a_{I2}, \cos\phi_R, \cos\phi_{I1}, \cos\phi_{I2}$
2	$P_{DCI1}, I_{DCI2}, V_{DCR}$ $\alpha_R, \gamma_{I1}, \gamma_{I2}$	$V_{DCI1}, V_{DCI2}, P_{DCI2}, P_{DCR}, I_{DCI1}, I_{DCR}, a_R, a_{I1}, a_{I2}, \cos\phi_R, \cos\phi_{I1}, \cos\phi_{I2}$
3	$P_{DCI2}, I_{DCI1}, V_{DCR}$ $\alpha_R, \gamma_{I1}, \gamma_{I2}$	$V_{DCI1}, V_{DCI2}, P_{DCI1}, P_{DCR}, I_{DCI2}, I_{DCR}, a_R, a_{I1}, a_{I2}, \cos\phi_R, \cos\phi_{I1}, \cos\phi_{I2}$
4	$I_{DCI1}, I_{DCI2}, V_{DCR}$ $\alpha_R, \gamma_{I1}, \gamma_{I2}$	$V_{DCI1}, V_{DCI2}, P_{DCI1}, P_{DCI2}, P_{DCR}, I_{DCR}, a_R, a_{I1}, a_{I2}, \cos\phi_R, \cos\phi_{I1}, \cos\phi_{I2}$
5	$P_{DCI1}, P_{DCI2}, V_{DCR}$ a_R, a_{I1}, a_{I2}	$V_{DCI1}, V_{DCI2}, P_{DCR}, I_{DCI1}, I_{DCI2}, I_{DCR}, \alpha_R, \gamma_{I1}, \gamma_{I2}, \cos\phi_R, \cos\phi_{I1}, \cos\phi_{I2}$
6	$I_{DCI1}, I_{DCI2}, V_{DCR}$ a_R, a_{I1}, a_{I2}	$V_{DCI1}, V_{DCI2}, P_{DCI1}, P_{DCI2}, P_{DCR}, I_{DCR}, \alpha_R, \gamma_{I1}, \gamma_{I2}, \cos\phi_R, \cos\phi_{I1}, \cos\phi_{I2}$

B. Control Strategy 2

In this control strategy, the DC voltage and the firing angle are specified for the rectifier. For inverter 1, the active power is specified while the DC current is specified for inverter 2. In addition, the extinction angles of both the inverters are specified.

C. Control Strategy 3

In this control strategy, at the rectifier terminal, the DC voltage and the firing angle are specified. The DC current is specified at inverter 1, while the active power is specified at inverter 2. Also, the extinction angles of both the inverters are specified.

D. Control Strategy 4

In this control strategy, the DC voltage and the firing angle are specified at the rectifier end, while the DC currents and the extinction angles are specified for both the inverters.

E. Control Strategy 5

In this control strategy, the DC voltage is specified for the rectifier while the active powers are specified for both the inverters. In addition, the tap settings for all the three-converter transformers are specified.

F. Control Strategy 6

In this control strategy, the DC voltage is specified for the rectifier while the DC currents are specified for both the inverters. The tap settings for all the three-converter transformers are also specified.

2.4 POWER-FLOW EQUATIONS OF INTEGRATED AC-MLDC SYSTEMS

Let us consider the integrated AC-DC system shown in Figure 2.1. From Figure 2.1, it can be observed that for any AC bus 'i', which is not connected to any DC link, the mismatches in the active and reactive power injections are given, respectively, by

$$\Delta P_i = P_i^{sp} - \sum_{p=1}^{n} V_i V_p Y_{ip} \cos\left(\theta_i - \theta_p - \phi_{ip}\right) \qquad (2.13)$$

$$\Delta Q_i = Q_i^{sp} - \sum_{p=1}^{n} V_i V_p Y_{ip} \sin\left(\theta_i - \theta_p - \phi_p\right) \tag{2.14}$$

Now, in the DC network shown in Figure 2.1, the three converters representing one rectifier and two inverters are connected to the AC system at buses 'i', 'j' and 'k', respectively, through their respective converter transformers. The DC network contains two HVDC links. The first link is connected in the branch 'i' and 'j' between any two AC buses 'i' and 'j' while the second one is connected in the branch 'i–k' between AC buses 'i' and 'k'. For solving the AC power flow, the effects of the DC links are included in the power-flow equations by injecting an equivalent amount of real and reactive power at the terminal AC buses connected to the converters. This results in appropriate modifications of the mismatch equations at the converter terminal AC buses. At the rectifier bus 'i', the effect of the DC link is incorporated in the AC power flow as equivalent active and reactive power injections 'P_{DCR}' and 'Q_{DCR}'. Similarly, at the inverter buses 'j' and 'k', the active and reactive power injections representing the effect of the DC network are 'P_{DCI1}', 'Q_{DCI1}', 'P_{DCI2}' and 'Q_{DCI2}', respectively.

Thus, for the AC buses 'i', 'j' and 'k', the mismatches in the active and reactive power injections can be written as

$$\Delta P_i = P_i^{sp} - \sum_{p=1}^{n} V_i V_p Y_{ip} \cos\left(\theta_i - \theta_p - \phi_p\right) - P_{DCR} \tag{2.15}$$

$$\Delta Q_i = Q_i^{sp} - \sum_{p=1}^{n} V_i V_p Y_{ip} \sin\left(\theta_i - \theta_p - \phi_p\right) - Q_{DCR} \tag{2.16}$$

$$\Delta P_j = P_j^{sp} - \sum_{p=1}^{n} V_j V_p Y_{jp} \cos\left(\theta_j - \theta_p - \phi_{jp}\right) + P_{DCI1} \tag{2.17}$$

$$\Delta Q_j = Q_j^{sp} - \sum_{p=1}^{n} V_j V_p Y_{jp} \sin\left(\theta_j - \theta_p - \phi_{jp}\right) - Q_{DCI1} \tag{2.18}$$

$$\Delta P_k = P_k^{sp} - \sum_{p=1}^{n} V_k V_p Y_{kp} \cos\left(\theta_k - \theta_p - \phi_{kp}\right) + P_{DCI2} \tag{2.19}$$

$$\Delta Q_k = Q_k^{sp} - \sum_{p=1}^{n} V_k V_p Y_{kp} \sin\left(\theta_k - \theta_p - \phi_{kp}\right) - Q_{DCI2} \tag{2.20}$$

where
$P_{DCR} = V_{DCR} I_{DCR}$, $Q_{DCR} = P_{DCR} \tan\phi_R$, $P_{DCI1} = V_{DCI1} I_{DCI1}$, $Q_{DCI1} = P_{DCI1} \tan\phi_{I1}$, $P_{DCI2} = V_{DCI2} I_{DCI2}$, $Q_{DCI2} = P_{DCI2} \tan\phi_{I2}$.

In the above equations, the equivalent active power injections 'P_{DCR}', 'P_{DCI1}' and 'P_{DCI2}' are usually specified or can be very easily computed by manipulation of the specified variables. However, for the equivalent reactive power injections Q_{DCR}, Q_{DCI1} and Q_{DCI2}, the case is different, depending on the control strategy adopted for the DC links. For control strategies 1, 2, and 3 and 4, ϕ_R, ϕ_{I1} and ϕ_{I2} (and hence Q_{DCR}, Q_{DCI1} and Q_{DCI2}) can be computed by manipulation of the specified variables. However, for control strategies 5 and 6, ϕ_R, ϕ_{I1} and ϕ_{I2} (and hence Q_{DCR}, Q_{DCI1} and Q_{DCI2}) are dependent on both the specified variables as well as the AC state variables.

It is important to note the conventions of the signs of the equivalent real and reactive power injections representing the DC link. It is assumed that the rectifier consumes both real and reactive powers from the AC grid, while the inverters supply real power and consume reactive power [6].

2.5 IMPLEMENTATION OF POWER-FLOW IN INTEGRATED AC-MLDC SYSTEMS

a. Unified Method

If the number of voltage controlled buses is (g–1), the unified AC-DC power-flow problem for a 'n' bus AC power system incorporating a three-terminal HVDC network employing control strategy 1 (Table 2.3) can be formulated as

$$\text{Solve: } \theta = \left[\theta_2 \cdots \theta_n\right]^T, \mathbf{V} = \left[V_{g+1} \cdots V_n\right]^T, \mathbf{X} = \left[V_{DCI1}\ V_{DCI2}\ a_R\ a_{I1}\ a_{I2}\ \phi_R\ \phi_{I1}\ \phi_{I2}\right]^T$$

$$\text{Specified: } \mathbf{P} = \left[P_2 \cdots P_n\right]^T, \mathbf{Q} = \left[Q_{g+1} \cdots Q_n\right]^T, \mathbf{f} = \left[f_{11} \cdots f_{18}\right]^T$$

where the individual functions 'f_{1q}' (q = 1, 2 ...8) comprising '$\mathbf{f}$' are derived from the basic converter equations, the DC network equations and the control specifications (corresponding to control strategy 1) and are detailed in Table 2.4.

For the above formulation, it has been assumed that the 'g' generators are connected at the first 'g' buses of the system with bus 1 being the slack bus. Thus the Newton power-flow equation can be written as

$$
\begin{bmatrix}
\dfrac{\partial \mathbf{P}}{\partial \theta} & \dfrac{\partial \mathbf{P}}{\partial \mathbf{V}} & \dfrac{\partial \mathbf{P}}{\partial \mathbf{X}} \\[2ex]
\dfrac{\partial \mathbf{Q}}{\partial \theta} & \dfrac{\partial \mathbf{Q}}{\partial \mathbf{V}} & \dfrac{\partial \mathbf{Q}}{\partial \mathbf{X}} \\[2ex]
\dfrac{\partial \mathbf{f}}{\partial \theta} & \dfrac{\partial \mathbf{f}}{\partial \mathbf{V}} & \dfrac{\partial \mathbf{f}}{\partial \mathbf{X}}
\end{bmatrix}
\begin{bmatrix}
\Delta\theta \\[1ex] \Delta\mathbf{V} \\[1ex] \Delta\mathbf{X}
\end{bmatrix}
=
\begin{bmatrix}
\Delta\mathbf{P} \\[1ex] \Delta\mathbf{Q} \\[1ex] \Delta\mathbf{f}
\end{bmatrix}
\qquad (2.21)
$$

The different Jacobian sub-matrices can be identified easily from Eq. (2.21). The details are given in Appendix A.

TABLE 2.4
Individual Functions Comprising 'f' for Control Strategy 1

Per-Unit System 1	Per-Unit System 2	Functions Related to Converter and Control
$V_{DCR} - a_R V_i \cos\alpha_R + X_c \left(\dfrac{2V_{DCR} - V_{DCI1} - V_{DCI2}}{R_{DC}} \right) = 0$	$V_{DCR} - \dfrac{3\sqrt{2}}{\pi} a_R n_b V_i \cos\alpha_R + \dfrac{3X_c}{\pi} n_b \left(\dfrac{2V_{DCR} - V_{DCI1} - V_{DCI2}}{R_{DC}} \right) = 0$	f_{11}
$V_{DCI1} - a_{I1} V_j \cos\gamma_{I1} + X_c \left(\dfrac{V_{DCR} - V_{DCI1}}{R_{DC}} \right) = 0$	$V_{DCI1} - \dfrac{3\sqrt{2}}{\pi} a_{I1} n_b V_j \cos\gamma_{I1} + \dfrac{3X_c}{\pi} n_b \left(\dfrac{V_{DCR} - V_{DCI1}}{R_d} \right) = 0$	f_{12}
$V_{DCI2} - a_{I2} V_k \cos\gamma_{I2} + X_c \left(\dfrac{V_{DCR} - V_{DCI2}}{R_{DC}} \right) = 0$	$V_{DCI2} - \dfrac{3\sqrt{2}}{\pi} a_{I2} n_b V_k \cos\gamma_{I2} + \dfrac{3X_c}{\pi} n_b \left(\dfrac{V_{DCR} - V_{DCI2}}{R_{DC}} \right) = 0$	f_{13}
$V_{DCR} - a_R V_i \cos\phi_R = 0$	$V_{DCR} - \dfrac{3\sqrt{2}}{\pi} a_R n_b V_i \cos\phi_R = 0$	f_{14}
$V_{DCI1} - a_{I1} V_j \cos\phi_{I1} = 0$	$V_{DCI1} - \dfrac{3\sqrt{2}}{\pi} a_{I1} n_b V_j \cos\phi_{I1} = 0$	f_{15}
$V_{DCI2} - a_{I2} V_k \cos\phi_{I2} = 0$	$V_{DCI2} - \dfrac{3\sqrt{2}}{\pi} a_{I2} n_b V_k \cos\phi_{I2} = 0$	f_{16}
$P_{DCI1} R_{DC} - V_{DCI1} V_{DCR} + V_{DCI1}^2 = 0$		f_{17}
$P_{DCI2} R_{DC} - V_{DCI2} V_{DCR} + V_{DCI2}^2 = 0$		f_{18}

In a similar manner, the Newton-Raphson power-flow formulations can be developed very easily for other control strategies.

b. Sequential Method

In this method, the AC and DC variables are calculated separately. First the DC network equations are solved to compute the DC voltages and/or currents. This is followed by the computation of the other DC variables (converter power factors, converter control angles or converter transformer tap ratios) from the basic converter equations. Subsequently, the equivalent active ('P_{DCR}', 'P_{DCI1}' and 'P_{DCI2}') and reactive (Q_{DCR}, Q_{DCI1} and Q_{DCI2}) power injections are computed for solving the AC power-flow equations. It is important to note that the computation of the reactive power injections is dependent on the control strategy employed for the DC link. For control strategies 5 and 6, the computation of the reactive power injections is dependent on the AC power-flow iterative process and is updated in every iteration, unlike control strategies 1, 2, 3 and 4. The steps involved in the computation of the active and reactive power injections for only two typical control strategies 1 and 5 are detailed in Table 2.5. In control strategy 5, the reactive power injections are dependent on the AC power-flow iterative process, while in control strategy 1, they are independent of it. Although the steps involved in the computation of the power injections pertaining to the rest of the control strategies are not shown, they can be computed in ways similar to control strategies 1 and 5.

If the number of voltage controlled buses is (g–1), the sequential AC-DC power-flow problem for an 'n' bus AC power system incorporating a three-terminal HVDC network can be formulated as

$$\text{Solve}: \ \theta = \begin{bmatrix} \theta_2 \cdots \theta_n \end{bmatrix}^T, \ \ V = \begin{bmatrix} V_{g+1} \cdots V_n \end{bmatrix}^T \tag{2.22}$$

$$\text{Specified}: \ P = \begin{bmatrix} P_2 \cdots P_n \end{bmatrix}^T, \ Q = \begin{bmatrix} Q_{g+1} \cdots Q_n \end{bmatrix}^T \tag{2.23}$$

The Newton Power-Flow equation would be represented as

$$\begin{bmatrix} \dfrac{\partial P}{\partial \theta} & \dfrac{\partial P}{\partial V} \\[2mm] \dfrac{\partial Q}{\partial \theta} & \dfrac{\partial Q}{\partial V} \end{bmatrix} \begin{bmatrix} \Delta \theta \\[1mm] \Delta V \end{bmatrix} = \begin{bmatrix} \Delta P \\[1mm] \Delta Q \end{bmatrix} \tag{2.24}$$

Flow charts for unified and sequential methods corresponding to control strategy 1 are shown in Figures 2.2 and 2.3, respectively.

TABLE 2.5

Steps to Compute Active and Reactive Power Injections in Control Strategies 1 and 5

Control Strategy 1		Control Strategy 5	
Specified Quantities	**Unknown Quantities**	**Specified Quantities**	**Unknown Quantities**
$P_{DC11}, P_{DC12}, V_{DCR}, \alpha_R, \gamma_{11}, \gamma_{12}$	$V_{DC11}, V_{DC12}, P_{DCR}, I_{DC11}, I_{DC12}, I_{DCR}, a_R, a_{I1}, a_{I2}, \cos\phi_R, \cos\phi_1, \cos\phi_2$	$P_{DC11}, P_{DC12}, V_{DCR}, a_R, a_{I1}, a_{I2}$	$V_{DC11}, V_{DC12}, P_{DCR}, I_{DC11}, I_{DC12}, I_{DCR}, a_R, \gamma_{11}, \gamma_{12}, \cos\phi_R, \cos\phi_1, \cos\phi_2$

Step 1: compute V_{DC11} and V_{DC12} using DC load flow

Step 2: compute $I_{DC11} = \dfrac{V_{DCR} - V_{DC11}}{R_{DC}}$

Step 3: compute $I_{DC12} = \dfrac{V_{DCR} - V_{DC12}}{R_{DC}}$

Step 4: compute $I_{DCR} = I_{DC11} + I_{DC12}$

Step 5: compute $P_{DCR} = V_{DCR} I_{DCR}$

Step 6: Compute $\cos\phi_R = \dfrac{V_{DCR}\cos\alpha_R}{V_{DCR} + X_c I_{DCR}}$

Step 7: Compute $Q_{DCR} = P_{DCR}\tan\phi_R$

Step 8: Compute $\cos\phi_1 = \dfrac{V_{DC11}\cos\gamma_{11}}{V_{DC11} + X_c I_{DC11}}$

Step 1: compute V_{DC11} and V_{DC12} using DC load flow

Step 2: compute $I_{DC11} = \dfrac{V_{DCR} - V_{DC11}}{R_{DC}}$

Step 3: compute $I_{DC12} = \dfrac{V_{DCR} - V_{DC12}}{R_{DC}}$

Step 4: compute $I_{DCR} = I_{DC11} + I_{DC12}$

Step 5: compute $P_{DCR} = V_{DCR} I_{DCR}$

Step 6: Compute $\cos\phi_R = \dfrac{V_{DCR}}{a_R V_i}$

Note 1: V_i is an AC power-flow variable and is updated every iteration. Hence, $\cos\phi_R$ changes in every iteration.

Step 7: Compute $Q_{DCR} = P_{DCR}\tan\phi_R$

Step 8: Compute $\cos\phi_1 = \dfrac{V_{DC11}}{a_{I1} V_j}$

(Continued)

TABLE 2.5 (Continued)
Steps to Compute Active and Reactive Power Injections in Control Strategies 1 and 5

Control Strategy 1		Control Strategy 5	
Specified Quantities	Unknown Quantities	Specified Quantities	Unknown Quantities
Step 9: Compute $\cos\phi_2 = \dfrac{V_{DCI2}\cos\gamma_{12}}{V_{DCI2} + X_c I_{DCI2}}$		Step 9: Compute $\cos\phi_2 = \dfrac{V_{DCI2}}{a_{12} V_k}$	
		Note 2: V_j and V_k are also AC power-flow variables and are updated every iteration, along with $\cos\phi_1$ and $\cos\phi_2$	
Step 10: Compute $Q_{DCI1} = P_{DCI1}\tan\phi_1$		Step 10: Compute $Q_{DCI1} = P_{DCI1}\tan\phi_1$	
Step 11: Compute $Q_{DCI2} = P_{DCI2}\tan\phi_2$		Step 11: Compute $Q_{DCI2} = P_{DCI2}\tan\phi_2$	
Note: P_{DCI1} are P_{DCI2} specified. P_{DCR}, Q_{DCR}, Q_{DCI1} and Q_{DCI2} can be computed prior to the AC power flow and hence, are independent of the iterative loop.		Note 3: P_{DCI1} and P_{DCI2} are specified. P_{DCR} can be computed prior to the AC power flow. However, Q_{DCR}, Q_{DCI1} and Q_{DCI2} depend upon ϕ_k, ϕ_1 and ϕ_2, respectively, and need to be updated every iteration.	

```
┌─────────────────────────────────────────┐
│ Input values of V_DCR, P_DCI1, P_DCI2, cosα_R, │
│        cos γ_I1, cos γ_I2, P^sp, Q^sp and f      │
└─────────────────────────────────────────┘
                    │
                    ▼
┌─────────────────────────────────────────┐
│   Initial estimates of unknown variables  │
│ θ, V, V_DCI1, V_DCI2, a_R, a_I1, a_I2, φ_R, Φ_I1 and φ_I2 │
└─────────────────────────────────────────┘
                    │
                    ▼
┌─────────────────────────────────────────┐
│  Calculate the parametric error for AC-DC │
│                 network                   │
│          Δ E = [ ΔP, ΔQ , Δf]             │
└─────────────────────────────────────────┘
                    │
              Max | Δ E | < = ε ──── Yes ───→ Output results
                    │                              │
                   No                              ▼
                    ▼                             End
┌─────────────────────────────────────────┐
│ Solve for θ, V, V_DCI1, V_DCI2, a_R, a_I1, a_I2, φ_R, φ_I1 │
│                 and φ_I2                   │
│            Δ X= J^-1 [Δ E]                 │
└─────────────────────────────────────────┘
                    │
                    ▼
┌─────────────────────────────────────────┐
│           Update the value of             │
│          X^(I+1) = X^(I) + Δ X            │
└─────────────────────────────────────────┘
```

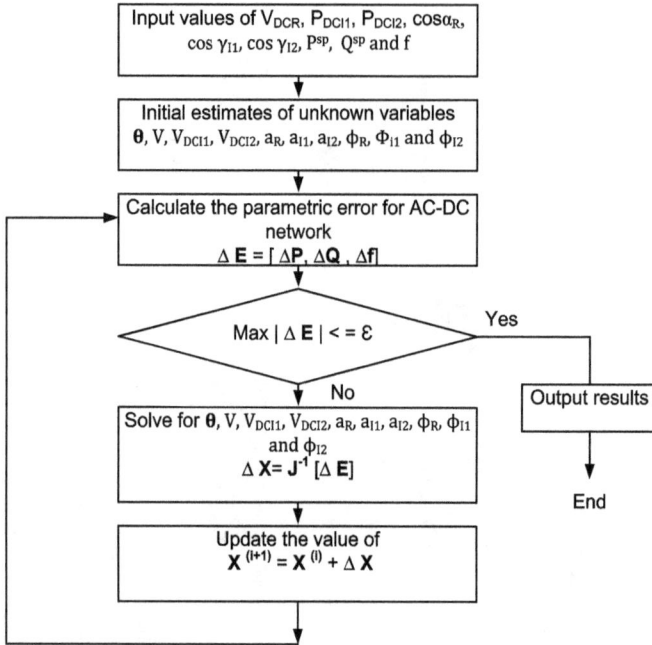

FIGURE 2.2 Flow chart of unified method corresponding to control strategy 1.

2.6   CASE STUDIES AND RESULTS

Numerous case studies were carried out in a three-terminal DC network incorporated in the IEEE 300-bus test system [115] and 12-terminal DC network incorporated in European 1354-bus test system [116]. All the converters are connected to their respective AC buses by converter transformers. Six typical control strategies are considered for the three-terminal DC network. In addition, two typical control strategies are considered for 12-terminal DC network. The effect of the different control strategies on the AC-DC power-flow convergence was studied. In addition, the base values chosen for the various DC quantities can be defined in several ways, giving rise to multiple per-unit HVDC system models. In this context, two different per-unit system models are considered in this chapter. It is observed that the adoption of different per-unit system models (depending on the selection of the base values chosen for the various DC quantities) affects the AC-DC power-flow convergence differently. For all the case studies, the commutating reactance and the DC link resistance were chosen as 0.1 and 0.01 p.u., respectively. The number of bridges 'n_b' for all the converters was taken to be equal to 2. The initial values of variables corresponding to LCC-based HVDC system were shown in Appendix A. A convergence tolerance of 10^{-8} p.u. is uniformly adopted for all the case studies. A three-terminal DC system is interfaced

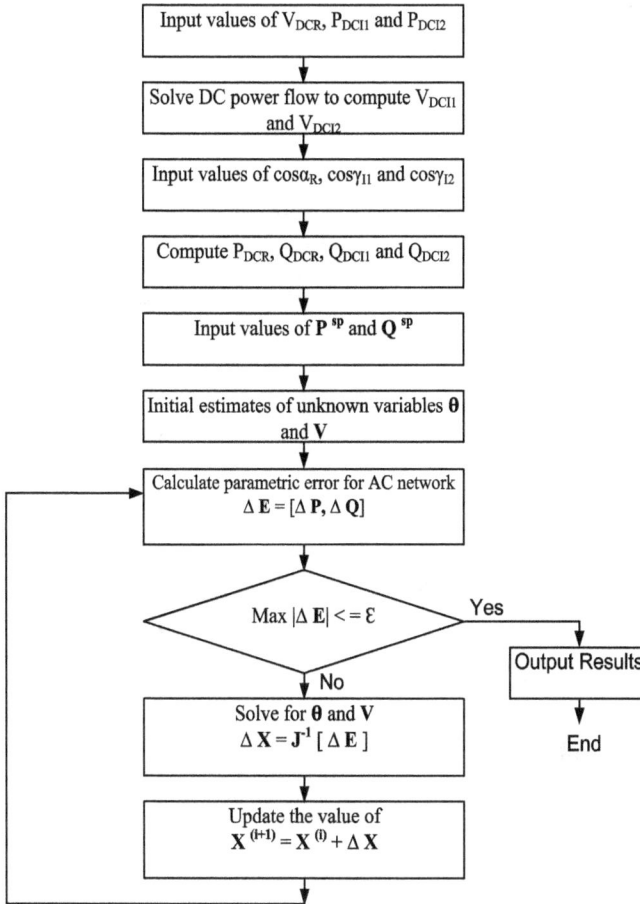

FIGURE 2.3 Flow chart of sequential method corresponding to control strategy 1.

with the IEEE 300-bus system and a 12-terminal DC system is interfaced with the 1354 bus European test system. In each of the case studies, 'NI' denotes the number of iterations for the algorithm to converge to the specified tolerance. All the case studies were implemented in MATLAB. Although a large number of case studies were conducted to validate the proposed model, a few sets of representative results are presented in this chapter.

2.6.1 STUDIES WITH UNIFIED POWER-FLOW MODEL OF IEEE 300-BUS TEST SYSTEM INTEGRATED WITH 3-TERMINAL LCC-HVDC GRID

Case I: Control strategy 1
In this case study, two HVDC links are considered. The first HVDC link is incorporated between AC buses '25' and '26' and the second one between buses '25'

and '232'. The converter connected to bus no. 25 is made to operate as a rectifier. On the other hand, both the converters connected to buses 26 and 232 are operated as inverters. As detailed in columns 1–5 of Table 2.6, the active power flows on the DC links 25–26 and 25–232 are set to values of 0.15 and 0.1 p.u., respectively. The DC voltages on the rectifier side for models 1 and 2 are set to values of 1 and 2.3 p.u., respectively. It may be noted that these values are different on account of the different constants being associated with the two models in p.u. (Table 2.2). The firing angles for the rectifier and the extinction angles for both the inverters are set to 5° and 15°, respectively. The power-flow solution is shown in columns 6–11 of Table 2.6. From Table 2.6, it is observed that although the final power-flow solutions corresponding to the two models are different, the power-flow convergence patterns are similar. Both the models require the same number of iterations to converge. The convergence characteristics corresponding to the base case power-flow and the unified AC-DC power-flow corresponding to models 1 and 2 are shown in Figures 2.4, 2.5 and 2.6, respectively. From Figures 2.4 to 2.6, it is observed that for control strategy 1, the proposed unified AC-DC power-flow algorithm does not demonstrate quadratic convergence characteristics, as in the base case power flow. The bus voltage profiles for models 1 and 2 are depicted in Figures 2.7 and 2.8, respectively. From Figures 2.7 and 2.8, it is observed that the bus voltage profile hardly changes except for the AC terminal buses at which the LCC-HVDC links are incorporated.

Case II: Control strategy 2
In this case study, two HVDC links are connected between AC buses '213' and '214' and '213' and '216'. The converter connected to bus no. 213 operates as a rectifier while both the converters connected to buses 214 and 216 operate as inverters. The rectifier end DC voltage is specified along with the firing angle. While the first inverter is operated in constant power mode, the second one is operated in the constant current mode. In addition, the extinction angles for both the inverters are specified. These values are detailed in columns 1–5 of Table 2.6. The power-flow solution is detailed in columns 6–11 of Table 2.6. From Table 2.6, it is observed that model 2 exhibits better convergence characteristics. The convergence characteristics of models 1 and 2, corresponding to control strategy 2, are shown in Figures 2.9 and 2.10, respectively. The bus voltage profiles with models 1 and 2 for control strategy 2 are shown in Figures 2.11 and 2.12, respectively.

Case III: Control strategy 3
In this case study, two HVDC links are connected between AC buses '109' and '146' and '109' and '147'. The converter connected to bus no. 109 operates as a rectifier, while those connected to buses 146 and 147 operate as inverters. The rectifier end DC voltage is specified along with the firing angle. While the first inverter acts on constant current mode, the second one operates on constant power mode. In addition, the extinction angles for both the inverters are specified. These values are detailed in columns 1–5 of Table 2.6. The power-flow solution is detailed in

LCC-Based Multi-Terminal DC Grids 43

TABLE 2.6

Study with Three Different Control Strategies 1, 2 and 3 of Unified Method

HVDC Links		HVDC Link Specification / Control Strategy 1			Power-Flow Solution					
					AC Terminal Buses			HVDC Variables		
Rectifier Bus	Inverter Buses	Spec. Values	Model 1	Model 2	ACSV	Model 1	Model 2	DCSV	Model 1	Model 2
25	26	P_{DCI1} (p.u.)	0.15	0.15	V_{25}	1.0075	0.9879	V_{DCI1}	0.9985	2.2993
								V_{DCI2}	0.999	2.2996
		P_{DCI2} (p.u.)	0.1	0.1	θ_{25}	−2.5588	−2.1461	P_{DCR}	0.2503	0.2501
								I_{DCI1}	0.1502	0.0652
		V_{DCR} (p.u.)	1	2.3	V_{26}	0.9892	0.9698	I_{DCI2}	0.1001	0.0435
								I_{DCR}	0.2503	0.1087
	232	α_R (deg.)	5	5	θ_{26}	−4.8856	−4.5598	a_R	1.0213	0.8731
								a_{I1}	1.0608	0.9137
								a_{I2}	1.0129	0.8805
		γ_{I1} (deg.)	15	15	V_{232}	1.00314	1.0047	$\cos\phi_R$	0.9719	0.9873
								$\cos\phi_{I1}$	0.9516	0.9607
		γ_{I2} (deg.)	15	15	θ_{232}	−2.4029	−1.836	$\cos\phi_{I2}$	0.9563	0.9624
								NI	9	9

(Continued)

TABLE 2.6 (Continued)
Study with Three Different Control Strategies 1, 2 and 3 of Unified Method

HVDC Links		HVDC Link Specification — Control Strategy 2			Power-Flow Solution — AC Terminal Buses			Power-Flow Solution — HVDC Variables		
Rectifier Bus	Inverter Buses	Spec. Values	Model 1	Model 2	ACSV	Model 1	Model 2	DC SV	Model 1	Model 2
213	214	P_{DCI1} (p.u.)	0.3	0.3	V_{213}	1.0409	1.0410	V_{DCI1}	0.997	2.2987
								V_{DCI2}	0.9901	2.2772
								P_{DCI2}	0.099	0.2277
		I_{DCI2} (p.u.)	0.1	0.1	θ_{213}	−20.9011	−20.999	P_{DCR}	0.4009	0.5302
								I_{DCI1}	0.3009	0.1305
								I_{DCR}	0.4009	0.2305
	216	V_{DCR} (p.u.)	1	2.3	V_{214}	1.0095	1.0095	a_R	1.0031	0.829
								a_{I1}	1.0533	0.8775
								a_{I2}	0.9801	0.8298
		α_R (deg.)	5	5	θ_{214}	−20.7274	−20.7806	$\cos\phi_R$	0.9578	0.9867
		γ_{I1} (deg.)	15	15	V_{216}	1.0564	1.0564	$\cos\phi_{I1}$	0.9376	0.9607
		γ_{I2} (deg.)	15	15	θ_{216}	−20.9194	−20.9184	$\cos\phi_{I2}$	0.9563	0.9619
								NI	8	8

(Continued)

TABLE 2.6 (Continued)
Study with Three Different Control Strategies 1, 2 and 3 of Unified Method

HVDC Links		HVDC Link Specification			Power-Flow Solution					
		Control Strategy 3			AC Terminal Buses			HVDC Variables		
Rectifier Bus	Inverter Buses	Spec. Values	Model 1	Model 2	ACSV	Model 1	Model 2	DC SV	Model 1	Model 2
109	146	P_{DCI2} (p.u.)	0.1	0.1	V_{109}	1.0195	1.0195	V_{DCI1}	0.9995	2.2995
								V_{DCI2}	1	2.3
								P_{DCR}	0.05	0.115
		I_{DCI1} (p.u.)	0.05	0.05	θ_{109}	5.6921	5.7010	P_{DCI1}	0.05	0.115
								I_{DCI2}	0	0
	147	V_{DCR} (p.u.)	1	2.3	V_{146}	0.9717	0.9716	I_{DCR}	0.05	0.05
								a_R	0.9896	0.842
		α_R (deg.)	5	5	θ_{146}	-6.6786	-6.5381	a_{I1}	1.0702	0.911
								a_{I2}	1.0325	0.8792
		γ_{I1} (deg.)	15	15	V_{147}	1.0027	1.0028	$\cos\phi_R$	0.9912	0.9921
								$\cos\phi_{I1}$	0.9611	0.9619
		γ_{I2} (deg.)	15	15	θ_{147}	-4.6434	-4.5892	$\cos\phi_{I2}$	0.9659	0.9659
								NI	7	7

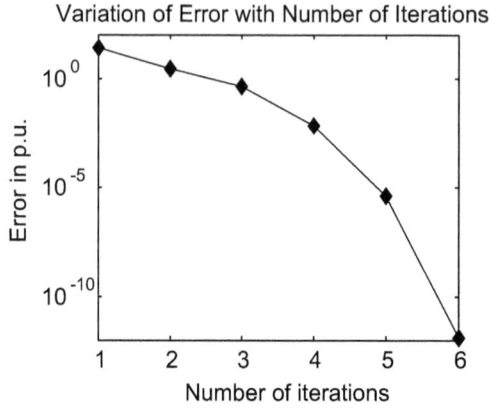

FIGURE 2.4 Convergence characteristic for the base case power flow in the IEEE 300-bus system.

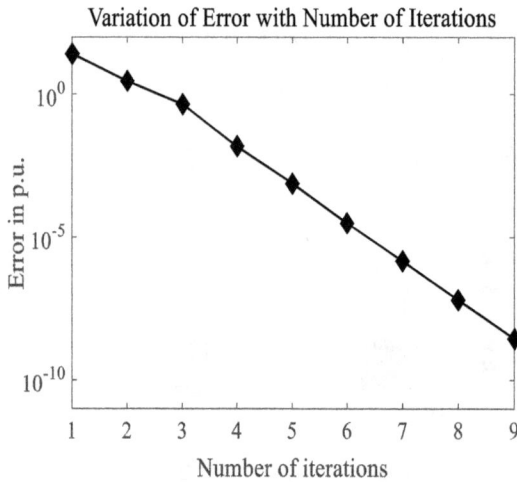

FIGURE 2.5 Convergence characteristic of Table 2.6 for model 1 employing control strategy 1.

columns 6–11 of Table 2.6. It is observed from the power-flow solution that both the models demonstrate almost similar convergence characteristics. The convergence characteristics of models 1 and 2 corresponding to control strategy 3 are shown in Figures 2.13 and 2.14, respectively. The bus voltage profiles with models 1 and 2 for control strategy 3 are shown in Figures 2.15 and 2.16, respectively.

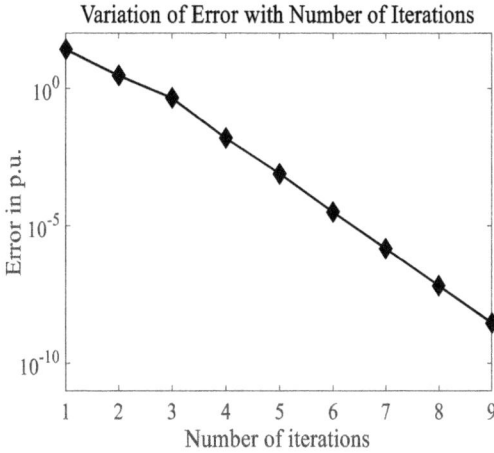

FIGURE 2.6 Convergence characteristic of Table 2.6 for model 2 employing control strategy 1.

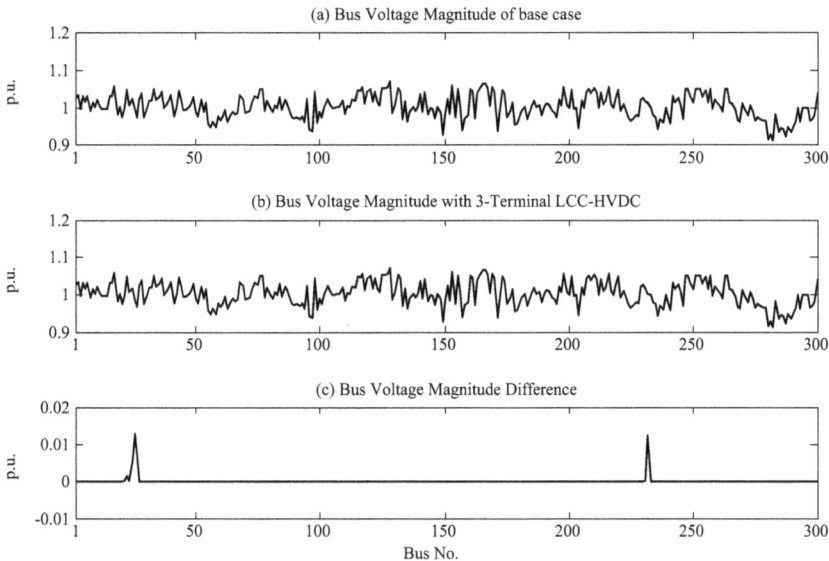

FIGURE 2.7 Bus voltage profile of Table 2.6 for model 1 employing control strategy 1.

Case IV: Control strategy 4

In this case study, the converter connected to AC bus no. 101 acts as a rectifier while those connected to AC buses 102 and 105 act as inverters. The rectifier end DC voltage is specified along with the firing angle. Both the inverters are operated at constant current and constant extinction angle. The values of the specified quantities are detailed in columns 1–5 of Table 2.7. The power-flow solution is

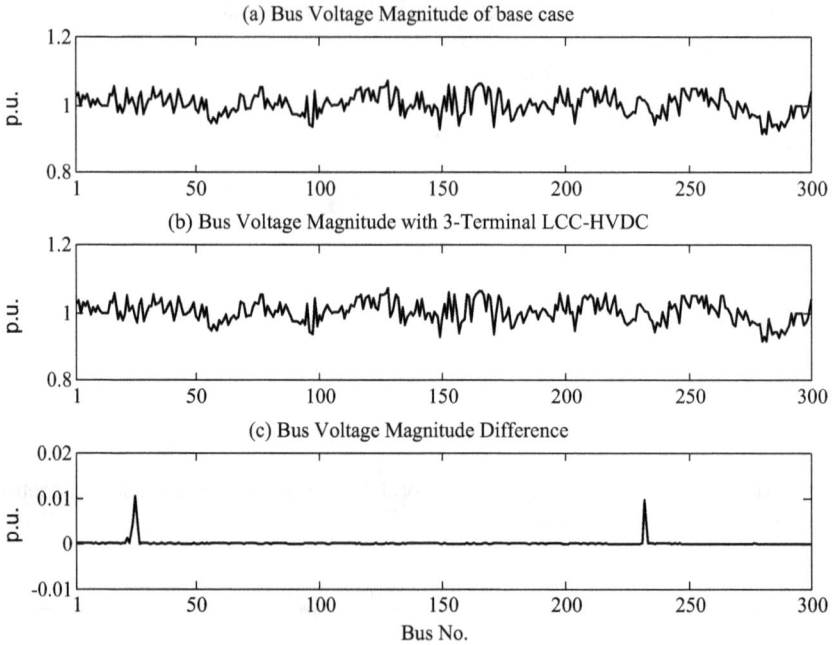

FIGURE 2.8   Bus voltage profile of Table 2.6 for model 2 employing control strategy 1.

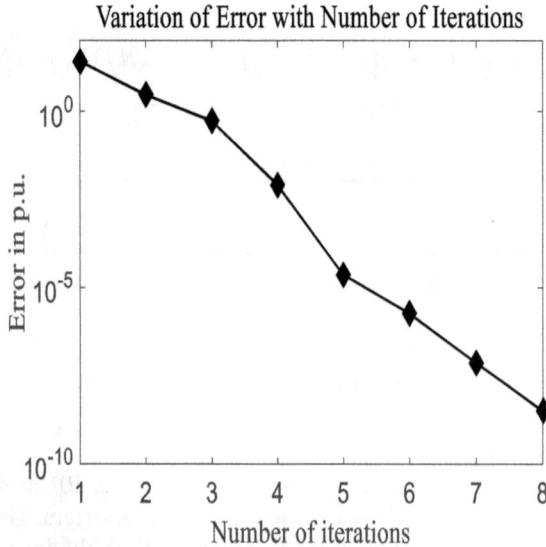

FIGURE 2.9   Convergence characteristic of Table 2.6 for model 1 employing control strategy 2.

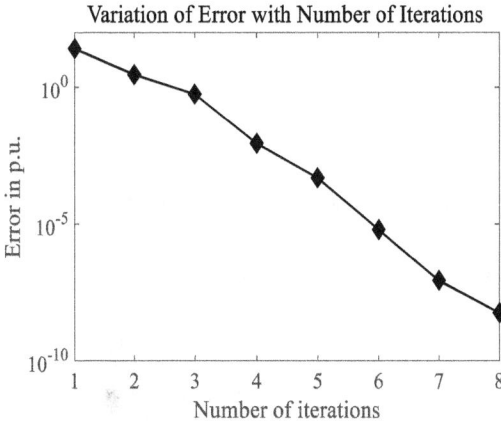

FIGURE 2.10   Convergence characteristic of Table 2.6 for model 2 employing control strategy 2.

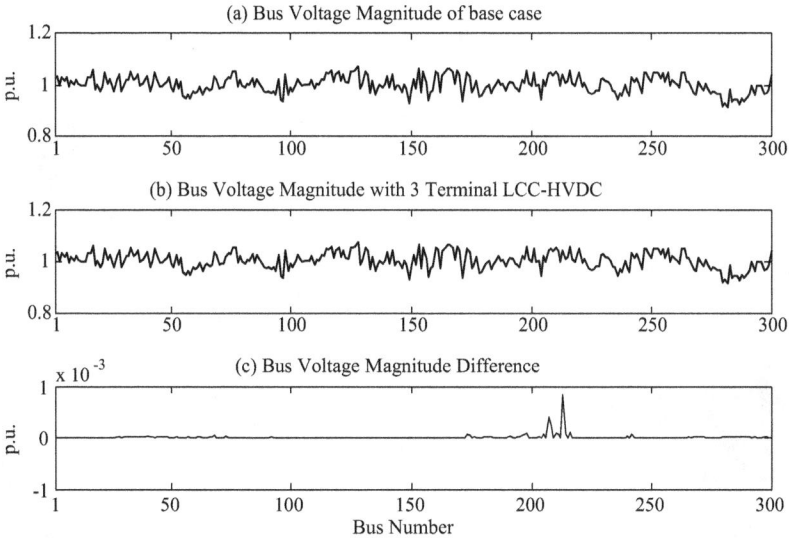

FIGURE 2.11   Bus voltage profile of Table 2.6 for model 1 employing control strategy 2.

detailed in columns 6–11 of Table 2.7. The convergence characteristics of models 1 and 2 corresponding to control strategy 4 are shown in Figures 2.17 and 2.18, respectively. It is again observed that both the models demonstrate similar convergence characteristics with model 1 taking less 'NI' to converge. The bus voltage profiles with models 1 and 2 for control strategy 4 are shown in Figures 2.19 and 2.20, respectively.

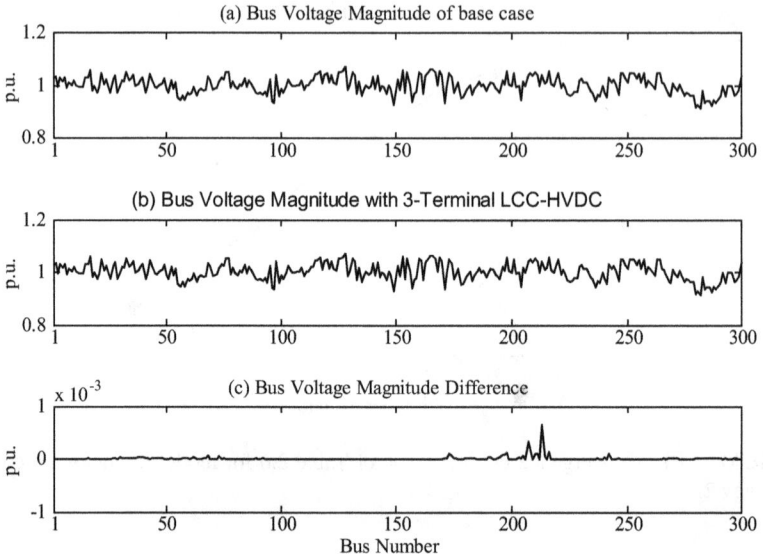

FIGURE 2.12 Bus voltage profile of Table 2.6 for model 2 employing control strategy 2.

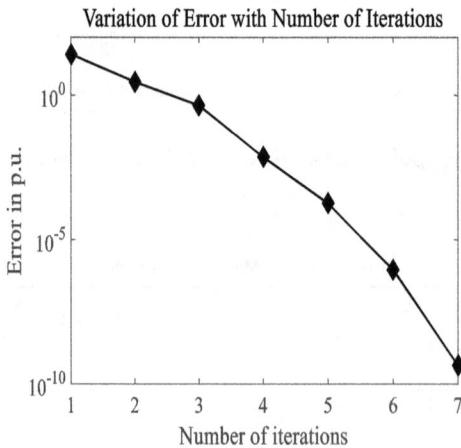

FIGURE 2.13 Convergence characteristic of Table 2.6 for model 1 employing control strategy 3.

Case V: Control strategy 5

In this case, two HVDC links are incorporated between AC buses '86–87' and '86–90'. The converter connected to bus no. 86 operates as a rectifier while both the converters connected to buses 87 and 90 operate as inverters. In this control strategy, the rectifier end DC voltage is specified and both the inverters are operated in the constant power mode. In addition, the tap ratios of all

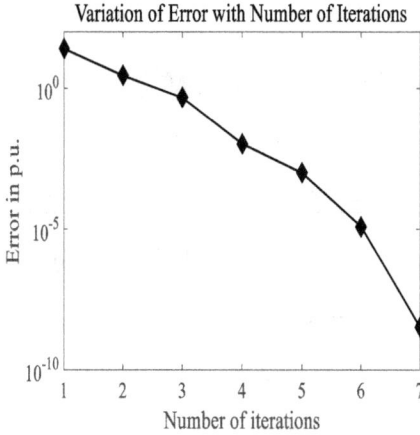

FIGURE 2.14 Convergence characteristic of Table 2.6 for model 2 employing control strategy 3.

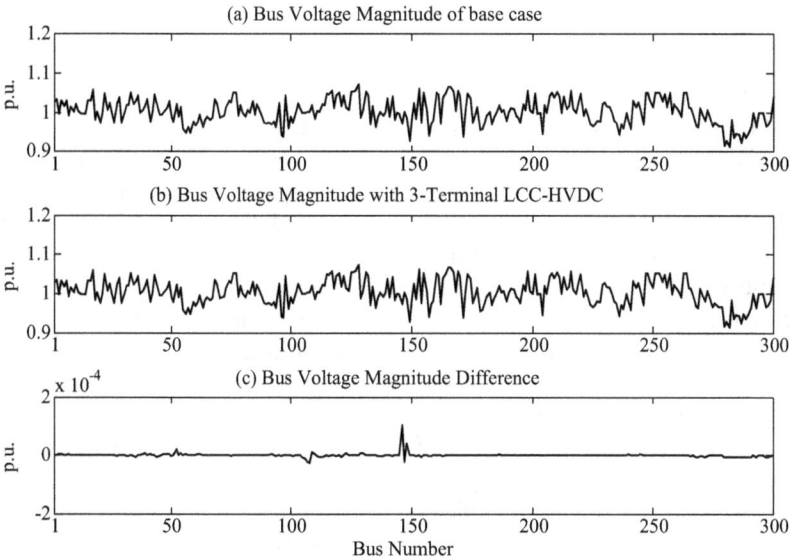

FIGURE 2.15 Bus voltage profile of Table 2.6 for model 1 employing control strategy 3.

the three-converter transformers are specified. Columns 1–5 of Table 2.7 show these specified values. The power-flow solution is shown in columns 6–11 of Table 2.7. The convergence characteristics of models 1 and 2 corresponding to control strategy 5 are shown in Figures 2.21 and 2.22, respectively. The bus voltage profiles with models 1 and 2 corresponding to control strategy 5 are shown in Figures 2.23 and 2.24, respectively.

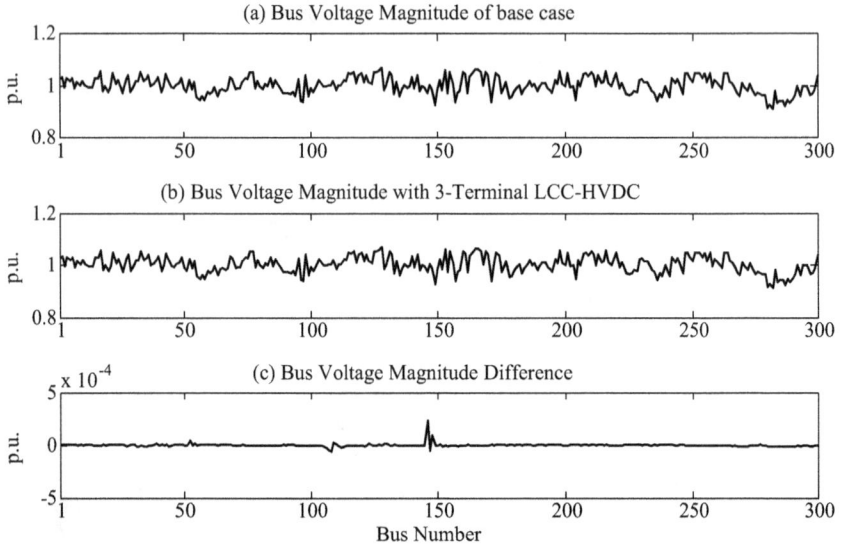

FIGURE 2.16 Bus voltage profile of Table 2.6 for model 2 employing control strategy 3.

Case VI: Control strategy 6

In this case, the two HVDC links are connected between AC buses '78–84' and '78–86'. While the converter connected to bus no. 78 operates as a rectifier, the converters connected to buses 84 and 86 operate as inverters. In this control strategy, the rectifier end DC voltage is specified. On the other hand, both the inverters are operated in the constant current mode. Also, the tap ratios of all the three-converter transformers are specified. While columns '1–5' of Table 2.7 show the values of the specified quantities, columns 6–11 detail the power-flow solution. Again, from the values of 'NI', it is observed that model 1 fares better than model 2. The convergence characteristics of models 1 and 2 corresponding to control strategy 6 are shown in Figures 2.25 and 2.26, respectively. The bus voltage profiles with models 1 and 2 corresponding to control strategy 6 are shown in Figures 2.27 and 2.28, respectively.

From the above studies of unified method, it is observed that the convergence characteristics corresponding to all the presented control strategies do not possess the quadratic convergence characteristics as in base case.

2.6.2 Studies with Unified Power-Flow Model of European 1354-Bus Test System Integrated with 12-Terminal LCC-HVDC Grid

Case I: Control strategy 1

In this case study, a 12-terminal DC grid as shown in Figure 2.29 is considered. Four rectifiers (1–4) are connected to AC buses 1281, 1286, 1292 and 1354,

TABLE 2.7

Study with Three Different Control Strategies 4, 5 and 6 of Unified Method

HVDC Link		HVDC Link Specification Control Strategy 4			Power-Flow Solution AC Terminal Buses			HVDC Variables		
Rectifier Bus	Inverter Buses	Spec. Val.	Model 1	Model 2	ACSV	Model 1	Model 2	DC SV	Model 1	Model 2
101	102	I_{DC11} (p.u.)	0.1	0.1	V_{101}	0.9745	0.9726	V_{DC11}	0.9985	2.2985
								V_{DC12}	0.999	2.299
								P_{DC11}	0.9999	0.2299
		I_{DC12} (p.u.)	0.05	0.05	θ_{101}	−14.5011	−14.7822	P_{DC12}	0.05	0.115
								P_{DCR}	0.15	0.345
								I_{DCR}	0.15	0.15
	105	V_{DCR} (p.u.)	1	2.3	V_{102}	1.0014	1.0015	a_R	1.0455	0.8898
								a_{I1}	1.0427	0.887
		α_R (deg.)	5	5	θ_{102}	−17.271	−16.8811	a_{I2}	1.0408	0.8863
								$\cos\theta_R$	0.9815	0.9839
		γ_{I1} (deg.)	15	15	V_{105}	0.9987	0.9984	$\cos\phi_{I1}$	0.9563	0.958
		γ_{I2} (deg.)	15	15	θ_{105}	−12.7184	−12.6926	$\cos\phi_{I2}$	0.9611	0.9619
								NI	7	9

(Continued)

TABLE 2.7 (Continued)
Study with Three Different Control Strategies 4, 5 and 6 of Unified Method

HVDC Link		HVDC Link Specification			Power-Flow Solution					
		Control Strategy 5			AC Terminal Buses			HVDC Variables		
Rectifier Bus	Inverter Buses	Spec. Values	Model 1	Model 2	ACSV	Model 1	Model 2	DC SV	Model 1	Model 2
86	87	P_{DCl1} (p.u.)	0.2	0.2	V_{86}	1.013	1.0154	V_{DCl1}	0.998	2.2991
								V_{DCl2}	0.998	2.2991
		P_{DCl2} (p.u.)	0.2	0.2	θ_{86}	-13.6274	-13.6533	P_{DCR}	0.4008	0.4002
								I_{DCl1}	0.2004	0.087
								I_{DCl2}	0.2004	0.087
		V_{DCR} (p.u.)	1	2.3	V_{87}	1.0025	1.0044	I_{DCR}	0.4008	0.174
								α_R	12.0888	8.405
					θ_{87}	-15.6906	-15.7093	γ_{I1}	21.0307	18.4729
	90	a_R	1.05	0.86				γ_{I2}	17.7966	21.7883
					V_{90}	0.972	0.9719	$\cos\phi_R$	0.9401	0.9752
		a_{I1}	1.09	0.9				$\cos\phi_{I1}$	0.9133	0.9417
		a_{I2}	1.1	0.95	θ_{90}	-23.6159	-23.5863	$\cos\phi_{I2}$	0.9334	0.9219
								NI	8	8

(Continued)

TABLE 2.7 (Continued)
Study with Three Different Control Strategies 4, 5 and 6 of Unified Method

HVDC Link		HVDC Link Specification Control Strategy 6			Power-Flow Solution AC Terminal Buses			HVDC Variables		
Rectifier Bus	Inverter Buses	Spec. Values	Model 1	Model 2	ACSV	Model 1	Model 2	DC SV	Model 1	Model 2
78	84	I_{DC11} (p.u.)	0.1	0.1	V_{78}	0.9769	0.9769	V_{DC11}	0.999	2.299
								V_{DC12}	0.999	2.299
								P_{DC11}	0.0999	0.2299
		I_{DC12} (p.u.)	0.1	0.1	θ_{78}	−11.43	−11.4228	P_{DC12}	0.0999	0.2299
								P_{DCR}	0.2	0.46
								I_{DCR}	0.2	0.2
	86	V_{DCR} (p.u.)	1	2.3	V_{84}	1.0262	1.0263	α_R	6.2257	10.0717
								γ_{I1}	22.1221	21.6945
								γ_{I2}	22.5174	18.9575
		a_R	1.04	0.9	θ_{84}	−10.77	−10.7698	$\cos\phi_R$	0.9746	0.9685
								$\cos\phi_{I1}$	0.9172	0.9215
		a_{I1}	1.06	0.9	V_{86}	1.0195	1.0195	$\cos\phi_{I2}$	0.9146	0.938
		a_{I2}	1.07	0.89	θ_{86}	−11.6496	−11.6497	NI	8	9

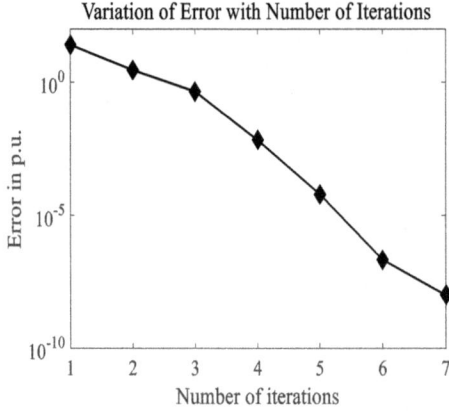

FIGURE 2.17 Convergence characteristic of Table 2.7 for model 1 employing control strategy 4.

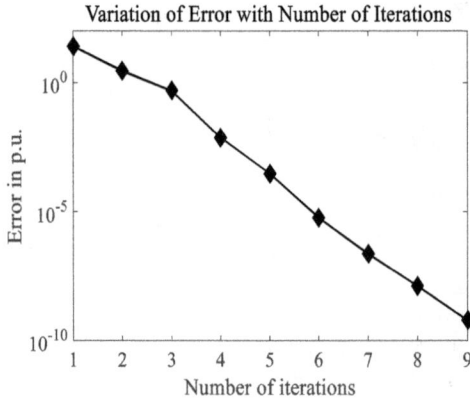

FIGURE 2.18 Convergence characteristic of Table 2.7 for model 2 employing control strategy 4.

respectively. Six inverters (1–6) are connected to AC buses 464, 792, 1009, 1313, 137 and 786, respectively. The base case power-flow results of the 1354-bus European test network are shown in the first row of Table 2.8. The specified values for both the models are given in the fifth row of Table 2.8. The DC voltage and firing angle of rectifier 1 are specified. Also, the DC powers and the firing angles are specified for the rest of the rectifiers. All the inverters are working on constant DC powers and constant extinction angles. For both the models, the firing angles (in degrees) for rectifiers 1–4 are set to values of 5, 6, 6 and 5, respectively. The extinction angles (in degrees) for inverters 1–6 are set to values of 15, 16, 17, 15, 16 and 17, respectively, in both the models. The DC voltages of

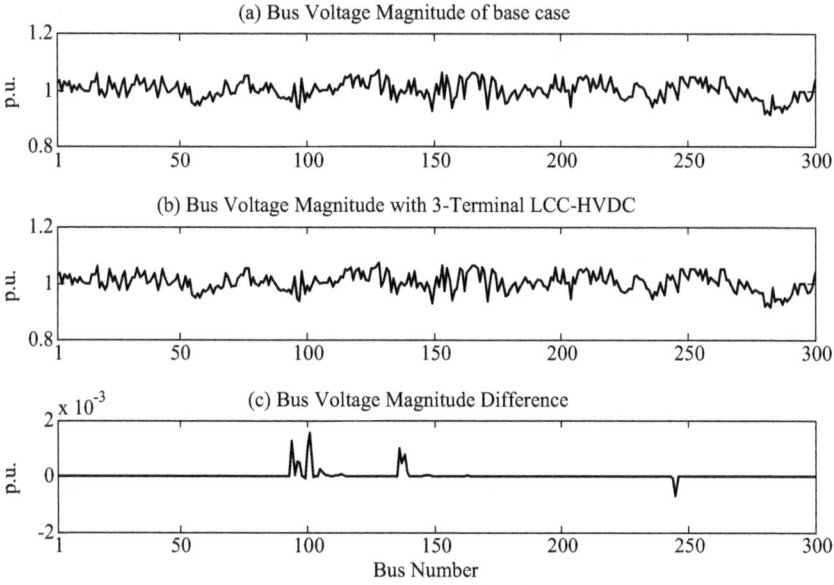

FIGURE 2.19   Bus voltage profile of Table 2.7 for model 1 employing control strategy 4.

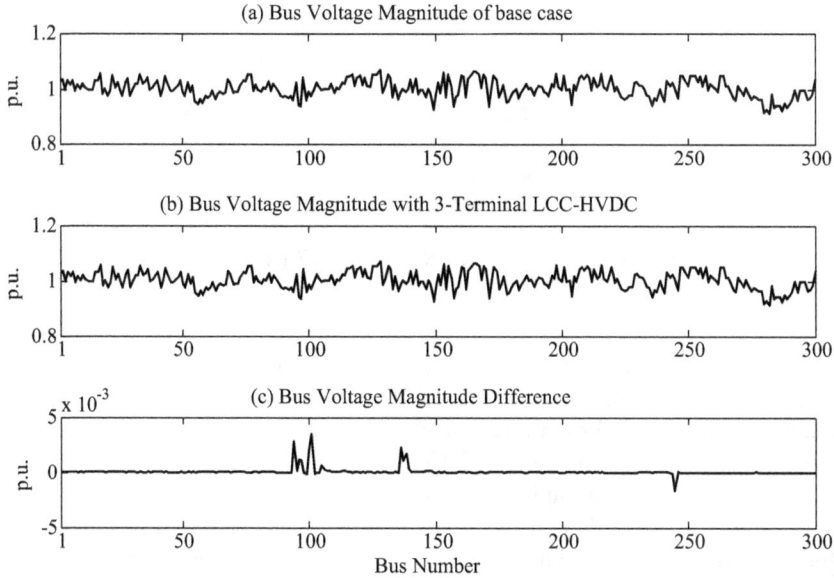

FIGURE 2.20   Bus voltage profile of Table 2.7 for model 2 employing control strategy 4.

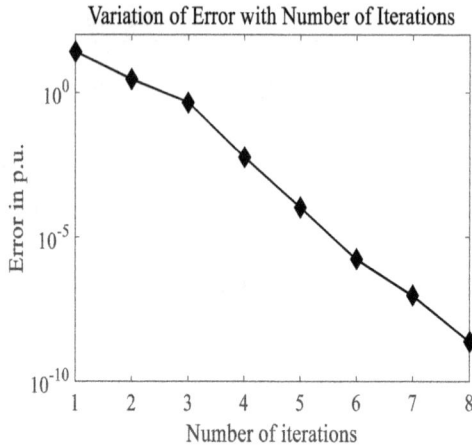

FIGURE 2.21 Convergence characteristic of Table 2.7 for model 1 employing control strategy 5.

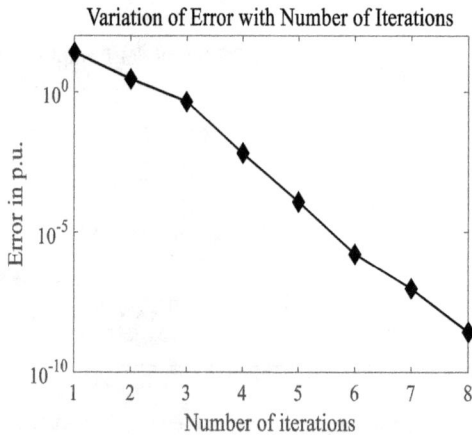

FIGURE 2.22 Convergence characteristic of Table 2.7 for model 2 employing control strategy 5.

rectifier 1 are specified to a value of 1 and 2.3 p.u. for models 1 and 2, respectively. The DC powers of rectifiers 2, 3 and 4 are specified to values of 0.4, 0.3 and 0.2 p.u., respectively, for both the models. The DC powers of inverters 1–6 have specified values of 0.2, 0.3, 0.2, 0.2, 0.3 and 0.2 p.u., respectively.

The power-flow solution is shown in rows 6–9 of Table 2.8. From Table 2.8, it is observed that although the final power-flow solutions corresponding to the two models are different, the power-flow convergence patterns are similar. Both the models require the same number of iterations (NI = 8) to converge. The convergence characteristics corresponding to the base case power-flow and

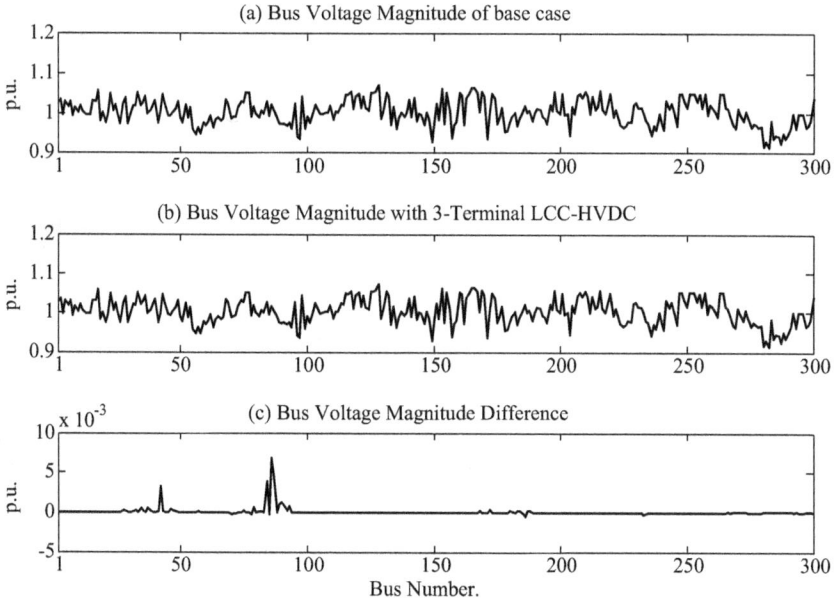

FIGURE 2.23 Bus voltage profile of Table 2.7 for model 1 employing control strategy 5.

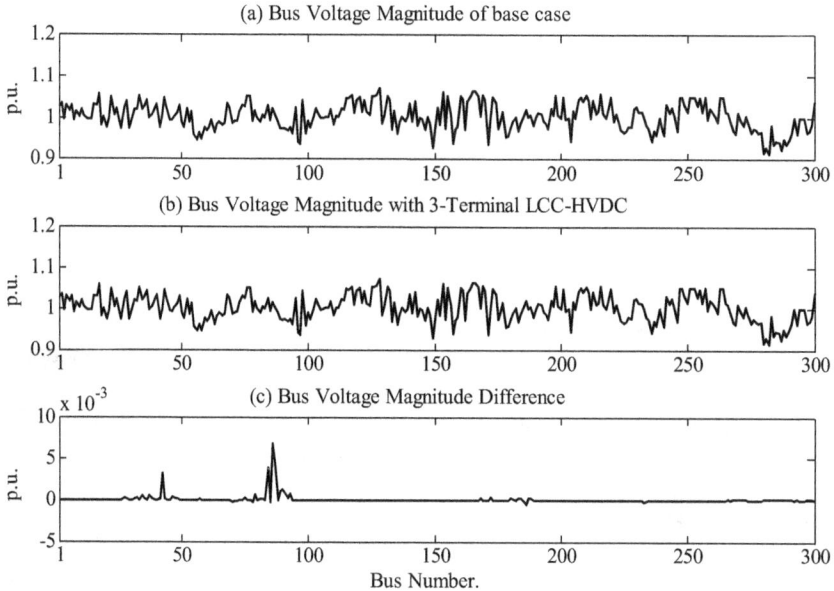

FIGURE 2.24 Bus voltage profile of Table 2.7 for model 2 employing control strategy 5.

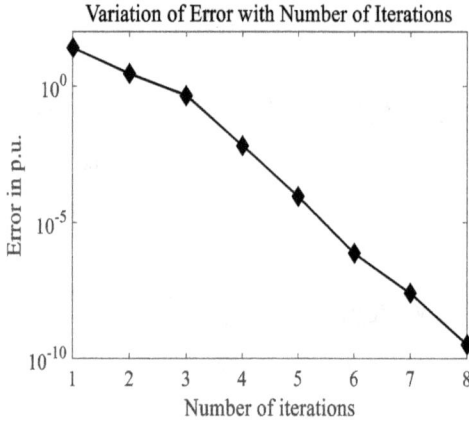

FIGURE 2.25   Convergence characteristic of Table 2.7 for model 1 employing control strategy 6.

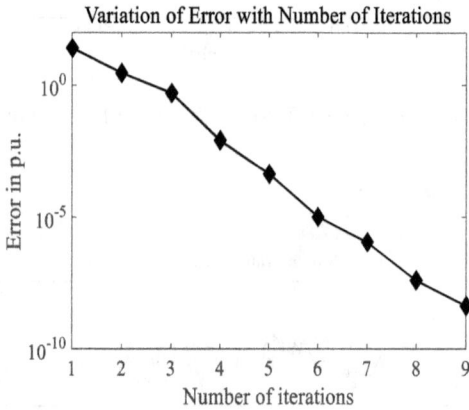

FIGURE 2.26   Convergence characteristic of Table 2.7 for model 2 employing control strategy 6.

the unified AC-DC power-flow corresponding to models 1 and 2 are shown in Figures 2.30, 2.31 and 2.32, respectively. The bus voltage profiles for models 1 and 2 are depicted in Figures 2.33 and 2.34, respectively. From Figures 2.33 and 2.34, it is observed that the bus voltage profile hardly changes except for the AC terminal buses at which the LCC-HVDC links are incorporated.

Case II: Control strategy 5
The test system used in this study is similar to previous study of case I. The specified values for both the models are given in the fifth row of Table 2.9. The DC voltage and tap setting of rectifier 1 are specified. The DC powers and tap setting are specified for the rest of the rectifiers. All inverters are working on constant

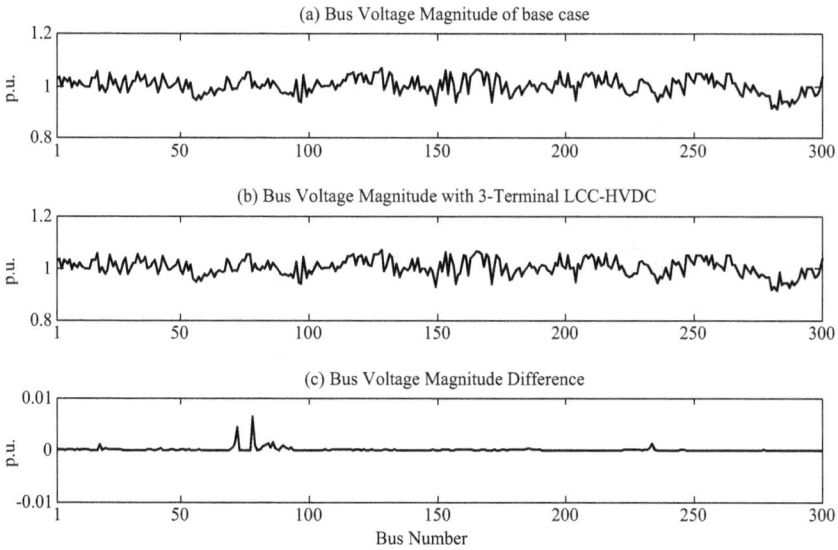

FIGURE 2.27   Bus voltage profile of Table 2.7 for model 1 employing control strategy 6.

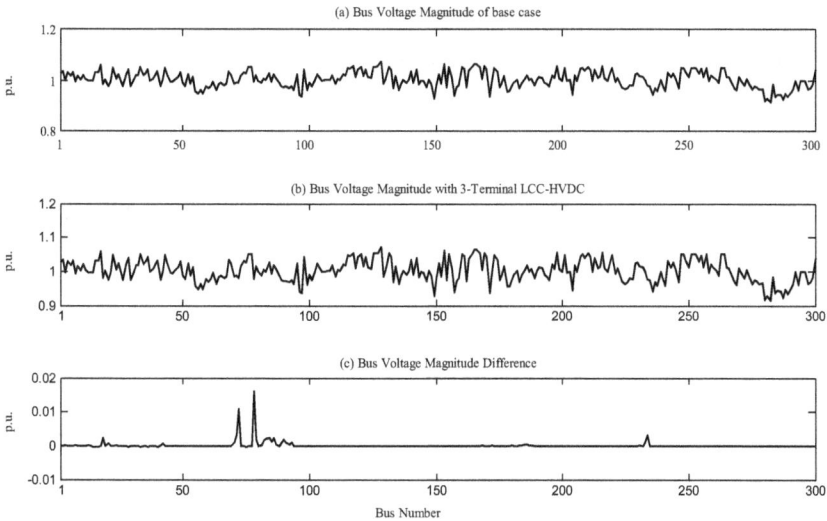

FIGURE 2.28   Bus voltage profile of Table 2.7 for model 2 employing control strategy 6.

DC power and constant tap setting. For model 1, the tap ratios of rectifiers 1–4 are set to values of 1.0219, 1.0368, 1.008 and 1.0001, respectively. The tap ratios for inverters 1–6 are set to values of 1, 1.01, 1.001, 1.0011, 1.0011 and 1.0012, respectively. In case of model 2, the tap settings for rectifiers 1–4 are set to values of 0.852, 0.858, 0.851 and 0.852, respectively. The tap ratios for inverters 1, 2, 3, 4, 5 and 6 are set to a value of 0.85, 0.86, 0.85, 0.86, 0.87 and 0.85, respectively. The DC

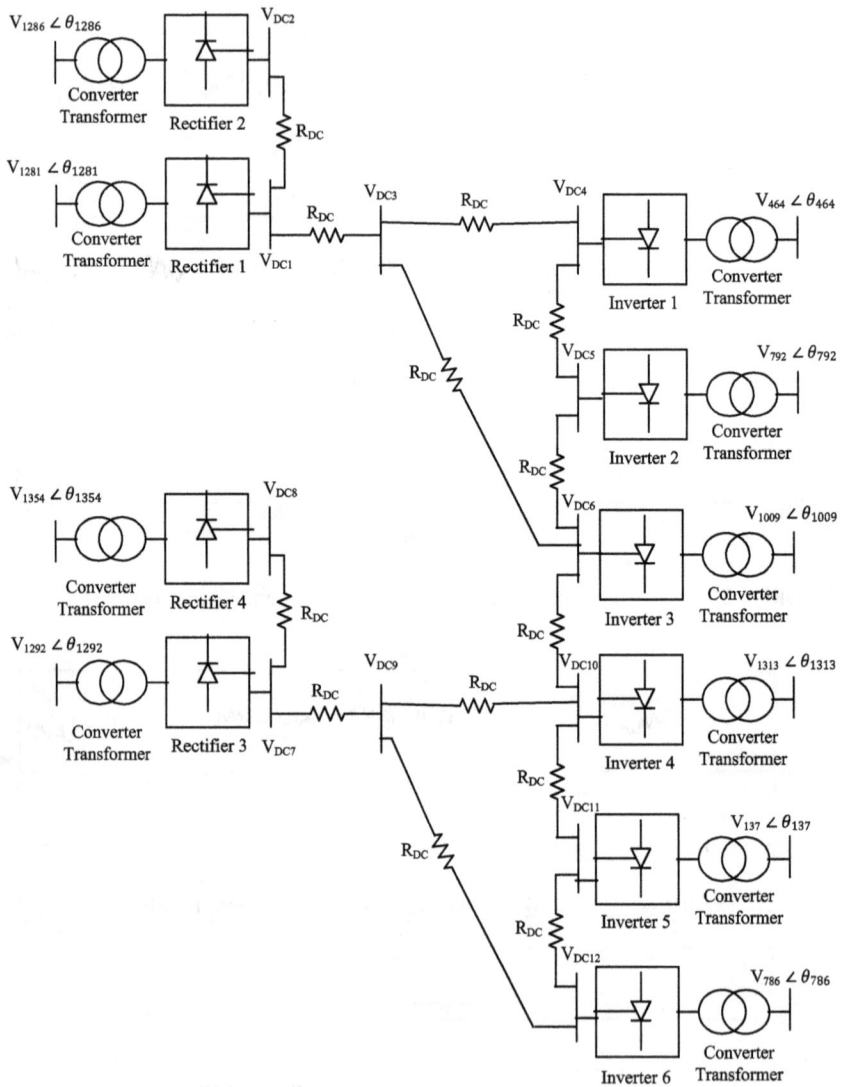

FIGURE 2.29   Twelve-terminal HVDC system incorporated in 1354 bus European test system.

voltage of rectifier 1 side has specified values of 1 and 2.3 p.u. for models 1 and 2, respectively. The DC powers of rectifiers 2–4 are specified as 0.4, 0.3 and 0.3 p.u., respectively, for both the models. For both models, the DC powers of inverters 1–6 are specified values of 0.3, 0.3, 0.2, 0.2, 0.3 and 0.2 p.u., respectively.

The power-flow solution is shown in rows 6–9 of Table 2.9. From Table 2.9, it is observed that although the final power-flow solutions corresponding to the two models are different, the power-flow convergence patterns are similar.

TABLE 2.8

Study of 12-Terminal DC System Incorporated in 1354-Bus European Test System with Control Strategy 1 of Unified Method

Base case power flow (NI = 7)

$V_{1281} = 1.0354 \angle -37.3082$; $V_{1286} = 1.0156 \angle -12.3760$; $V_{1292} = 1.0289 \angle -35.2306$; $V_{1354} = 1.0581 \angle -21.7988$; $V_{464} = 1.0641 \angle 6.4670$; $V_{792} = 1.0656 \angle -33.2984$; $V_{1009} = 1.0424 \angle -23.5338$; $V_{1313} = 1.0351 \angle -37.4755$; $V_{137} = 1.0539 \angle -30.2000$; $V_{786} = 1.0533 \angle -23.8588$;

HVDC link

Rectifier buses: 1281, 1286, 1292, 1354; Inverter buses: 464, 792, 1009, 1313, 137, 786;

HVDC Link Specification

Model 1	Model 2
Rectifiers	Rectifiers
$V_{DCR1} = 1$; $\alpha_{R1} = 5$; $P_{DCR2} = 0.4$; $\alpha_{R2} = 6$;	$V_{DCR1} = 1$; $\alpha_{R1} = 5$; $P_{DCR2} = 0.4$; $\alpha_{R2} = 6$;
$P_{DCR3} = 0.3$; $\alpha_{R3} = 6$; $P_{DCR4} = 0.2$; $\alpha_{R4} = 5$;	$P_{DCR3} = 0.3$; $\alpha_{R3} = 6$; $P_{DCR4} = 0.2$; $\alpha_{R4} = 5$;
Inverters	Inverters
$P_{DCI1} = -0.2$; $\gamma_{I1} = 15$; $P_{DCI2} = -0.3$; $\gamma_{I2} = 16$; $P_{DCI3} = -0.2$; $\gamma_{I3} = 17$;	$P_{DCI1} = -0.2$; $\gamma_{I1} = 15$; $P_{DCI2} = -0.3$; $\gamma_{I2} = 16$; $P_{DCI3} = -0.2$; $\gamma_{I3} = 17$;
$P_{DCI4} = -0.2$; $\gamma_{I4} = 15$; $P_{DCI5} = -0.3$; $\gamma_{I5} = 16$; $P_{DCI6} = -0.2$; $\gamma_{I6} = 17$;	$P_{DCI4} = -0.2$; $\gamma_{I4} = 15$; $P_{DCI5} = -0.3$; $\gamma_{I5} = 16$; $P_{DCI6} = -0.2$; $\gamma_{I6} = 17$;

(Continued)

TABLE 2.8 (Continued)
Study of 12-Terminal DC System Incorporated in 1354-Bus European Test System with Control Strategy 1 of Unified Method

Power-Flow Solution

Model 1	Model 2
AC terminal buses	AC terminal buses
$V_{1281} = 1.0340 \angle -39.7442$; $V_{1286} = 1.0124 \angle -13.0975$;	$V_{1281} = 1.0344 \angle -39.7098$; $V_{1286} = 1.0131 \angle -13.0951$;
$V_{1292} = 1.0253 \angle -37.4127$; $V_{1354} = 1.0574 \angle -23.6539$;	$V_{1292} = 1.0261 \angle -37.4013$; $V_{1354} = 1.0575 \angle -23.6403$;
$V_{464} = 1.0644 \angle 5.6337$; $V_{792} = 1.0655 \angle -34.6668$;	$V_{464} = 1.0646 \angle 5.6361$; $V_{792} = 1.0656 \angle -34.6636$;
$V_{1009} = 1.0426 \angle -25.2922$; $V_{1313} = 1.0342 \angle -39.8895$;	$V_{1009} = 1.0430 \angle -25.2822$; $V_{1313} = 1.0345 \angle -39.8554$;
$V_{137} = 1.0537 \angle -31.9810$; $V_{786} = 1.0532 \angle -25.9255$;	$V_{137} = 1.0539 \angle -31.9749$; $V_{786} = 1.0534 \angle -25.9103$;
HVDC variables	HVDC variables
Rectifiers	Rectifiers
$P_{DCR1} = 0.52$; $V_{DCR2} = 1.004$; $V_{DCR3} = 0.9907$; $V_{DCR4} = 0.9927$;	$P_{DCR1} = 0.5037$; $V_{DCR2} = 2.3017$; $V_{DCR3} = 2.2961$; $V_{DCR4} = 2.2969$;
$I_{DCR1} = 0.52$; $I_{DCR2} = 0.3984$; $I_{DCR3} = 0.3028$; $I_{DCR4} = 0.2015$;	$I_{DCR1} = 0.2190$; $I_{DCR2} = 0.1738$; $I_{DCR3} = 0.1307$; $I_{DCR4} = 0.0871$;
$a_{R1} = 1.0213$; $a_{R2} = 1.0368$; $a_{R3} = 1.0012$; $a_{R4} = 0.9615$;	$a_{R1} = 0.8414$; $a_{R2} = 0.8581$; $a_{R3} = 0.8421$; $a_{R4} = 0.8131$;
$\cos\phi_{R1} = 0.947$; $\cos\phi_{R2} = 0.9565$;	$\cos\phi_{R1} = 0.9784$; $\cos\phi_{R2} = 0.9804$;
$\cos\phi_{R3} = 0.9650$; $\cos\phi_{R4} = 0.9764$;	$\cos\phi_{R3} = 0.9838$; $\cos\phi_{R4} = 0.9890$;
Inverters	Inverters
$V_{DCI1} = 0.9867$; $V_{DCI2} = 0.9847$; $V_{DCI3} = 0.9857$; $V_{DCI4} = 0.9836$;	$V_{DCI1} = 2.2943$; $V_{DCI2} = 2.2935$; $V_{DCI3} = 2.2939$; $V_{DCI4} = 2.2930$;
$V_{DCI5} = 0.9816$; $V_{DCI6} = 0.9826$; $I_{DCI1} = -0.2027$; $I_{DCI2} = -0.3047$;	$V_{DCI5} = 2.2921$; $V_{DCI6} = 2.2926$;
$I_{DCI3} = -0.2029$;	$I_{DCI1} = -0.0872$; $I_{DCI2} = -0.1308$; $I_{DCI3} = -0.0872$;
$I_{DCI4} = -0.2033$; $I_{DCI5} = -0.3056$; $I_{DCI6} = -0.2035$;	$I_{DCI4} = -0.0872$; $I_{DCI5} = -0.1309$; $I_{DCI6} = -0.0872$;
$a_{I1} = 0.9401$; $a_{I2} = 0.9316$; $a_{I3} = 0.9683$; $a_{I4} = 0.9643$;	$a_{I1} = 0.8201$; $a_{I2} = 0.8199$; $a_{I3} = 0.8453$; $a_{I4} = 0.8435$;
$a_{I5} = 0.9389$; $a_{I6} = 0.9554$;	$a_{I5} = 0.8245$; $a_{I6} = 0.8283$;
$\cos\phi_{I1} = 0.9862$; $\cos\phi_{I2} = 0.9920$; $\cos\phi_{I3} = 0.9764$;	$\cos\phi_{I1} = 0.9730$; $\cos\phi_{I2} = 0.9719$; $\cos\phi_{I3} = 0.9633$;
$\cos\phi_{I4} = 0.9863$; $\cos\phi_{I5} = 0.9922$; $\cos\phi_{I6} = 0.9765$;	$\cos\phi_{I4} = 0.9730$; $\cos\phi_{I5} = 0.9766$; $\cos\phi_{I6} = 0.9730$;
NI = 8	NI = 8

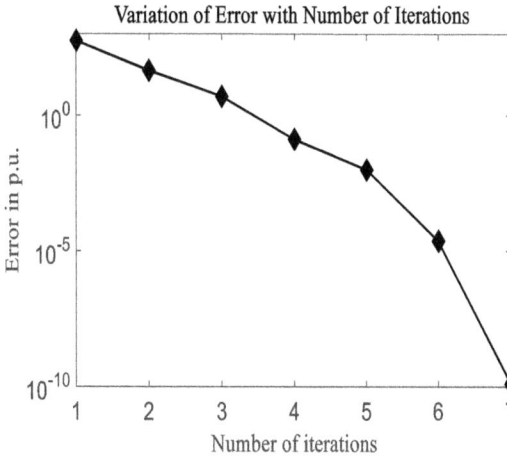

FIGURE 2.30 Convergence characteristic for the base case power flow in the European 1354-bus system.

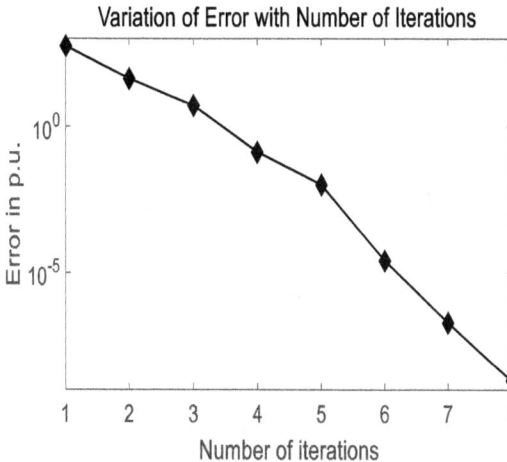

FIGURE 2.31 Convergence characteristic of Table 2.8 for model 1 employing control strategy 1.

Both the models require the same eight number of iterations to converge. The convergence characteristics corresponding to Table 2.9 for models 1 and 2 are shown in Figures 2.35 and 2.36, respectively. The bus voltage profiles for models 1 and 2 are depicted in Figures 2.37 and 2.38, respectively. From Figures 2.37 and 2.38, it is observed that the bus voltage profile hardly changes except for the AC terminal buses at which the LCC-HVDC links are incorporated.

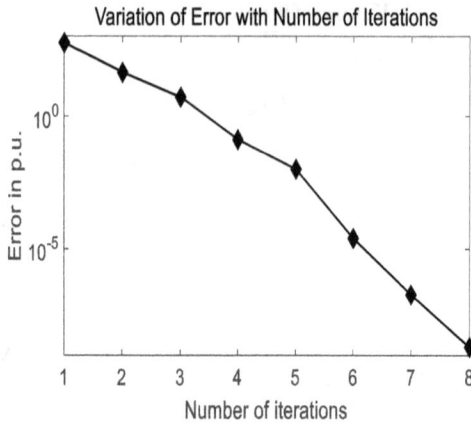

FIGURE 2.32   Convergence characteristic of Table 2.8 for model 2 employing control strategy 1.

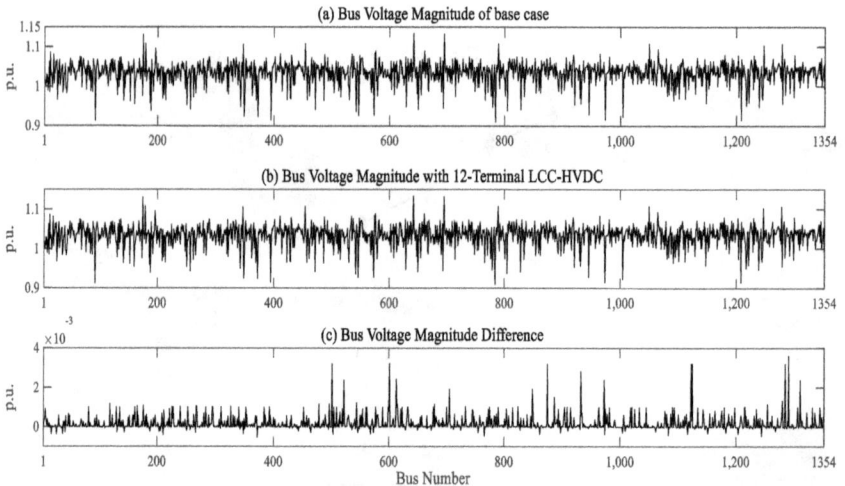

FIGURE 2.33   Bus voltage profile of Table 2.8 for model 1 employing control strategy 1.

2.6.3   STUDIES WITH SEQUENTIAL POWER-FLOW MODEL OF IEEE 300-BUS TEST SYSTEM INTEGRATED WITH 3-TERMINAL LCC-HVDC GRID

Case I: Control strategy 1

In this case study, two HVDC links are considered. The first link is incorporated between AC buses '25–26' and the second one between buses '25 and 232'. The converter connected to bus no. 25 operates as a rectifier. On the other hand, both the converters connected to buses 26 and 232 operate as inverters. The active power flows on the links 25–26 and 25–232 are set to 0.15 and 0.1 p.u.,

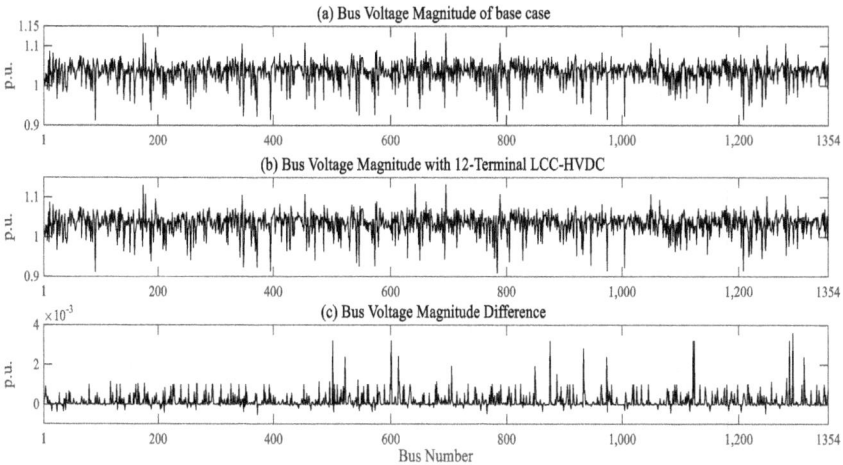

FIGURE 2.34 Bus voltage profile of Table 2.8 for model 2 employing control strategy 1.

respectively. The DC voltages on the rectifier side for models 1 and 2 are set to values of 1 and 2.3 p.u., respectively. It may be noted that these values are different on account of the different constants being associated with the two models (Table 2.2). The firing angles for the rectifier and the extinction angles for both the inverters are set to 5° and 15°, respectively. The power-flow solution is shown in columns 6–11 of Table 2.10. From Table 2.10, it is observed that although the final power-flow solutions corresponding to two models are different, the power-flow convergence patterns are similar. Both the models require the same 'NI' to converge. The convergence characteristics with models 1 and 2 corresponding to control strategy 1 are shown in Figures 2.39 and 2.40, respectively. The bus voltage profiles with models 1 and 2 for control strategy 1 are shown in Figures 2.41 and 2.42, respectively.

Case II: Control strategy 2
In this case study, two HVDC links are connected between AC buses '213–214' and '213–216'. The converter connected to bus no. 213 operates as a rectifier while both the converters connected to buses 214 and 216 operate as inverters. The rectifier end DC voltage is specified along with the firing angle. While the first inverter acts on constant power mode, the second one acts on constant current mode. In addition, the extinction angles for both the inverters are specified. These values are detailed in columns 1–5 of Table 2.10. The power-flow solution is detailed in columns 6–11 of Table 2.10. From Table 2.10, it is observed that both models 1 and 2 demonstrate similar convergence characteristics. The convergence characteristics with models 1 and 2 corresponding to control strategy 2 are shown in Figures 2.43 and 2.44, respectively. The bus voltage profiles of models 1 and 2 for control strategy 2 are shown in Figures 2.45 and 2.46, respectively.

TABLE 2.9
Study of 12-Terminal DC System Incorporated in 1354-Bus European Test System with Control Strategy 5 of Unified Method

Base case power flow (NI = 7)

$V_{1281} = 1.0354 \angle -37.3082$; $V_{1286} = 1.0156 \angle -12.3760$; $V_{1292} = 1.0289 \angle -35.2306$; $V_{1354} = 1.0581 \angle -21.7988$;

$V_{464} = 1.0641 \angle 6.4670$; $V_{792} = 1.0656 \angle -33.2984$

$V_{137} = 1.0539 \angle -30.2000$; $V_{786} = 1.0533 \angle -23.8588$

HVDC link

Rectifier buses: 1281, 1286, 1292, 1354; Inverter buses: 464, 792, 1009, 1313, 137, 786;

HVDC Link Specification

Model 1	Model 2
Rectifiers	Rectifiers
$V_{DCR1} = 1$; $a_{R1} = 1.0219$; $P_{DCR2} = 0.4$; $a_{R2} = 1.0368$; $P_{DCR3} = 0.3$; $a_{R3} = 1.008$; $P_{DCR4} = 0.3$; $a_{R4} = 1.0001$;	$V_{DCR1} = 2.3$; $a_{R1} = 0.852$; $P_{DCR2} = 0.4$; $a_{R2} = 0.858$; $P_{DCR3} = 0.3$; $a_{R3} = 0.851$; $P_{DCR4} = 0.3$; $a_{R4} = 0.852$;
Inverters	Inverters
$P_{DCI1} = -0.3$; $a_{I1} = 1$; $P_{DCI2} = -0.3$; $a_{I2} = 1.01$; $P_{DCI3} = -0.2$; $a_{I3} = 1.001$; $P_{DCI4} = -0.2$; $a_{I4} = 1.0011$; $P_{DCI5} = -0.3$; $a_{I5} = 1.0011$; $P_{DCI6} = -0.2$; $a_{I6} = 1.0012$	$P_{DCI1} = -0.3$; $a_{I1} = 0.85$; $P_{DCI2} = -0.3$; $a_{I2} = 0.86$; $P_{DCI3} = -0.2$; $a_{I3} = 0.85$; $P_{DCI4} = -0.2$; $a_{I4} = 0.86$; $P_{DCI5} = -0.3$; $a_{I5} = 0.87$; $P_{DCI6} = -0.2$; $a_{I6} = 0.85$

(Continued)

TABLE 2.9 (Continued)

Study of 12-Terminal DC System Incorporated in 1354-Bus European Test System with Control Strategy 5 of Unified Method

Power-Flow Solution

Model 1	Model 2
AC terminal buses	AC terminal buses
$V_{1281} = 1.0341 \angle -39.8415$; $V_{1286} = 1.0124 \angle -13.1210$; $V_{1292} = 1.0252 \angle -37.5066$; $V_{1354} = 1.0566 \angle -23.8707$; $V_{464} = 1.0658 \angle 5.4666$; $V_{792} = 1.0656 \angle -34.7351$; $V_{1009} = 1.0434 \angle -25.3786$; $V_{1313} = 1.0343 \angle -39.9868$; $V_{137} = 1.0544 \angle -32.0616$; $V_{786} = 1.0538 \angle -26.0221$;	$V_{1281} = 1.0343 \angle -39.8053$; $V_{1286} = 1.0131 \angle -13.1186$; $V_{1292} = 1.0257 \angle -37.4929$; $V_{1354} = 1.0566 \angle -23.8556$; $V_{464} = 1.0656 \angle 5.4697$; $V_{792} = 1.0656 \angle -34.7318$; $V_{1009} = 1.0431 \angle -25.3656$; $V_{1313} = 1.0345 \angle -39.9513$; $V_{137} = 1.0544 \angle -32.0526$; $V_{786} = 1.0536 \angle -26.0072$;
HVDC variables	HVDC variables
Rectifiers	Rectifiers
$P_{DCR1} = 0.5217$; $V_{DCR2} = 1.004$; $V_{DCR3} = 0.9939$; $V_{DCR4} = 0.9969$; $I_{DCR1} = 0.5217$; $I_{DCR2} = 0.3984$; $I_{DCR3} = 0.3019$; $I_{DCR4} = 0.3009$; $\alpha_{R1} = 5.3207$; $\alpha_{R2} = 6.0285$; $\alpha_{R3} = 7.7236$; $\alpha_{R4} = 13.623$; $\cos\phi_{R1} = 0.9463$; $\cos\phi_{R2} = 0.9565$; $\cos\phi_{R3} = 0.9617$; $\cos\phi_{R4} = 0.9434$;	$P_{DCR1} = 0.5040$; $V_{DCR2} = 2.3017$; $V_{DCR3} = 2.2975$; $V_{DCR4} = 2.2988$; $I_{DCR1} = 0.2191$; $I_{DCR2} = 0.1738$; $I_{DCR3} = 0.1306$; $I_{DCR4} = 0.1305$; $\alpha_{R1} = 10.2842$; $\alpha_{R2} = 5.9718$; $\alpha_{R3} = 9.9001$; $\alpha_{R4} = 17.124$; $\cos\phi_{R1} = 0.9664$; $\cos\phi_{R2} = 0.9804$; $\cos\phi_{R3} = 0.9745$; $\cos\phi_{R4} = 0.9454$;
Inverters	Inverters
$V_{DCI1} = 0.9862$; $V_{DCI2} = 0.9847$; $V_{DCI3} = 0.9862$; $V_{DCI4} = 0.9851$; $V_{DCI5} = 0.9833$; $V_{DCI6} = 0.9846$; $I_{DCI1} = -0.3042$; $I_{DCI2} = -0.3047$; $I_{DCI3} = -0.2028$; $I_{DCI4} = -0.2030$; $I_{DCI5} = -0.3051$; $I_{DCI6} = -0.2031$; $\gamma_{I1} = 26.2602$; $\gamma_{I2} = 27.5551$; $\gamma_{I3} = 22.3655$; $\gamma_{I4} = 21.2935$; $\gamma_{I5} = 25.4917$; $\gamma_{I6} = 23.9469$; $\cos\phi_{I1} = 0.9253$; $\cos\phi_{I2} = 0.9149$; $\cos\phi_{I3} = 0.9442$; $\cos\phi_{I4} = 0.9513$; $\cos\phi_{I5} = 0.9316$; $\cos\phi_{I6} = 0.9332$; NI = 8	$V_{DCI1} = 2.2941$; $V_{DCI2} = 2.2935$; $V_{DCI3} = 2.2941$; $V_{DCI4} = 2.2937$; $V_{DCI5} = 2.2929$; $V_{DCI6} = 2.2934$; $I_{DCI1} = -0.1308$; $I_{DCI2} = -0.1308$; $I_{DCI3} = -0.0872$; $I_{DCI4} = -0.0872$; $I_{DCI5} = -0.1308$; $I_{DCI6} = -0.0872$; $\gamma_{I1} = 21.9468$; $\gamma_{I2} = 23.5864$; $\gamma_{I3} = 18.0090$; $\gamma_{I4} = 18.6264$; $\gamma_{I5} = 23.7505$; $\gamma_{I6} = 19.7367$; $\cos\phi_{I1} = 0.9377$; $\cos\phi_{I2} = 0.9266$; $\cos\phi_{I3} = 0.9580$; $\cos\phi_{I4} = 0.9546$;$\cos\phi_{I5} = 0.9254$; $\cos\phi_{I6} = 0.9481$; NI = 8

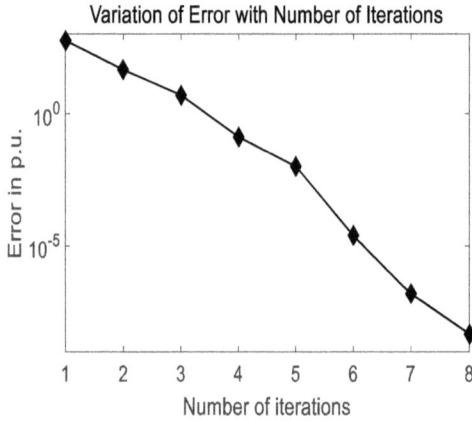

FIGURE 2.35 Convergence characteristic of Table 2.9 for model 1 employing control strategy 5.

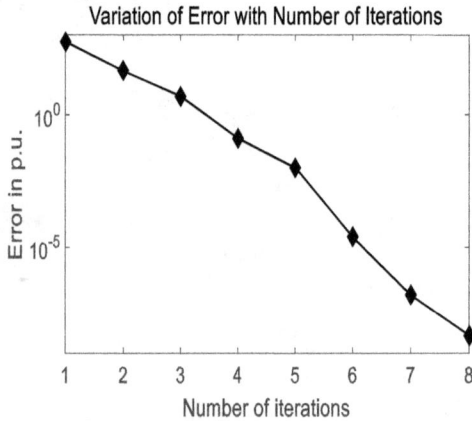

FIGURE 2.36 Convergence characteristic of Table 2.9 for model 2 employing control strategy 5.

Case III: Control strategy 3

In this case, two HVDC links are connected between AC buses '109–146' and '109–147'. While the converter connected to bus no. 109 acts as a rectifier, both the converters connected to buses 146 and 147 operate as inverters. The rectifier end DC voltage is specified along with the firing angle. While the first inverter operates in constant current mode, the second one operates in the constant power mode. In addition, the extinction angles for both the inverters are specified. These values are detailed in columns 1–5 of Table 2.10. The power-flow solution is detailed in columns 6–11 of Table 2.10. It is observed from the power-flow solution that both the models demonstrate almost similar convergence characteristics. The convergence characteristics with models 1 and 2 corresponding to control strategy 3 are

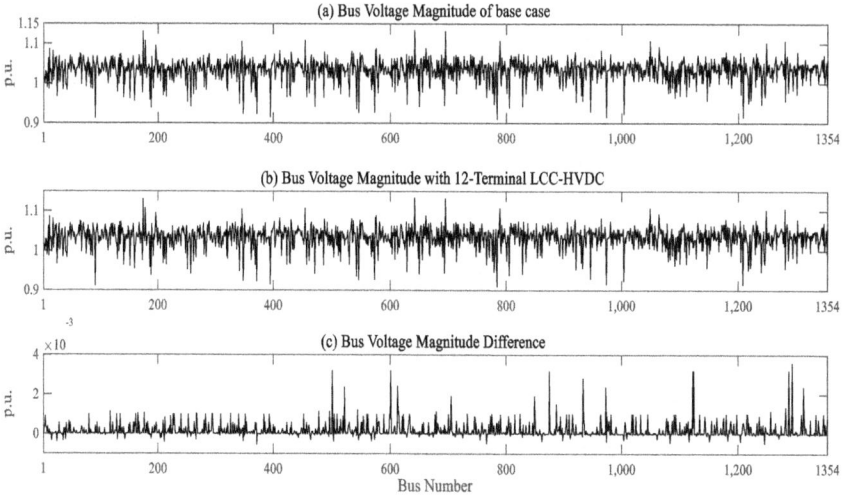

FIGURE 2.37 Bus voltage profile of Table 2.9 for model 1 employing control strategy 5.

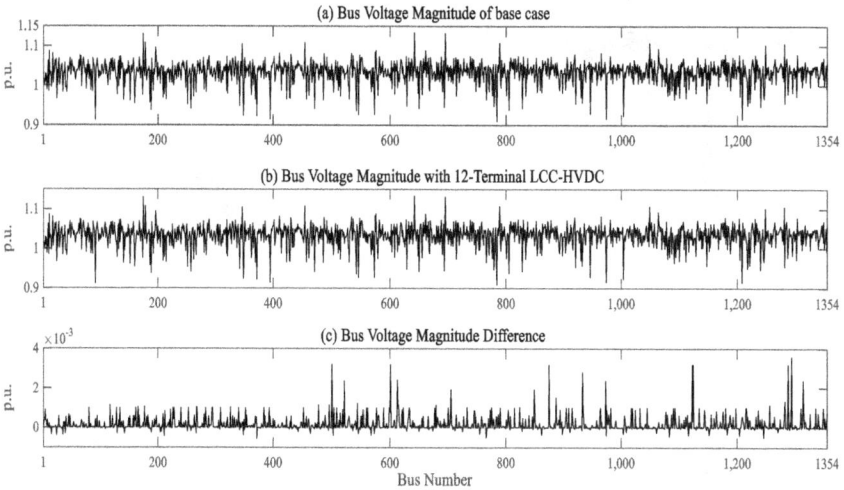

FIGURE 2.38 Bus voltage profile of Table 2.9 for model 2 employing control strategy 5.

shown in Figures 2.47 and 2.48, respectively. The bus voltage profiles with models 1 and 2 for control strategy 3 are shown in Figures 2.49 and 2.50, respectively.

Case IV: Control strategy 4

In this case study, the rectifier is connected to AC bus no. 101 while the inverters are connected to buses 102 and 105, respectively. The rectifier end DC voltage is specified along with the firing angle. Both the inverters are operated at constant current and constant extinction angle mode. The values of the specified quantities

TABLE 2.10

Study with Three Different Control Strategies 1, 2 and 3 of Sequential Method

HVDC Links		HVDC Link Specification			Power-Flow Solution					
		Control Strategy 1			AC Terminal Buses			HVDC Variables		
Rectifier Bus	Inverter Buses	Spec. Values	Model 1	Model 2	ACSV	Model 1	Model 2	DC SV	Model 1	Model 2
25	26	P_{DCI1} (p.u.)	0.15	0.15	V_{25}	0.9938	0.9879	V_{DCI1}	0.9985	2.2993
								V_{DCI2}	0.999	2.2996
								0.2501		
		P_{DCI2} (p.u.)	0.1	0.1	θ_{25}	0.2585	−2.1461	I_{DCI1}	0.1502	0.0652
								I_{DCI2}	0.1001	0.0435
								I_{DCR}	0.2503	0.1087
		V_{DCR} (p.u.)	1	2.3	V_{26}	0.9732	0.9698	a_R	1.0237	0.8731
								a_{I1}	1.0904	0.9137
								a_{I2}	1.034	0.8805
	232	(deg.)	5	5	θ_{26}	−2.0488	−4.5599	$\cos\phi_{I1}$	0.9814	0.9873
									0.9423	0.9607
								$\cos\phi_{I2}$	0.9563	0.9624
		γ_{I1} (deg.)	15	15	V_{232}	1.0103	1.0047	NI	6	6
		γ_{I2} (deg.)	15	15	θ_{232}	0.5706	−1.836			

(Continued)

TABLE 2.10 (Coinitnued)

Study with Three Different Control Strategies 1, 2 and 3 of Sequential Method

HVDC Links		HVDC Link Specification			Power-Flow Solution					
		Control Strategy 2			AC Terminal Buses			HVDC Variables		
Rectifier Bus	Inverter Buses	Spec. Values	Model 1	Model 2	ACSV	Model 1	Model 2	DC SV	Model 1	Model 2
213	214	(p.u.)	0.3	0.3	V_{215}	1.0409	1.0409	V_{DCI1}	0.997	2.2987
								V_{DCI2}	0.999	2.299
								P_{DCI2}	0.0999	0.2299
		I_{DC12} (p.u.)	0.1	0.1	θ_{213}	−20.8941	−20.9826	P_{DCR}	0.4009	0.5302
									0.3009	0.1305
		V_{DCR} (p.u.)	1	2.3	V_{214}	1.0095	1.0095	I_{DCR}	0.4009	0.2305
	216							a_R	1.0031	0.837
		α_R (deg.)	5	5	θ_{214}	−20.7203	−20.7635	a_{I1}	1.0533	0.8822
								a_{I2}	0.9888	0.8411
		γ_{I1} (deg.)	15	15	V_{216}	1.0564	1.0563	$\cos\phi_R$	0.9578	0.9775
								$\cos\phi_{I1}$	0.9376	0.9556
		γ_{I2} (deg.)	15	15	θ_{216}	−20.9123	−20.9012	$\cos\phi_{I2}$	0.9564	0.958
								NI	6	6

(Continued)

TABLE 2.10 (Continued)
Study with Three Different Control Strategies 1, 2 and 3 of Sequential Method

HVDC Links		HVDC Link Specification			Power-Flow Solution					
		Control Strategy 3			AC Terminal Buses			HVDC Variables		
Rectifier Bus	Inverter Buses	Spec. Values	Model 1	Model 2	ACSV	Model 1	Model 2	DC SV	Model 1	Model 2
109	146	(p.u.)	0.1	0.1	V_{109}	1.0190	1.0194	V_{DCI1}	0.9995	2.995
								V_{DCI2}	0.999	2.996
		I_{DCI1} (p.u.)	0.05	0.05	θ_{109}	5.6989	5.7082	P_{DCR}	0.1501	0.215
								P_{DCI1}	0.05	0.115
	147	V_{DCR} (p.u.)	1	2.3	V_{146}	0.9713	0.9716	I_{DCI2}	0.1001	0.0435
								I_{DCR}	0.1501	0.0935
		α_R (deg.)	5	5	θ_{146}	-6.6123	-6.4623	a_R	1.133	0.8451
								a_{I1}	1.1186	0.9109
		γ_{I1} (deg.)	15	15	V_{147}	1.002	1.0026	a_{I2}	1.1356	0.8823
								$\cos\phi_R$	0.8662	0.9885
		γ_{I2} (deg.)	15	15	θ_{147}	-4.4854	-4.4246	$\cos\phi_{I1}$	0.9199	0.9619
								$\cos\phi_{I2}$	0.878	0.9624
								NI	6	6

Variation of Error with Number of Iterations

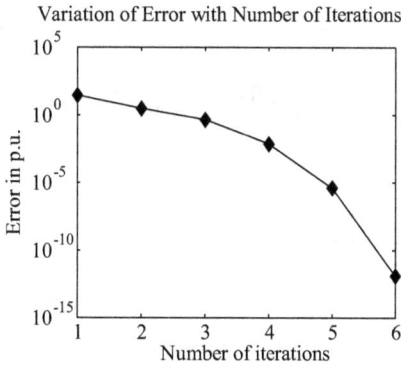

FIGURE 2.39 Convergence characteristic of Table 2.10 for model 1 employing control strategy 1.

Variation of Error with Number of Iterations

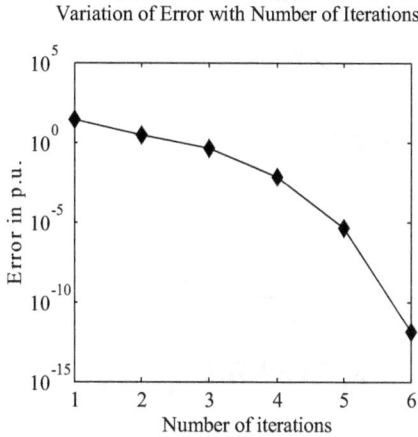

FIGURE 2.40 Convergence characteristic of Table 2.10 for model 2 employing control strategy 1.

are detailed in columns 1–5 of Table 2.11. The power-flow solution is detailed in columns 6–11 of Table 2.11. It is again observed that both the models demonstrate similar convergence characteristics. The convergence characteristics with models 1 and 2 corresponding to control strategy 4 are shown in Figures 2.51 and 2.52, respectively. The bus voltage profiles with models 1 and 2 for control strategy 4 are shown in Figures 2.53 and 2.54, respectively.

Case V: Control strategy 5

In this case, two HVDC links are incorporated between AC buses '86–87' and '86–90'. The converter connected to bus no. 86 operates as a rectifier while both the converters connected to buses 87 and 90 operate as inverters. In this strategy, the

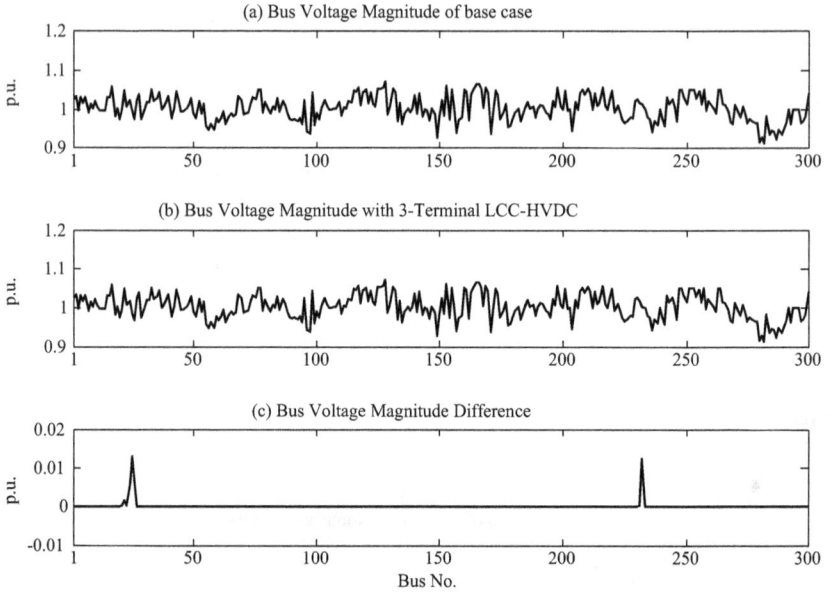

FIGURE 2.41   Bus voltage profile of Table 2.10 for model 1 employing control strategy 1.

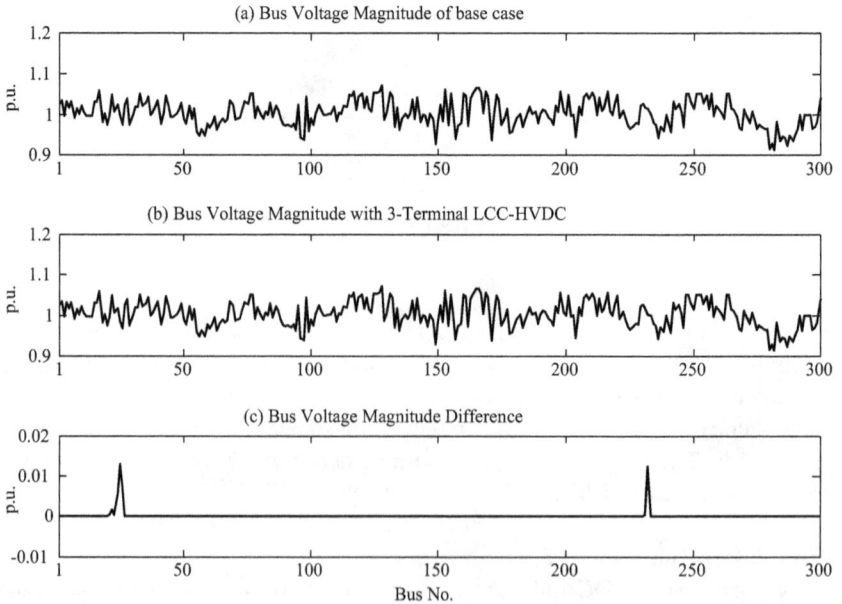

FIGURE 2.42   Bus voltage profile of Table 2.10 for model 2 employing control strategy 1.

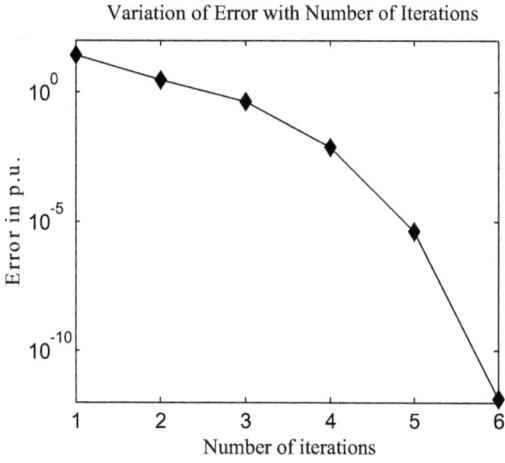

FIGURE 2.43 Convergence characteristic of Table 2.10 for model 1 employing control strategy 2.

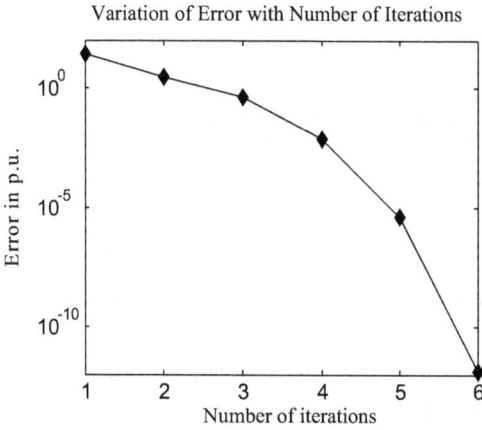

FIGURE 2.44 Convergence characteristic of Table 2.10 for model 2 employing control strategy 2.

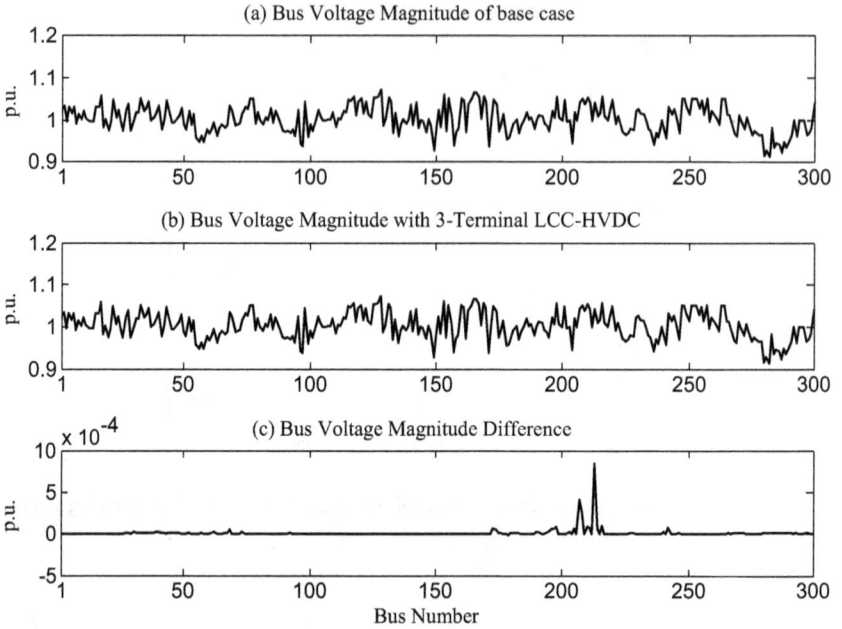

FIGURE 2.45   Bus voltage profile of Table 2.10 for model 1 employing control strategy 2.

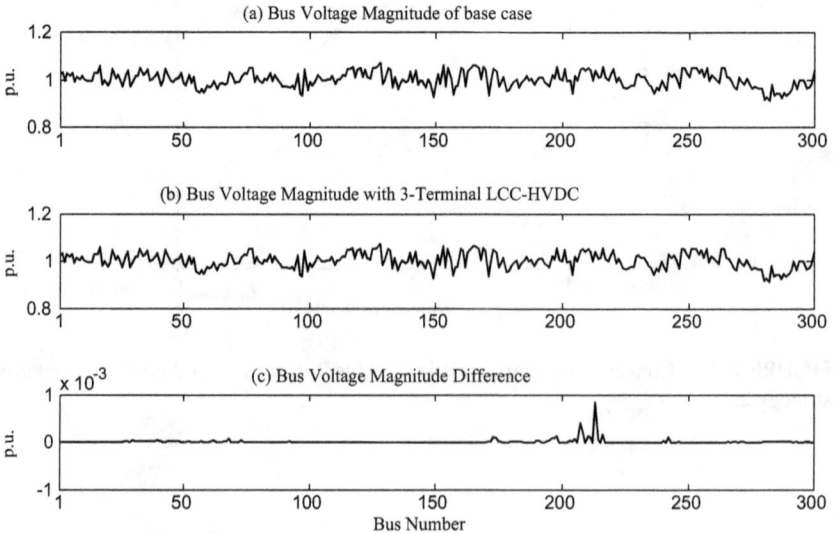

FIGURE 2.46   Bus voltage profile of Table 2.10 for model 2 employing control strategy 2.

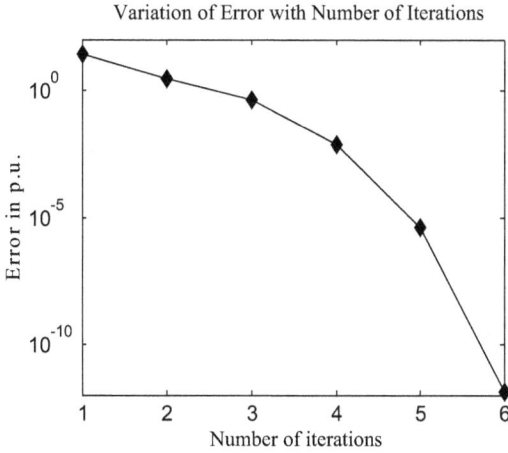

FIGURE 2.47 Convergence characteristic of Table 2.10 for model 1 employing control strategy 3.

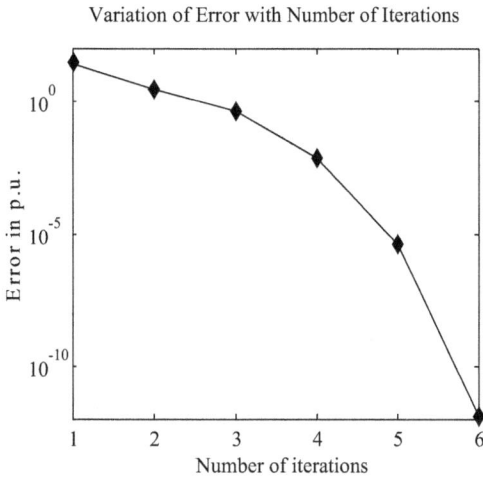

FIGURE 2.48 Convergence characteristic of Table 2.10 for model 2 employing control strategy 3.

rectifier end DC voltage is specified. Both the inverters are operated in the constant power mode. In addition, the tap ratios of all the three-converter transformers are specified. Columns 1–5 of Table 2.11 show these specified values. The power-flow solution is shown in columns 6–11 of Table 2.11. From the power-flow solution, it is observed that the number of iterations 'NI' taken to converge has increased for both the models. This is because for this control strategy, the equivalent reactive power injections (at the converter terminal buses) cannot be computed a priori and need to be updated in every iteration. This affects the convergence pattern adversely.

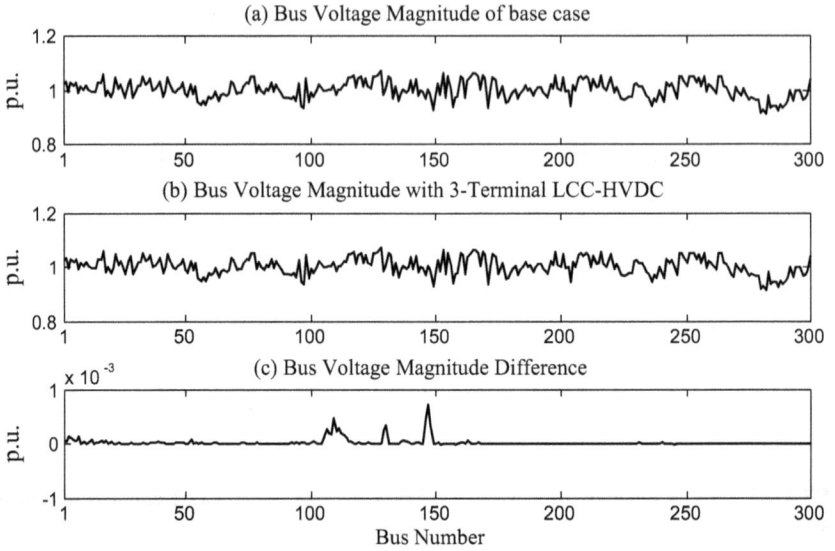

FIGURE 2.49   Bus voltage profile of Table 2.10 for model 1 employing control strategy 3.

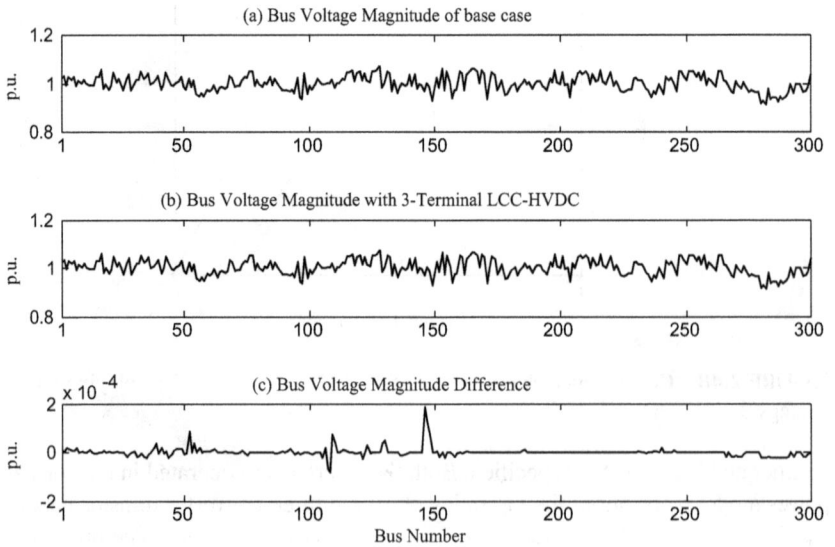

FIGURE 2.50   Bus voltage profile of Table 2.10 for model 2 employing control strategy 3.

TABLE 2.11
Study with Three Different Control Strategies 4, 5 and 6 of Sequential Method

HVDC Link		Control Strategy 4			HVDC Variables					
Rectifier Bus	Inverter Buses	Spec. Values	Model 1	Model 2	ACSV	Model 1	Model 2	DC SV	Model 1	Model 2
101	102	I_{DCI1} (p.u.)	0.1	0.1	V_{101}	0.9745	0.9384	V_{DCI1}	0.999	2.295
θ_{101}								V_{DCI2}	0.9995	2.296
		I_{DCI2} (p.u.)	0.05	0.05	θ_{101}	−14.5005	−18.0564		1.1475	
								P_{DCI2}	0.05	0.9184
		V_{DCR} (p.u.)	1	2.3	V_{102}	1.0014	0.9901	P_{DCR}	0.15	2.07
								I_{DCR}	0.15	0.9
	105				θ_{102}	−17.2702	−14.7398	a_R	1.0455	0.979
		α_R (deg.)	5	5				a_{I1}	1.0432	0.9255
								a_{I2}	1.0413	0.9174
		γ_{I1} (deg.)	15	15	V_{105}	0.9987	0.9912	$\cos\phi_R$	0.9815	0.9269
								$\cos\phi_{I1}$	0.9564	0.9273
		γ_{I2} (deg.)	15	15	θ_{105}	−12.7177	−12.8546	$\cos\phi_{I2}$	0.9611	0.9348
								NI	6	6

(Continued)

TABLE 2.11 (Continued)
Study with Three Different Control Strategies 4, 5 and 6 of Sequential Method

HVDC Link		HVDC Link Specification			Power-Flow Solution					
		Control Strategy 5			AC Terminal Buses			HVDC Variables		
Rectifier Bus	Inverter Buses	Spec. Val.	Model 1	Model 2	ACSV	Model 1	Model 2	DC SV	Model 1	Model 2
86	87	P_{DCI1} (p.u.)	0.2	0.2	V_{86}	1.013	1.0141	V_{DCI1}	0.998	2.991
								V_{DCI2}	0.998	2.991
		P_{DCI2} (p.u.)	0.2	0.2	θ_{86}	−13.6274	−13.6383	P_{DCR}	0.4008	0.4002
								I_{DCI1}	0.2004	0.087
		V_{DCR} (p.u.)	1	2.3	V_{87}	1.0025	1.0035	I_{DCI2}	0.2004	0.087
								I_{DCR}	0.4008	0.174
								α_R	12.0888	14.5339
								γ_{I1}	21.037	20.1385
								γ_{I2}	17.7966	16.703
	90	a_R	1.05	0.88	θ_{87}	−15.6906	−15.7001	$\cos\phi_R$	0.9401	0.9542
		a_{I1}	1.09	0.91	V_{90}	0.972	0.9733	$\cos\phi_{I1}$	0.9133	0.9321
		a_{I2}	1.1	0.92	θ_{90}	−23.6159	−23.6212	$\cos\phi_{I2}$	0.9334	0.9506
								NI	9	9

(Continued)

TABLE 2.11 (Continued)
Study with Three Different Control Strategies 4, 5 and 6 of Sequential Method

HVDC Link		HVDC Link Specification			Power-Flow Solution					
		Control Strategy 6			AC Terminal Buses			HVDC Variables		
Rectifier Bus	Inverter Buses	Spec. Val.	Model 1	Model 2	ACSV	Model 1	Model 2	DC SV	Model 1	Model 2
78	84	I_{DCI1} (p.u.)	0.1	0.1	V_{78}	0.9861	0.9757	V_{DCI1}	0.999	2.299
								V_{DCI2}	0.999	2.299
		I_{DCI2} (p.u.)	0.1	0.1	θ_{78}	−10.2544	−11.4187	P_{DCI1}	0.0999	0.2299
								P_{DCI2}	0.0999	0.2299
	86	V_{DCR} (p.u.)	1	2.3	V_{84}	1.0276	1.0263	P_{DCR}	0.2	0.46
								I_{DCR}	0.2	0.2
					θ_{84}	−11.5247	−10.776	α_R	9.8874	12.8457
		a_R	1.05	0.91				γ_{I1}	20.7534	21.6895
					V_{86}	1.0209	1.0196	γ_{I2}	22.5238	18.8897
		a_{I1}	1.05	0.9				$\cos\phi_R$	0.9658	0.959
					θ_{86}	−12.4037	−11.6552	$\cos\phi_{I1}$	0.9258	0.9215
		a_{I2}	1.07	0.89				$\cos\phi_{I2}$	0.9146	0.938
								NI	8	10

Variation of Error with Number of Iterations

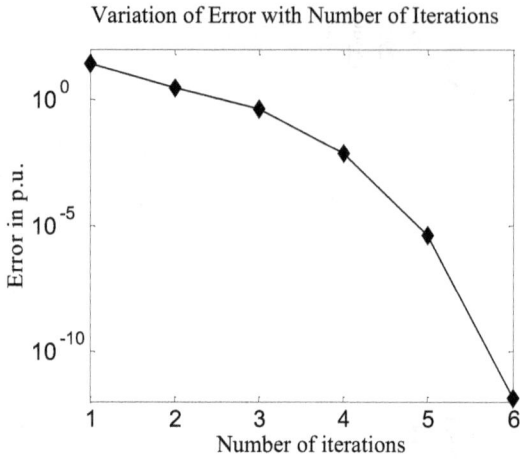

FIGURE 2.51   Convergence characteristic of Table 2.11 for model 1 employing control strategy 4.

Variation of Error with Number of Iterations

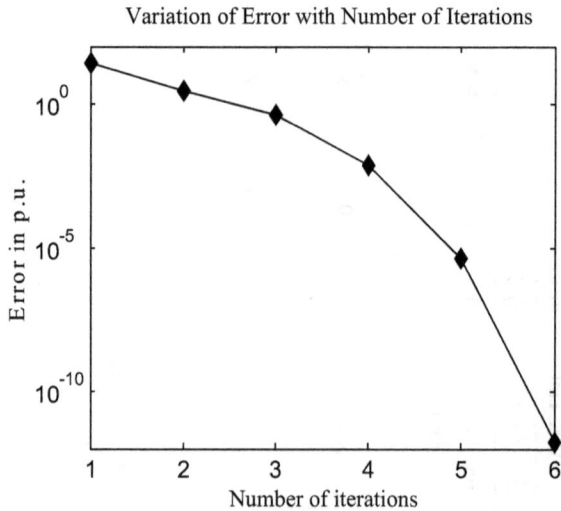

FIGURE 2.52   Convergence characteristic of Table 2.11 for model 2 employing control strategy 4.

The convergence characteristics with models 1 and 2 corresponding to control strategy 5 are shown in Figures 2.55 and 2.56, respectively. The bus voltage profiles of models 1 and 2 for control strategy 5 are shown in Figures 2.57 and 2.58, respectively.

Case VI: Control strategy 6

In this case, the two HVDC links are connected between AC buses '78–84' and '78–86'. While the converter connected to bus no. 78 operates as a rectifier,

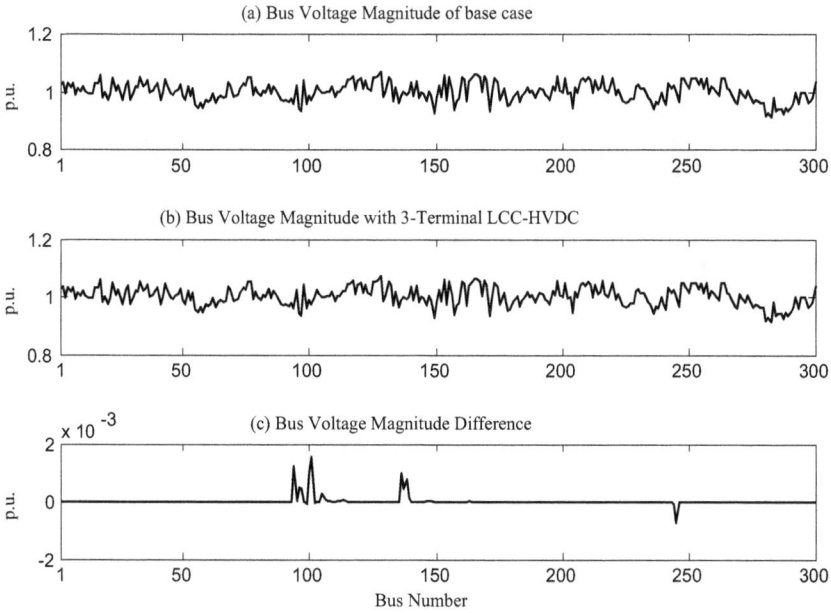

FIGURE 2.53   Bus voltage profile of Table 2.11 for model 1 employing control strategy 4.

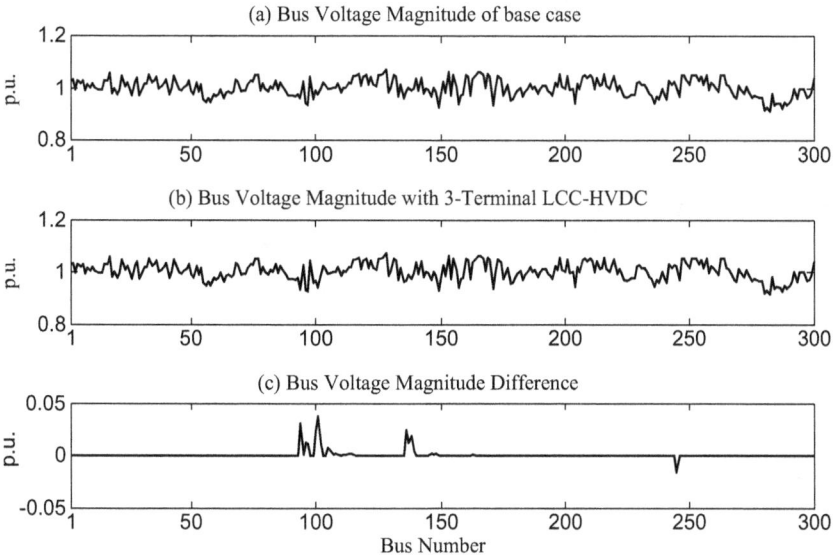

FIGURE 2.54   Bus voltage profile of Table 2.11 for model 2 employing control strategy 4.

the converters connected to buses 84 and 86 operate as inverters. In this control strategy, the rectifier end DC voltage is specified. On the other hand, both the inverters are operated in the constant current mode. Also, the tap ratios of all the

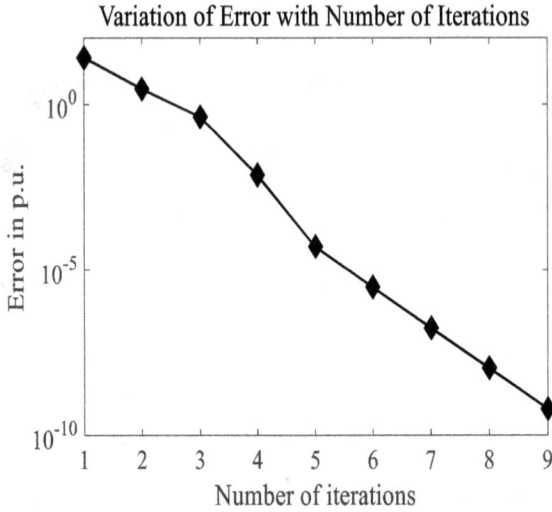

FIGURE 2.55 Convergence characteristic of Table 2.11 for model 1 employing control strategy 5.

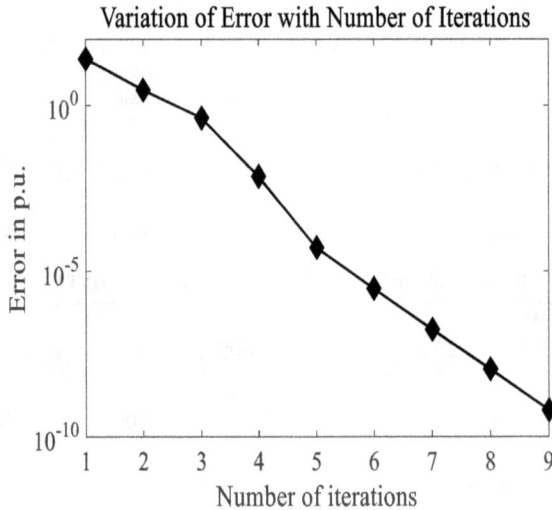

FIGURE 2.56 Convergence characteristic of Table 2.11 for model 2 employing control strategy 5.

three-converter transformers are specified. While columns '1–5' of Table 2.11 show the values of the specified quantities, columns 6–11 detail the power-flow solution. From the power-flow solution, it is observed that again the convergence pattern is adversely affected by this control strategy, as the converter equivalent reactive power injections are updated in each iteration. Again, from the values of 'NI', it is observed that model 1 fares better than model 2. The convergence

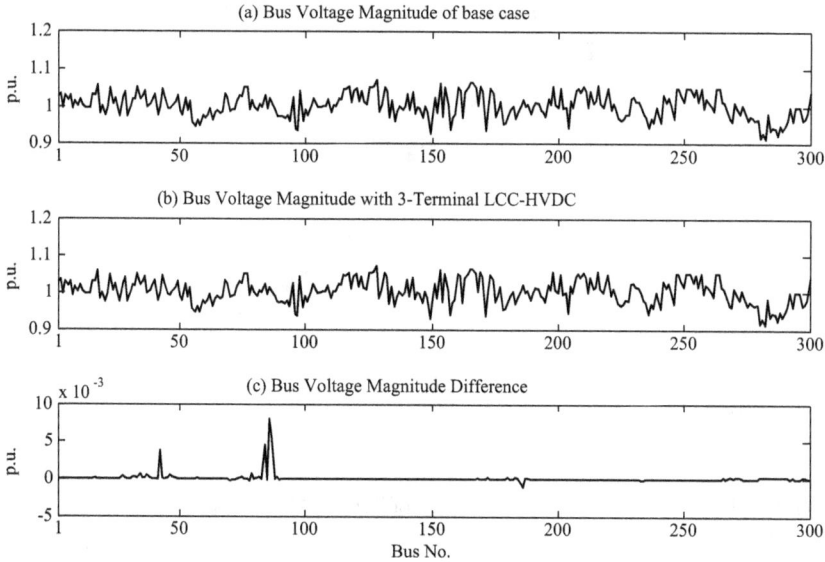

FIGURE 2.57   Bus voltage profile of Table 2.11 for model 1 employing control strategy 5.

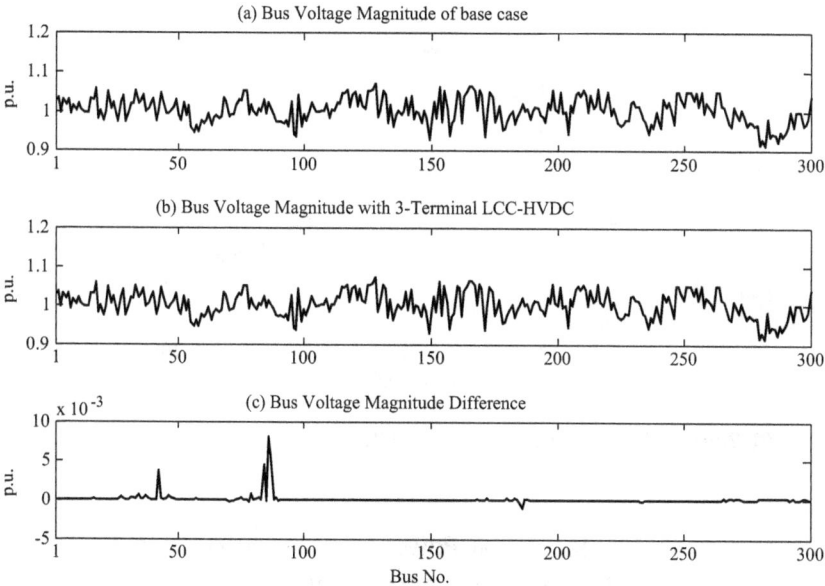

FIGURE 2.58   Bus voltage profile of Table 2.11 for model 2 employing control strategy 5.

characteristics with models 1 and 2 corresponding to control strategy 6 are shown in Figures 2.59 and 2.60, respectively. The bus voltage profiles with models 1 and 2 for control strategy 6 are shown in Figures 2.61 and 2.62, respectively.

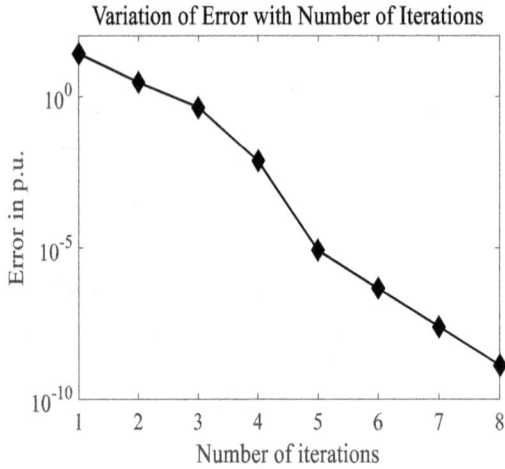

FIGURE 2.59 Convergence characteristic of Table 2.11 for model 1 employing control strategy 6.

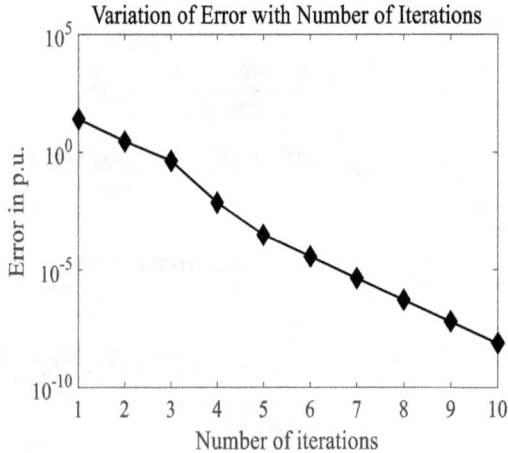

FIGURE 2.60 Convergence characteristic of Table 2.11 for model 2 employing control strategy 6.

2.6.4 STUDIES WITH SEQUENTIAL POWER-FLOW MODEL OF EUROPEAN 1354-BUS TEST SYSTEM INTEGRATED WITH 12-TERMINAL LCC-HVDC GRID

Case I: Control strategy 1

This study is similar to the study of Section 2.6.2 corresponding to case I, except the power-flow method. In this study, sequential method is used to solve the power-flow equations. The results are shown in Table 2.12. It is shown that the results

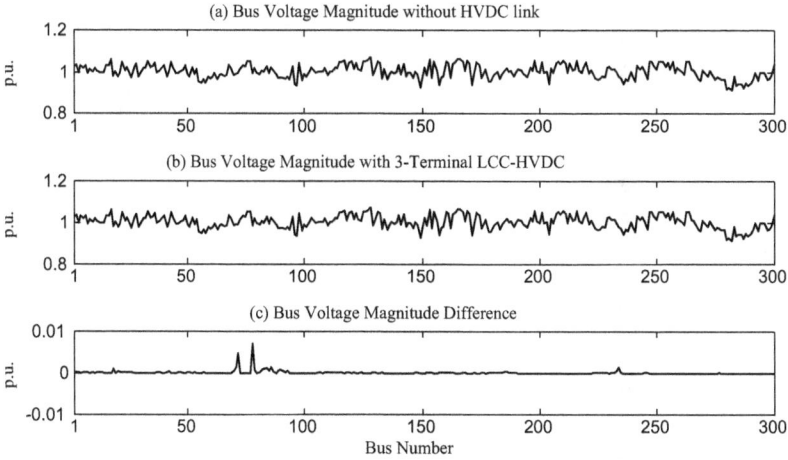

FIGURE 2.61   Bus voltage profile of Table 2.11 for model 1 employing control strategy 6.

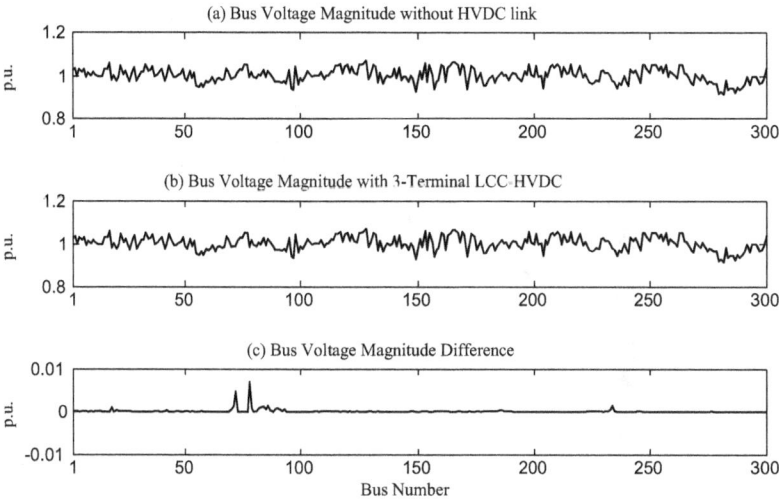

FIGURE 2.62   Bus voltage profile of Table 2.11 for model 2 employing control strategy≈6.

of Table 2.12 are identical to the Table 2.8. The convergence characteristics with models 1 and 2 corresponding to control strategy 1 are shown in Figures 2.63 and 2.64, respectively. From Figures 2.63 and 2.64, it is observed that both models demonstrate the same convergence characteristics as in base case. Both the models require the same seven number of iterations to converge, with model 1 taking slightly less computational time than model 2. The bus voltage profiles with models 1 and 2 for control strategy 1 are shown in Figures 2.65 and 2.66, respectively.

TABLE 2.12

Study of 12-Terminal DC System Incorporated in 1354-Bus European Test System with Control Strategy 1 of Sequential Method

Base case power flow (NI = 7)

$V_{1281} = 1.0354 \angle -37.3082$; $V_{1286} = 1.0156 \angle -12.3760$; $V_{1292} = 1.0289 \angle -35.2306$; $V_{1354} = 1.0581 \angle -21.7988$;
$V_{464} = 1.0641 \angle 6.4670$; $V_{792} = 1.0656 \angle -33.2984$; $V_{1009} = 1.0424 \angle -23.5338$; $V_{1313} = 1.0351 \angle -37.4755$;
$V_{137} = 1.0539 \angle -30.2000$; $V_{786} = 1.0533 \angle -23.8588$;

HVDC link

Rectifier buses: 1281, 1286, 1292, 1354; Inverter Buses: 464, 792, 1009, 1313, 137, 786;

HVDC Link Specification

Model 1	Model 2
Rectifiers	Rectifiers
$V_{DCR1} = 1$; $\alpha_{R1} = 5$; $P_{DCR2} = 0.4$; $\alpha_{R2} = 6$;	$V_{DCR1} = 1$; $\alpha_{R1} = 5$; $P_{DCR2} = 0.4$; $\alpha_{R2} = 6$;
$P_{DCR3} = 0.3$; $\alpha_{R3} = 6$; $P_{DCR4} = 0.2$; $\alpha_{R4} = 5$;	$P_{DCR3} = 0.3$; $\alpha_{R3} = 6$; $P_{DCR4} = 0.2$; $\alpha_{R4} = 5$;
Inverters	Inverters
$P_{DCI1} = -0.2$; $\gamma_{I1} = 15$; $P_{DCI2} = -0.3$; $\gamma_{I2} = 16$; $P_{DCI3} = -0.2$; $\gamma_{I3} = 17$; $P_{DCI4} = -0.2$; $\gamma_{I4} = 15$; $P_{DCI5} = -0.3$; $\gamma_{I5} = 16$;	$P_{DCI1} = -0.2$; $\gamma_{I1} = 15$; $P_{DCI2} = -0.3$; $\gamma_{I2} = 16$; $P_{DCI3} = -0.2$; $\gamma_{I3} = 17$; $P_{DCI4} = -0.2$; $\gamma_{I4} = 15$; $P_{DCI5} = -0.3$; $\gamma_{I5} = 16$;
$P_{DCI6} = -0.2$; $\gamma_{I6} = 17$;	$P_{DCI6} = -0.2$; $\gamma_{I6} = 17$;

(Continued)

TABLE 2.12 (*Continued*)

Study of 12-Terminal DC System Incorporated in 1354-Bus European Test System with Control Strategy 1 of Sequential Method

Power-Flow Solution

Model 1	Model 2
AC terminal buses	AC terminal buses
$V_{1281} = 1.0340 \angle -39.7442$; $V_{1286} = 1.0124 \angle -13.0975$;	$V_{1281} = 1.0344 \angle -39.7097$; $V_{1286} = 1.0131 \angle -13.0951$;
$V_{1292} = 1.0253 \angle -37.4127$; $V_{1354} = 1.0574 \angle -23.6539$;	$V_{1292} = 1.0261 \angle -37.4012$; $V_{1354} = 1.0575 \angle -23.6403$;
$V_{464} = 1.0644 \angle 5.6337$; $V_{792} = 1.0655 \angle -34.6668$;	$V_{464} = 1.0646 \angle 5.6361$; $V_{792} = 1.0656 \angle -34.6636$;
$V_{1009} = 1.0426 \angle -25.2922$; $V_{1313} = 1.0342 \angle -39.8895$;	$V_{1009} = 1.0430 \angle -25.2822$; $V_{1313} = 1.0345 \angle -39.8553$;
$V_{137} = 1.0537 \angle -31.9810$; $V_{786} = 1.0532$;	$V_{137} = 1.0540 \angle -31.9756$; $V_{786} = 1.0534 \angle -25.9095$
HVDC variables	HVDC variables
Rectifiers	Rectifiers
$P_{DCR1} = 0.52$; $V_{DCR2} = 1.004$; $V_{DCR3} = 0.9907$; $V_{DCR4} = 0.9927$;	$P_{DCR1} = 0.5037$; $V_{DCR2} = 2.3017$; $V_{DCR3} = 2.2961$; $= 2.2969$;
$I_{DCR1} = 0.52$; $I_{DCR2} = 0.3984$; $I_{DCR3} = 0.3028$; $I_{DCR4} = 0.2015$;	$I_{DCR1} = 0.2190$; $I_{DCR2} = 0.1738$; $I_{DCR3} = 0.1307$; $I_{DCR4} = 0.0871$;
$a_{R1} = 1.0213$; $a_{R2} = 1.0367$; $a_{R3} = 1.0012$; $a_{R4} = 0.9615$;	$a_{R1} = 0.8418$; $a_{R2} = 0.8580$; $a_{R3} = 0.8421$; $a_{R4} = 0.8131$;
$\cos\phi_{R1} = 0.9469$; $\cos\phi_{R2} = 0.9566$;	$\cos\phi_{R1} = 0.9784$; $\cos\phi_{R2} = 0.9804$;
$\cos\phi_{R3} = 0.9650$; $\cos\phi_{R4} = 0.9764$;	$\cos\phi_{R3} = 0.9838$; $\cos\phi_{R4} = 0.9890$;
Inverters	Inverters
$V_{DCI1} = 0.9867$; $V_{DCI2} = 0.9847$; $V_{DCI3} = 0.9857$; $V_{DCI4} = 0.9836$;	$V_{DCI1} = 2.2943$; $V_{DCI2} = 2.2935$; $V_{DCI3} = 2.2939$; $V_{DCI4} = 2.2930$;
$V_{DCI5} = 0.9816$; $V_{DCI6} = 0.9826$;	$V_{DCI5} = 2.2921$; $V_{DCI6} = 2.2926$;
$I_{DCI1} = -0.2027$; $I_{DCI2} = -0.3047$; $I_{DCI3} = -0.2029$;	$I_{DCI1} = -0.0872$; $I_{DCI2} = -0.1308$; $I_{DCI3} = -0.0872$;
$I_{DCI4} = -0.2033$; $I_{DCI5} = -0.3056$; $I_{DCI6} = -0.2035$;	$I_{DCI4} = -0.0872$; $I_{DCI5} = -0.1309$; $I_{DCI6} = -0.0872$;
$a_{I1} = 0.9401$; $a_{I2} = 0.9316$; $a_{I3} = 0.9683$; $a_{I4} = 0.9643$;	$a_{I1} = 0.8201$; $a_{I2} = 0.8200$; $a_{I3} = 0.8453$; $a_{I4} = 0.8434$;
$a_{I5} = 0.9390$; $a_{I6} = 0.9554$;	$a_{I5} = 0.8285$; $a_{I6} = 0.8365$;
$\cos\phi_{I1} = 0.9862$; $\cos\phi_{I2} = 0.9920$; $\cos\phi_{I3} = 0.9764$;	$\cos\phi_{I1} = 0.9730$; $\cos\phi_{I2} = 0.9718$; $\cos\phi_{I3} = 0.9633$;
$\cos\phi_{I4} = 0.9863$; $\cos\phi_{I5} = 0.9922$; $\cos\phi_{I6} = 0.9765$;	$\cos\phi_{I4} = 0.9730$; $\cos\phi_{I5} = 0.9719$; $\cos\phi_{I6} = 0.9633$;
NI = 7	NI = 7

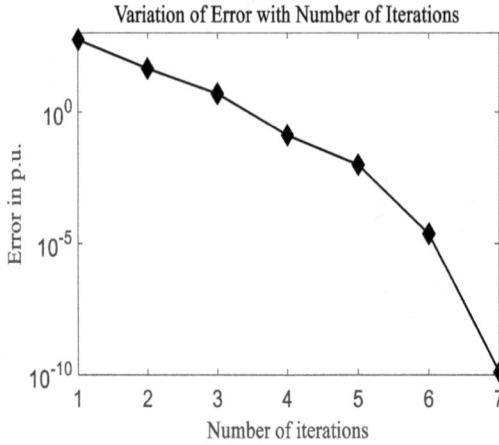

FIGURE 2.63 Convergence characteristic of Table 2.12 for model 1 employing control strategy 1.

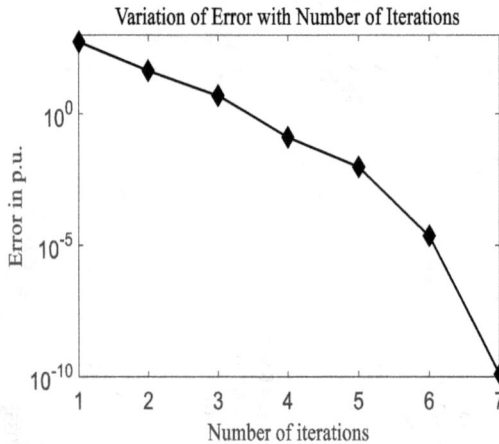

FIGURE 2.64 Convergence characteristic of Table 2.12 for model 2 employing control strategy 1.

Case II: Control strategy 5

This study is similar to the study of Section 2.6.2 corresponding to case II, except the power-flow method. In this study sequential method is used to solve the power-flow equations. The results are shown in Table 2.13. It is shown that the results of Table 2.13 are identical to the Table 2.9. The convergence characteristics with models 1 and 2 corresponding to control strategy 1 are shown in Figures 2.67 and 2.68, respectively. Both the models require the same number of iterations (8) to converge. The bus voltage profiles with models 1 and 2 for control strategy 1 are shown in Figures 2.69 and 2.70, respectively.

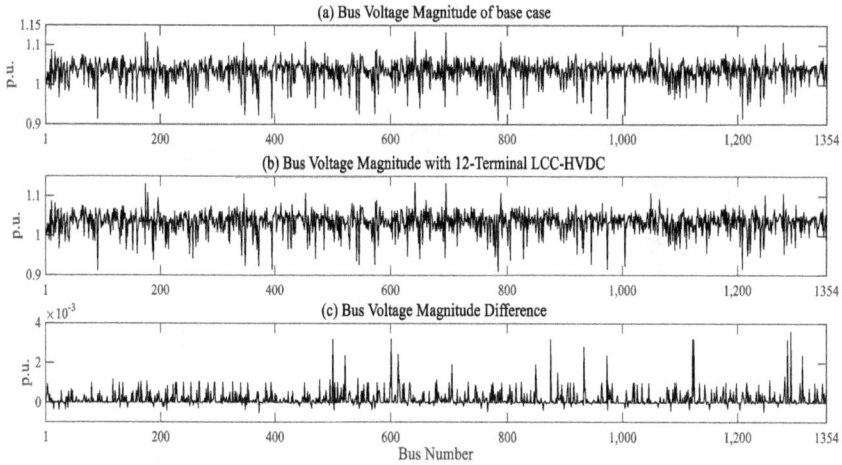

FIGURE 2.65 Bus voltage profile of Table 2.12 for model 1 employing control strategy 1.

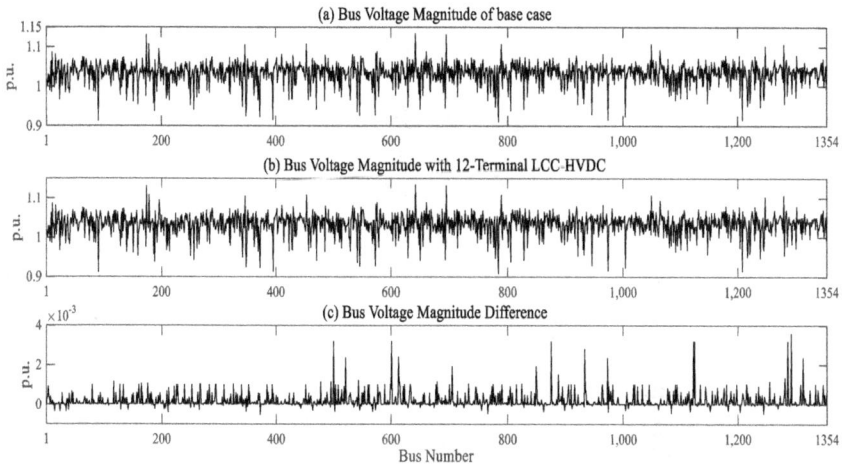

FIGURE 2.66 Bus voltage profile of Table 2.12 for model 2 employing control strategy 1.

The convergence characteristics corresponding to some of the control strategies in the sequential method retain the quadratic convergence characteristics similar to the base case. However, for other control strategies, the quadratic convergence characteristics are not retained. In addition, the bus voltage profiles for these studies do not change except the terminal AC buses at which the LCC-HVDC converters are placed.

TABLE 2.13
Study of 12-Terminal DC System Incorporated in 1354-Bus European Test System with Control Strategy 5 of Sequential Method

Base case power flow (NI = 7)

$V_{1281} = 1.0354 \angle -37.3082$; $V_{1286} = 1.0156 \angle -12.3760$; $V_{1292} = 1.0289 \angle -35.2306$; $V_{1354} = 1.0581 \angle -21.7988$;

$V_{464} = 1.0641 \angle 6.4670$; $V_{792} = 1.0656 \angle -33.2984$;

$V_{137} = 1.0539 \angle -30.2000$; $V_{786} = 1.0533 \angle -23.8588$;

HVDC link

Rectifier buses: 1281, 1286, 1292, 1354; Inverter Buses: 464, 792, 1009, 1313, 137, 786;

HVDC Link Specification

Model 1	Model 2
Rectifiers	Rectifiers
$V_{DCR1} = 1$; $a_{R1} = 1.0219$; $P_{DCR2} = 0.4$; $a_{R2} = 1.0368$;	$V_{DCR1} = 2.3$; $a_{R1} = 0.852$; $P_{DCR2} = 0.4$; $a_{R2} = 0.858$;
$P_{DCR3} = 0.3$; $a_{R3} = 1.008$; $P_{DCR4} = 0.3$; $a_{R4} = 1.0001$;	$P_{DCR3} = 0.3$; $a_{R3} = 0.851$; $P_{DCR4} = 0.3$; $a_{R4} = 0.852$;
Inverters	Inverters
$P_{DCI1} = -0.3$; $a_{I1} = 1$; $P_{DCI2} = -0.3$; $a_{I2} = 1.01$; $P_{DCI3} = -0.3$; $a_{I3} = 1.001$; $P_{DCI4} = -0.2$; $a_{I4} = 1.0011$;	$P_{DCI1} = -0.3$; $a_{I1} = 0.85$; $P_{DCI2} = -0.3$; $a_{I2} = 0.86$; $P_{DCI3} = -0.2$;
$P_{DCI5} = -0.3$; $a_{I5} = 1.0011$; $P_{DCI6} = -0.2$; $a_{I6} = 1.0012$;	$a_{I3} = 0.85$; $P_{DCI4} = -0.2$; $a_{I4} = 0.86$; $P_{DCI5} = -0.3$; $a_{I5} = 0.87$;
	$P_{DCI6} = -0.2$; $a_{I6} = 0.85$;

(Continued)

TABLE 2.13 (Continued)

Study of 12-Terminal DC System Incorporated in 1354-Bus European Test System with Control Strategy 5 of Sequential Method

Power-Flow Solution

Model 1	Model 2
AC terminal buses $V_{1281} = 1.0341 \angle -39.8415$; $V_{1286} = 1.0124 \angle -13.1210$; $V_{1292} = 1.0252 \angle -37.5066$; $V_{1354} = 1.0566 \angle -23.8707$; $V_{464} = 1.0658 \angle 5.4666$; $V_{792} = 1.0656 \angle -34.7351$; $V_{1009} = 1.0434 \angle -25.3786$; $V_{1313} = 1.0343 \angle -39.9868$; $V_{137} = 1.0544 \angle -32.0616$; $V_{786} = 1.0538 \angle -26.0221$; HVDC variables Rectifiers $P_{DCR1} = 0.5217$; $V_{DCR2} = 1.004$; $V_{DCR3} = 0.9939$; $V_{DCR4} = 0.9969$; $I_{DCR1} = 0.5217$; $I_{DCR2} = 0.3984$; $I_{DCR3} = 0.3019$; $I_{DCR4} = 0.3009$; $\alpha_{R1} = 5.3207$; $\alpha_{R2} = 6.0285$; $\alpha_{R3} = 7.7236$; $\alpha_{R4} = 13.623$; $\cos\phi_{R1} = 0.9463$; $\cos\phi_{R2} = 0.9565$; $\cos\phi_{R3} = 0.9617$; $\cos\phi_{R4} = 0.9434$; Inverters $V_{DCI1} = 0.9862$; $V_{DCI2} = 0.9847$; $V_{DCI3} = 0.9862$; $- V_{DCI4} = 0.9851$; $V_{DCI5} = 0.9833$; $V_{DCI6} = 0.9846$; $I_{DCI1} = -0.3042$; $I_{DCI2} = -0.3047$; $I_{DCI3} = -0.2028$; $I_{DCI4} = -0.2030$; $I_{DCI5} = -0.3051$; $I_{DCI6} = -0.2031$; $\gamma_{I1} = 26.2602$; $\gamma_{I2} = 27.5551$; $\gamma_{I3} = 22.3665$; $\gamma_{I4} = 21.2935$; $\gamma_{I5} = 25.4917$; $\gamma_{I6} = 23.9469$; $\cos\phi_{I1} = 0.9253$; $\cos\phi_{I2} = 0.9149$; $\cos\phi_{I3} = 0.9442$; $\cos\phi_{I4} = 0.9513$; $\cos\phi_{I5} = 0.9316$; $\cos\phi_{I6} = 0.9332$; NI = 8	AC terminal $V_{1281} = 1.0343 \angle -39.8053$; $V_{1286} = 1.0131 \angle -13.1186$; $V_{1292} = 1.0257 \angle -37.4929$; $V_{1354} = 1.0566 \angle -23.8556$; $V_{464} = 1.0656 \angle 5.4697$; $V_{792} = 1.0656 \angle -34.7318$; $V_{1009} = 1.0431 \angle -25.3656$; $V_{1313} = 1.0345 \angle -39.9513$; $V_{137} = 1.0544 \angle -32.0526$; $V_{786} = 1.0536 \angle -26.0072$; HVDC variables Rectifiers $P_{DCR1} = 0.5040$; $V_{DCR2} = 2.3017$; $V_{DCR3} = 2.2975$; $V_{DCR4} = 2.2988$; $I_{DCR1} = 0.2191$; $I_{DCR2} = 0.1738$; $I_{DCR3} = 0.1306$; $I_{DCR4} = 0.1305$; $\alpha_{R1} = 10.2842$; $\alpha_{R2} = 5.9718$; $\alpha_{R3} = 9.9001$; $\alpha_{R4} = 17.124$; $= 0.9664$; $\cos\phi_{R2} = 0.9804$; $\cos\phi_{R3} = 0.9745$; $\cos\phi_{R4} = 0.9454$; Inverters $V_{DCI1} = 2.2941$; $V_{DCI2} = 2.2935$; $V_{DCI3} = 2.2941$; $V_{DCI4} = 2.2937$; $V_{DCI5} = 2.2929$; $V_{DCI6} = 2.2934$; $I_{DCI1} = -0.1308$; $I_{DCI2} = -0.1308$; $I_{DCI3} = -0.0872$; $I_{DCI4} = -0.0872$; $I_{DCI5} = -0.1308$; $I_{DCI6} = -0.0872$; $\gamma_{I1} = 21.9468$; $\gamma_{I2} = 23.5864$; $\gamma_{I3} = 18.0090$; $\gamma_{I4} = 18.6264$; $\gamma_{I5} = 23.7505$; $\gamma_{I6} = 19.7367$; $\cos\phi_{I1} = 0.9377$; $\cos\phi_{I2} = 0.9266$; $\cos\phi_{I3} = 0.9580$; $\cos\phi_{I4} = 0.9546$; $\cos\phi_{I5} = 0.9254$; $\cos\phi_{I6} = 0.9481$; NI = 8

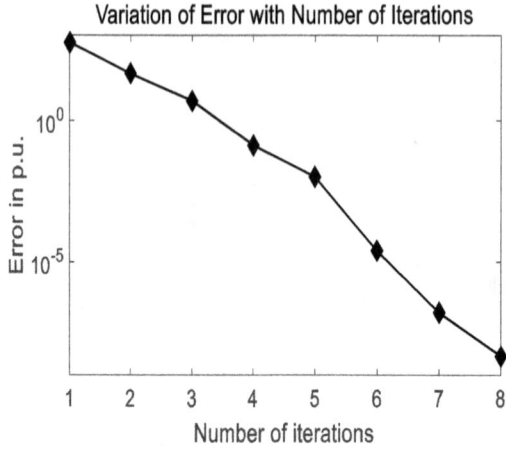

FIGURE 2.67 Convergence characteristic of Table 2.13 for model 1 employing control strategy 5.

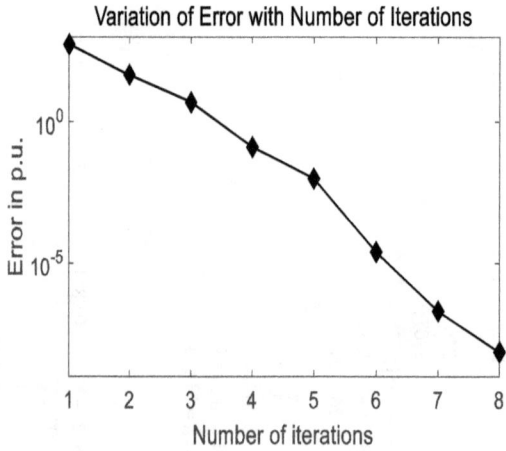

FIGURE 2.68 Convergence characteristic of Table 2.13 for model 2 employing control strategy 5.

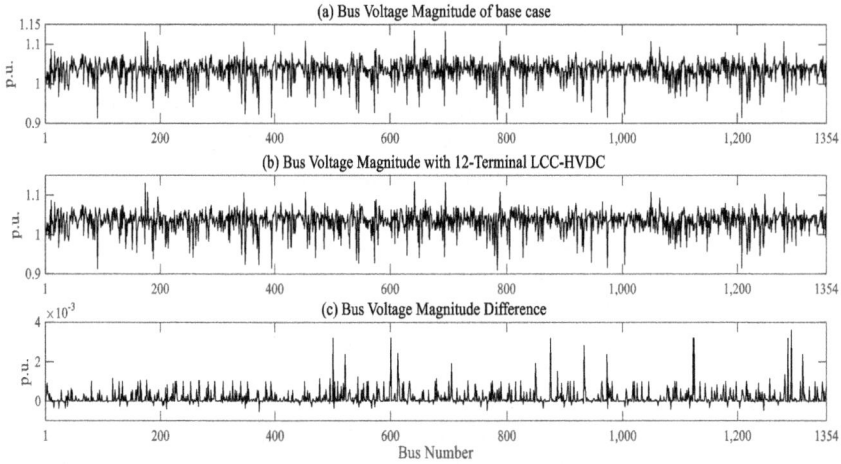

FIGURE 2.69 Bus voltage profile of Table 2.13 for model 1 employing control strategy 5.

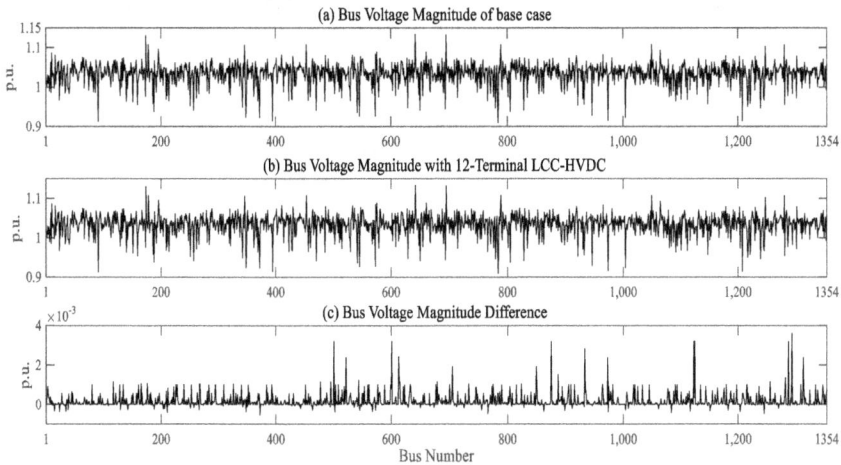

FIGURE 2.70 Bus voltage profile of Table 2.13 for model 2 employing control strategy 5.

2.7 SUMMARY

In this chapter, both unified and sequential power-flow models of LCC-based integrated AC-DC systems have been presented. Based on the selection of the base values of the various DC quantities, two different per-unit AC-DC system models have been considered. The convergence characteristics of both the unified and the sequential Newton power-flow algorithms have been investigated in light of these two per-unit AC-DC system models and diverse DC link control strategies employed. All the power-flow models were implemented on multi-terminal MLDC networks incorporated in the IEEE 300-bus and the European 1354-bus test systems [115,116]. Six different control strategies have been considered for control of the HVDC links.

It is observed that corresponding to the unified power-flow model, the convergence characteristics with both the per-unit system models (models 1 and 2) are similar, independent of the control strategy adopted. However, the convergence characteristics vary slightly with the location of the MLDC network, i.e. the AC system buses between which the link is incorporated and the values of the electrical quantities specified in the control strategy adopted for the DC links.

Corresponding to the sequential power-flow model, it is observed that the convergence characteristics are strongly dependent on the control strategy employed for the DC links. For some of the control strategies, the converter equivalent reactive power injections need to be computed in every power-flow iteration. This affects the convergence of the algorithm. Similar to the unified model, the convergence characteristics are also observed to be affected by the location of the MLDC network in the AC system and the values of the electrical quantities specified in the DC link control strategy. It is also observed that in most of the cases, the convergence characteristics are almost similar for both the per-unit models.

In the next chapter, the power-flow modelling of VSC-based integrated AC-DC systems is presented.

3 Power-Flow Modelling of AC Power Systems Integrated with VSC-Based Multi-Terminal DC (AC-MVDC) Grids Employing DC Slack-Bus Control

3.1 INTRODUCTION

The development of the insulated gate bipolar transistor (IGBT) paved the way for VSC-based HVDC technology. VSC-HVDC based on pulse-width modulation (PWM) scheme has the advantage of independent active and reactive power control, reversible power flow without the change of voltage polarity along with reduction in filter size [8–18].

Unlike a two-terminal VSC-HVDC interconnection, a multi-terminal VSC-based HVDC (MVDC) interconnection is able to exploit the economic and technical advantages of the VSC-HVDC technology in a superior way. It is also better suited if futuristic integration of renewable energy sources is planned [19–21]. In a MVDC system, the converters stations can be located closely, in the same sub-station or remotely, at different locations. The corresponding configurations are known as back-to-back (BTB) or point-to-point (PTP), respectively. Most of the MVDC systems installed worldwide are in the PTP configuration; their DC sides being interconnected through DC links or cables [4, 10, 12].

In a MVDC system, one of the VSCs acts as a master converter while the rest act as slave converters [10, 12]. The master converter controls the voltage magnitude of its AC terminal bus while the slave converters control the active and reactive powers at the terminal end of the lines connecting them to the AC system buses [10, 12]. Some comprehensive research works on the control of VSC-based HVDC systems are presented in Refs. [24–37].

For planning, operation and control of AC power systems incorporating VSC-HVDC networks, power-flow solution of the integrated AC–DC systems is an essential requirement. In this respect, [86–90, 95, 98–104] present some

DOI: 10.1201/9781003252078-3

comprehensive research works on the development of efficient power-flow algorithms for VSC-based integrated AC–DC systems.

However, it is observed that in none of the above works, the modulation index of the converter has been considered as an unknown. The converter modulation index 'm' is an important parameter for VSC operation. Operational considerations limit the minimum and maximum values of the modulation index 'm'. It has been reported that [11] while a low 'm' limits the maximum fundamental AC side voltage of the VSC, over-modulation (m > 1) may result in low-order harmonics in the AC voltage spectrum. Practical ranges of 'm' have been reported in Ref. [18].

Thus, a power-flow model should yield the value of 'm' and 'V_{DC}' directly, for a given operating condition, so that it can be checked whether 'm' lies within its specified limits (with sufficient margin for a dynamic response), along with 'V_{DC}'.

This chapter presents the development of unified and sequential Newton power-flow models of integrated AC-MVDC systems. For control of the MVDC grid, DC slack-bus control has been employed. In all the models, the converter modulation indices appear as unknowns along with the DC side voltages and the phase angles of the AC side voltage phasors of the VSCs. In addition, both the number of VSCs and the MVDC network topology can be arbitrarily chosen in the model. All the models account for the converter losses.

3.2   MODELLING OF INTEGRATED AC-MVDC SYSTEMS EMPLOYING DC SLACK-BUS CONTROL

For the power-flow modelling of an integrated AC-MVDC system, the following assumptions have been adopted [4, 8, 10, 12]:

- The supply voltages are sinusoidal and balanced (contain only fundamental frequency and positive sequence components).
- All the transmission lines are represented by their equivalent-pi models.
- The harmonics generated by the VSCs are neglected.
- The switches are assumed to be ideal.

Now, based on the locations of the VSCs, a MVDC system can have a back-to-back (BTB) or a point-to-point (PTP) configuration. In the back-to-back (BTB) scheme, the converters usually exist at the same location. On the other hand, in a point-to-point (PTP) scheme, the DC links are used to transmit the bulk power between converters which are at different locations. Depending upon the MVDC configuration (PTP or BTB), the power-flow equations and their implementation are slightly different. These are elaborated below.

3.2.1   MODELLING OF INTEGRATED AC-MVDC SYSTEMS IN THE PTP CONFIGURATION

Figure 3.1 shows an 'n' bus AC power system network incorporating a MVDC grid, which is interconnected in the PTP configuration. The MVDC grid

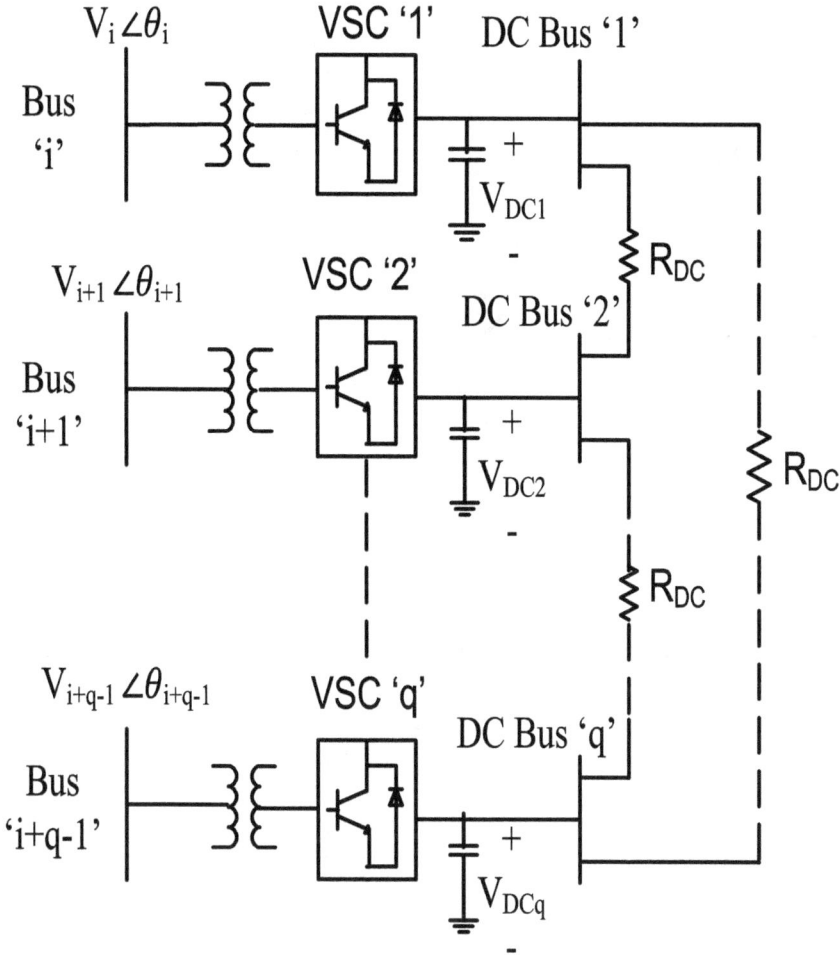

FIGURE 3.1   Schematic diagram of a 'q' terminal PTP VSC-HVDC system.

comprises 'q' VSCs which are connected to 'q' AC buses through their respective coupling transformers. Without loss of generality, it is assumed that the 'q' VSC converters are connected to AC buses 'i', '(i + 1)', and so on, up to bus '(i + q–1)'. The equivalent circuit of Figure 3.1 is shown in Figure 3.2.

In Figure 3.2, the 'q' VSCs are represented by 'q' fundamental frequency, positive sequence voltage sources. The ath ($1 \leq a \leq q$) voltage source V_{sha} (not shown) is connected to AC bus '(i + a–1)' through the leakage impedance $Z_{sha} = R_{sha} + j\, X_{sha}$ of the ath coupling transformer.

Now, let $y_{sha} = 1/Z_{sha}$. Then, from Figure 3.2, the current through the ath ($1 \leq a \leq q$) coupling transformer can be written as

$$I_{sha} = y_{sha}\left(V_{sha} - V_{i+a-1}\right) \tag{3.1}$$

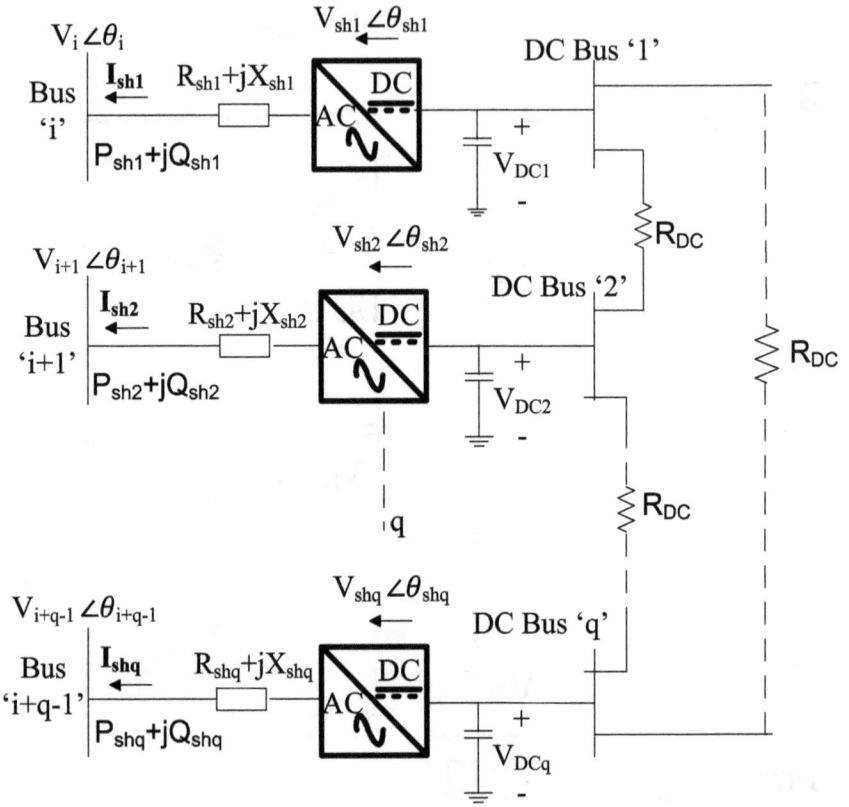

FIGURE 3.2 Equivalent circuit of the 'q' terminal PTP VSC-HVDC system.

where V_{sha} is the voltage phasor representing the output of the ath VSC and is given by $V_{sha} = V_{sha} \angle\theta_{sha} = m_a c V_{DCa} \angle \theta_{sha}$, where '$m_a$' and '$V_{DCa}$' are the modulation index and the DC side voltage of the ath ($1 \le a \le q$) VSC, with 'c' being a constant which depends on the type of converter [11,12]. The ath VSC is connected to the AC terminal bus '(i + a − 1)' whose voltage is represented by the phasor $V_{i+a-1} = V_{i+a-1} \angle \theta_{i+a-1}$.

Hence from Figure 3.2, the net current injection at the AC bus '(i + a−1)' connected to the ath ($1 \le a \le q$) VSC can be written as

$$I_{i+a-1} = \left[Y^{old}_{(i+a-1)(i+a-1)} + y_{sha} \right] V_{i+a-1} + \sum_{k=1,\ k\ne i+a-1}^{n} Y_{(i+a-1)k} V_k - y_{sha} V_{sha} \quad (3.2)$$

$$\text{or,} \quad I_{i+a-1} = \sum_{k=1}^{n} Y_{(i+a-1)k} V_k - y_{sha} V_{sha} \quad (3.3)$$

where $Y^{old}_{(i+a-1)(i+a-1)} = y_{(i+a-1)0} + \sum\limits_{k=1,\,k \neq i+a-1}^{n} y_{(i+a-1)k}$ and $Y_{(i+a-1)(i+a-1)} = Y^{old}_{(i+a-1)(i+a-1)} + y_{sha}$

are the self-admittances of bus '(i + a−1)' for the original 'n' bus AC system without any VSC and with the ath VSC connected, respectively. '$y_{(i+a-1)0}$' accounts for the shunt capacitances of all the transmission lines connected to bus '(i + a−1)'.

3.2.2 Power-Flow Equations of Integrated AC-MVDC System in the PTP Configuration

With the ath $(1 \leq a \leq q)$ VSC connected, the net injected active power at the corresponding AC terminal bus '(i + a−1)' can be written as

$$P_{i+a-1} = \text{Re} \left\{ V_{i+a-1} I^*_{i+a-1} \right\}$$

$$= \sum\limits_{k=1}^{n} V_{i+a-1} V_k Y_{(i+a-1)k} \cos\left[\theta_{i+a-1} - \theta_k - \phi_{(i+a-1)k} \right]$$

$$- m_a c V_{DCa} V_{i+a-1} y_{sha} \cos\left(\theta_{i+a-1} - \theta_{sha} - \phi_{sha} \right) \tag{3.4}$$

since $V_{sha} = m_a c\, V_{DCa}$, as already discussed.

In a similar manner, the net injected reactive power at bus '(i + a−1)' can be written as

$$Q_{i+a-1} = \sum\limits_{k=1}^{n} V_{i+a-1} V_k Y_{(i+a-1)k} \sin\left[\theta_{i+a-1} - \theta_k - \phi_{(i+a-1)k} \right]$$

$$- m_a c\, V_{DCa} V_{i+a-1} y_{sha} \sin\left(\theta_{i+a-1} - \theta_{sha} - \phi_{sha} \right) \tag{3.5}$$

Also, from Figure 3.2, the active and reactive power flows at the terminal end of the line connecting the ath VSC to AC bus '(i + a−1)' can be written as

$$P_{sha} = \text{Re} \left[V_{i+a-1}\, I^*_{sha} \right]$$

$$= m_a c V_{DCa} V_{i+a-1} y_{sha} \cos\left(\theta_{i+a-1} - \theta_{sha} - \phi_{sha} \right) - V^2_{i+a-1} y_{sha} \cos\phi_{sha} \tag{3.6}$$

$$Q_{sha} = m_a c\, V_{DCa} V_{i+a-1} y_{sha} \sin\left(\theta_{i+a-1} - \theta_{sha} - \phi_{sha} \right) + V^2_{i+a-1} y_{sha} \sin\phi_{sha} \tag{3.7}$$

From Eqs. (3.6) and (3.7), the apparent power flow at the terminal end of the line connecting the ath VSC to AC bus '(i + a−1)' can be calculated as

$$S_{sha} = V_{i+a-1} I_{sha} = \sqrt{P^2_{sha} + Q^2_{sha}}$$

$$= y_{sha} \left[V^4_{i+a-1} + m_a^2 c^2 V^2_{DCa} V^2_{i+a-1} - 2 m_a c V_{DCa} V^3_{i+a-1} \cos\left(\theta_{i+a-1} - \theta_{sha} \right) \right]^{1/2} \tag{3.8}$$

The derivation of Eq. (3.8) is given in Appendix A.

Now, for the 'q' terminal DC system shown in Figure 3.2, the net current injection at the uth $(1 \le u \le q)$ DC bus, i.e., at the DC terminal of the uth VSC, is given as

$$I_{DCu} = \sum_{v=1}^{q} Y_{DCuv} V_{DCv} \tag{3.9}$$

where $Y_{DCuv} = -\dfrac{1}{R_{DCuv}}$, '$R_{DCuv}$' being the DC link resistance between the DC buses 'u' and 'v'.

Now, from Figure 3.2, the active power delivered by the ath $(1 \le a \le q)$ VSC at its AC terminal can be written as $P_{ACa} = \mathrm{Re}\left(V_{sha}\, I^*_{sha}\right)$. In a similar manner, the net power injection at the ath DC terminal is given by

$$P_{DCa} = V_{DCa} I_{DCa} = \sum_{v=1}^{q} V_{DCa} V_{DCv} Y_{DCav}.$$

Now, for the ath VSC, the converter losses [18,102] are

$$P_{lossa} = a_1 + b_1\, I_{sha} + c_1\, I^2_{sha} \tag{3.10}$$

where 'a_1' indicates the VSC losses no load, and 'b_1' and 'c_1' are constants representative of the linear and quadratic dependency of the VSC losses on the converter current magnitude (I_{sha}), respectively. The derivation of the converter current magnitude (I_{sha}) is given in Appendix A.

As already discussed earlier, for the master converter, the line active and reactive powers are not specified. Hence, Eq. (3.10) can be written as

$$P_{lossa} = a_1 + b_1\, \frac{S^{cal}_{sha}}{V_{i+a-1}} + c_1 \left(\frac{S^{cal}_{sha}}{V_{i+a-1}}\right)^2 \tag{3.11}$$

where S^{cal}_{sha} is the calculated value of apparent power.

For any slave converter, Eq. (3.10) can be written as

$$P_{lossa} = a_1 + b_1\, \frac{\sqrt{P^{sp\,2}_{sha} + Q^{sp\,2}_{sha}}}{V_{i+a-1}} + c_1 \left(\frac{\sqrt{P^{sp\,2}_{sha} + Q^{sp\,2}_{sha}}}{V_{i+a-1}}\right)^2 \tag{3.12}$$

where P^{sp}_{sha} and Q^{sp}_{sha} are the specified values of the active and reactive powers at the terminal end of the line connecting the ath VSC to the AC bus '$(i+a-1)$', respectively.

Thus, from Figure 3.2, for the ath VSC,

$$\mathrm{Re}\left(V_{sha}\, I^*_{sha}\right) + \sum_{v=1}^{q} V_{DCa} V_{DCv} Y_{DCav} = -P_{lossa} \tag{3.13}$$

Substituting Eqs. (3.1) and (3.9) in Eq. (3.13) and manipulating, we get

$$\left(m_a c\, V_{DCa}\right)^2 y_{sha}\cos\phi_{sha} - m_a c\, V_{DCa} V_{i+a-1} y_{sha} \cos\left(\theta_{sha} - \theta_{i+a-1} - \phi_{sha}\right)$$

$$+ \sum_{v=1}^{q} V_{DCa} V_{DCv} Y_{DCav} + P_{lossa} = 0$$

$$\text{or,} \quad f_{1a} = 0 \tag{3.14}$$

where $1 \le a \le q$.

Thus, 'q' independent equations are obtained. The derivation of Eq. (3.14) is shown in Appendix A.

Again from Figure 3.2, for the 'q' terminal VSC-HVDC system, there are one master VSC and '(q–1)' slave VSCs. The master VSC is used to control the voltage magnitude of its AC terminal bus while the slave VSCs operate in the PQ or PV control modes. The slave VSCs control the active and reactive power flows P_{sh} and Q_{sh} {as given by Eqs. (3.6) and (3.7), respectively} at the terminal end of the lines connecting the VSCs to their respective AC system buses. Again, without loss of generality, if the rth ($1 \le r \le q$) VSC is chosen to be the master VSC, the additional equations obtained for the slave VSCs can be expressed as

$$P_{sha}^{sp} - P_{sha}^{cal} = 0 \tag{3.15}$$

$$Q_{sha}^{sp} - Q_{sha}^{cal} = 0 \tag{3.16}$$

$$\forall a,\ 1 \le a \le q, a \ne r$$

In the above equations, P_{sha}^{sp} and Q_{sha}^{sp} are the specified active and the reactive powers, respectively, in the line connecting the ath slave VSC {$1 \le a \le q, a \ne r$} to its AC terminal bus '$(i + a-1)$' while P_{sha}^{cal} and Q_{sha}^{cal} are their calculated values which can be obtained using Eqs. (3.6) and (3.7).

Thus, we get '(2q–2)' independent equations corresponding to the '(q–1)' slave VSCs.

Now, the master VSC 'r' controls the voltage magnitude at its AC terminal bus. Thus, for the AC terminal bus corresponding to the ath VSC, if V_{i+a-1}^{sp} is the bus voltage control reference and V_{i+a-1}^{cal} is the calculated value of the voltage magnitude at bus '$(i + a-1)$', this can be expressed as

$$V_{i+a-1}^{sp} - V_{i+a-1}^{cal} = 0 \tag{3.17}$$

$\forall a, 1 \le a \le q,\ a = r$, {as the rth ($1 \le r \le q$) VSC is chosen to be the master VSC}.

It may be noted that a slave converter may be used to control the AC bus voltage magnitude rather than the line reactive power, in which case Eq. (3.16) becomes

$$V_{i+a-1}^{sp} - V_{i+a-1}^{cal} = 0 \qquad (3.18)$$

$$\forall a, 1 \le a \le q, \ a \ne r.$$

Now, similar to the AC power flow, a slack bus is chosen for the DC power flow and its voltage is pre specified. It serves the dual role of providing the DC voltage control and balancing the active power exchange among the VSCs. From Figure 3.2, in the 'q' terminal DC system, the first terminal is chosen as the DC slack bus, by convention. This is represented as

$$V_{DC1}^{sp} - V_{DC1}^{cal} = 0 \qquad (3.19)$$

At this stage, it is worthwhile to take stock of the unknown and the specified quantities. Corresponding to each VSC, three new variables enter into the picture. These include the modulation index 'm', the DC side voltage 'V_{DC}' and the phase angle 'θ_{sh}' of the AC side output voltage (phasor) of the VSC. Also, as discussed earlier, the DC side voltage 'V_{DC1}' of the first VSC is chosen as the slack bus. Thus, due to incorporation of the 'q' terminal VSC-HVDC network, '$(3q-1)$' additional variables need to be solved.

Against this, we have 'q' independent equations corresponding to the function 'f_1' {Eq. (3.14)} along with '$(2q-2)$' independent equations for the line active and reactive powers {Eqs. (3.6) and (3.7)} corresponding to the '$(q-1)$' slave VSCs. This gives as $(3q-2)$ independent equations. Now as the master VSC 'r' controls the voltage magnitude of its AC terminal bus, the net reactive power injection at that bus is available as a specified quantity. This can be expressed as

$$Q_{i+a-1}^{sp} - Q_{i+a-1}^{cal} = 0 \qquad (3.20)$$

$\forall a, 1 \le a \le q, \ a = r$. This completes the formulation.

3.2.3 MODELLING OF INTEGRATED AC-MVDC SYSTEMS IN THE BTB CONFIGURATION

Figure 3.3 shows the AC-MVDC network shown in Figure 3.1, now connected in the BTB configuration. The equivalent circuit of Figure 3.3 is shown in Figure 3.4.

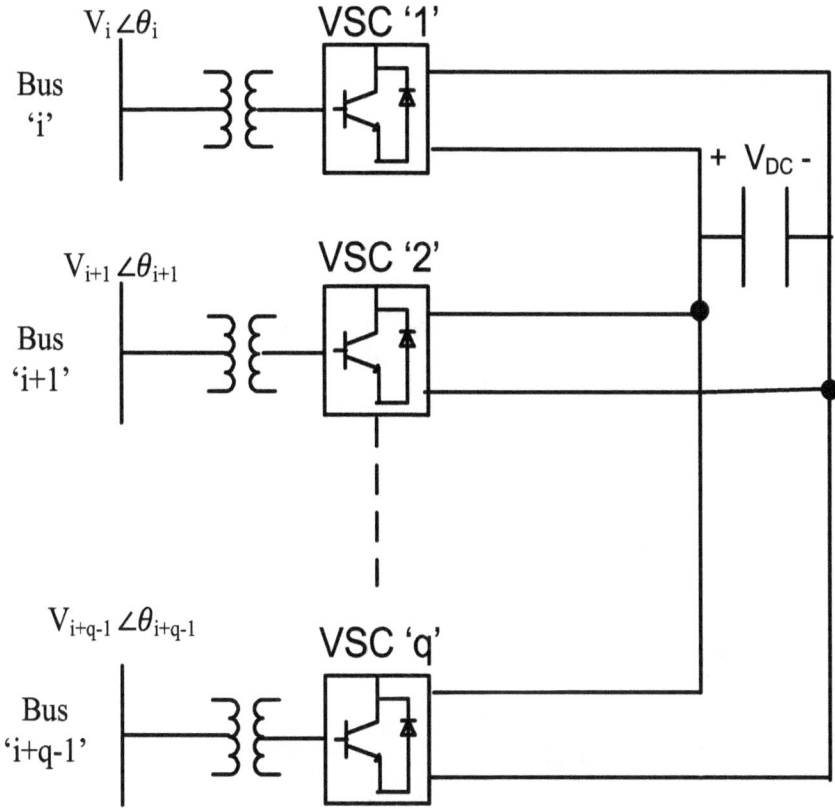

FIGURE 3.3   Schematic diagram of a 'q' terminal BTB VSC-HVDC system.

3.2.4   POWER-FLOW EQUATIONS OF INTEGRATED AC-MVDC SYSTEMS IN THE BTB CONFIGURATION

The equations for the AC-MVDC system with the VSC-HVDC network connected in the BTB configuration would be similar to those in Section 3.2.2, except some minor modifications. These are elaborated below.

It may be noted that in the BTB configuration of Figure 3.3, the DC side voltage 'V_{DC}' is common to all the 'q' VSCs. Thus, $V_{DCa} = V_{DC}$ $\forall a, 1 \leq a \leq q$. As a consequence, Eqs. (3.6) and (3.7) are retained with appropriate modifications ('V_{DCa}' replaced by 'V_{DC}'). However, as the DC network is now rendered lossless, the 'q' independent equations represented by Eq. (3.14) are now replaced by a single independent equation as shown below.

From Figures 3.3 and 3.4, it can be observed that

$$\sum_{a=1}^{q} \mathrm{Re}\left(\mathbf{V_{sha}}\ \mathbf{I_{sha}^{*}}\right) + \sum_{a=1}^{q} P_{lossa} = 0 \qquad (3.21)$$

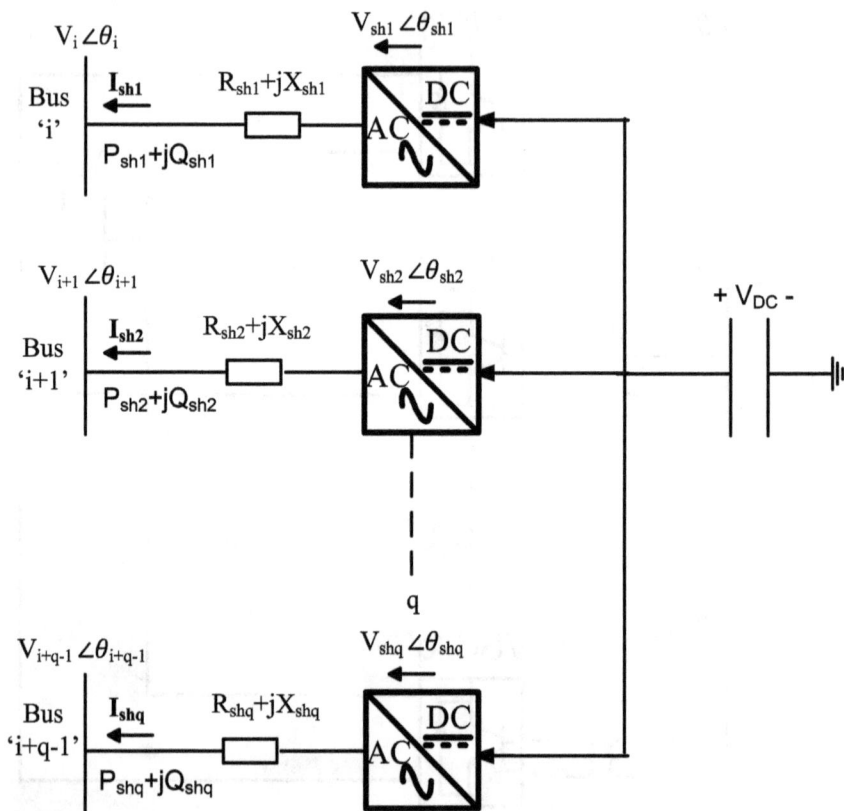

FIGURE 3.4 Equivalent circuit of a 'q' terminal BTB VSC-HVDC system.

Substituting Eqs. (3.1) and (3.9) in Eq. (3.21) and manipulating, we get

$$\sum_{a=1}^{q} \left[(m_a c\, V_{DC})^2\, y_{sha} \cos\phi_{sha} - m_a c\ _{DC} V_{i+a-1} y_{sha} \cos\left(\theta_{sha} - \theta_{i+a-1} - \phi_{sha}\right) + P_{lossa} \right] = 0$$

$$f_1 = 0 \qquad\qquad\qquad (3.22)$$

Thus, we get a single independent equation.

It may also be noted that Eq. (3.22) will be retained, without any modification.

3.3   IMPLEMENTATION OF POWER-FLOW IN INTEGRATED AC-MVDC SYSTEMS

In this section, the unified and the sequential AC–DC power-flow algorithms are used for solving the power-flow equations developed in Section 3.2. The unified method is implemented first for both the PTP and the BTB configurations, followed by the sequential method.

3.3.1 Unified AC–DC Power-Flow Method

This method deals with the simultaneous solution of the AC and DC variables. The unified method is used to solve the AC–DC power-flow equations with the MVDC system connected in the PTP and the BTB configurations. The PTP configuration is discussed first.

3.3.1.1 Unified AC–DC Power-Flow Method for PTP Configuration

In Figure 3.1, without any loss of generality, if it is assumed that there are 'g' generators connected at the first 'g' buses of the 'n' bus AC system with bus 1 being the slack bus, then the Newton power-flow equation for the AC power system incorporated with the 'q' terminal VSC-HVDC network connected in the PTP configuration can be written as

Solve: $\boldsymbol{\theta}, \mathbf{V}, \mathbf{X}$
Specified: $\mathbf{P}, \mathbf{Q}, \mathbf{R}$

where

$$\boldsymbol{\theta} = \left[\theta_2 \cdots \theta_n\right]^T, \mathbf{V} = \left[V_{g+1} \cdots V_n\right]^T, \boldsymbol{\theta}_{sh} = \left[\theta_{sh1} \cdots \theta_q\right]^T, \mathbf{m} = \left[m_1 \cdots m_q\right]^T$$

$$\mathbf{V_{DC}} = \left[V_{DC2} \cdots V_{DCq}\right]^T \text{ and } \mathbf{X} = \left[\boldsymbol{\theta}_{sh}^T \ \mathbf{m}^T \ \mathbf{V}_{DC}^T\right]^T$$

$$\mathbf{P} = \left[P_2 \cdots P_n\right]^T, \mathbf{Q} = \left[Q_{g+1} \cdots Q_n\right]^T, \mathbf{P}_{sh} = \left[P_{sh2}, \ldots P_{shq}\right], \mathbf{Q}_{sh} = \left[Q_{sh2}, \ldots Q_{shq}\right],$$

$$\mathbf{f}_1 = \left[f_{11} \cdots f_{1q}\right] \text{ and } \mathbf{R} = \left[\mathbf{P}_{sh}, \mathbf{Q}_{sh}, V_{i+r-1}, \mathbf{f}_1\right]^T$$

Thus the basic equation for the Newton power-flow analysis is given below:

$$
\begin{bmatrix}
\mathbf{J}_{old} & & \dfrac{\partial \mathbf{P}}{\partial \boldsymbol{\theta}_{sh}} & \dfrac{\partial \mathbf{P}}{\partial \mathbf{m}} & \dfrac{\partial \mathbf{P}}{\partial \mathbf{V}_{DC}} \\[2mm]
 & & \dfrac{\partial \mathbf{Q}}{\partial \boldsymbol{\theta}_{sh}} & \dfrac{\partial \mathbf{Q}}{\partial \mathbf{m}} & \dfrac{\partial \mathbf{Q}}{\partial \mathbf{V}_{DC}} \\[2mm]
\dfrac{\partial \mathbf{R}}{\partial \boldsymbol{\theta}} & \dfrac{\partial \mathbf{R}}{\partial \mathbf{V}} & \dfrac{\partial \mathbf{R}}{\partial \boldsymbol{\theta}_{sh}} & \dfrac{\partial \mathbf{R}}{\partial \mathbf{m}} & \dfrac{\partial \mathbf{R}}{\partial \mathbf{V}_{DC}}
\end{bmatrix}
\begin{bmatrix}
\Delta \boldsymbol{\theta} \\ \Delta \mathbf{V} \\ \Delta \boldsymbol{\theta}_{sh} \\ \Delta \mathbf{m} \\ \Delta \mathbf{V}_{DC}
\end{bmatrix}
=
\begin{bmatrix}
\Delta \mathbf{P} \\ \Delta \mathbf{Q} \\ \Delta \mathbf{R}
\end{bmatrix}
\quad (3.23)
$$

In Eq. (3.23), $\mathbf{J}_{old}$ is the conventional load flow (without incorporating HVDC link) Jacobian sub-block given as follows:

$$
\mathbf{J}_{old} =
\begin{bmatrix}
\dfrac{\partial \mathbf{P}}{\partial \boldsymbol{\theta}} & \dfrac{\partial \mathbf{P}}{\partial \mathbf{V}} \\[2mm]
\dfrac{\partial \mathbf{Q}}{\partial \boldsymbol{\theta}} & \dfrac{\partial \mathbf{Q}}{\partial \mathbf{V}}
\end{bmatrix}
$$

Also, in Eq. (3.23), '$\Delta \mathbf{P}$', '$\Delta \mathbf{Q}$' and '$\Delta \mathbf{R}$' represent the mismatch vectors. In addition, $\Delta \boldsymbol{\theta}, \Delta \mathbf{V}, \Delta \boldsymbol{\theta}_{sh}, \Delta \mathbf{m}$ and $\Delta \mathbf{V}_{DC}$ represent correction vectors. In the

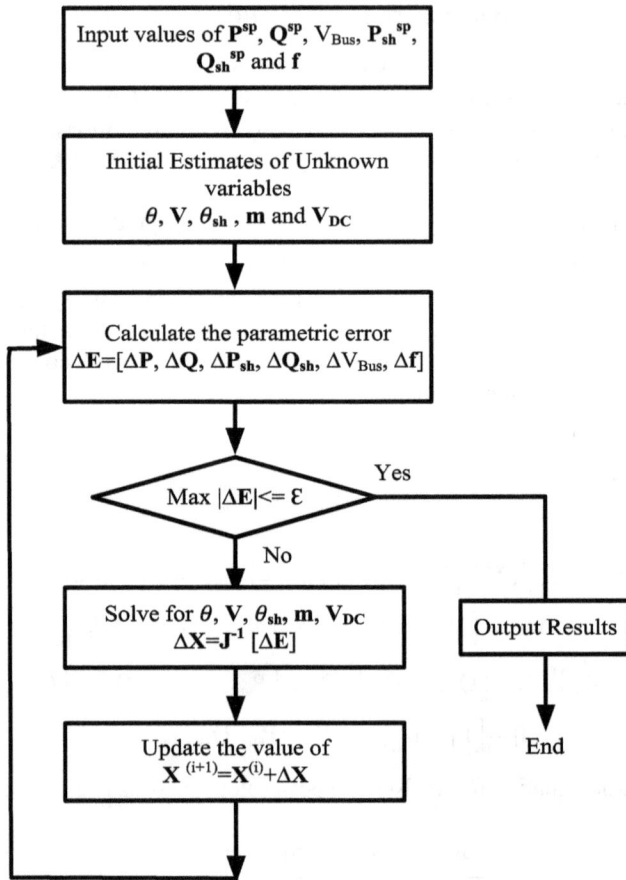

FIGURE 3.5   Flow chart of the unified NR AC-MVDC power-flow algorithm for the PTP connection.

above formulation, it is assumed that all the slave converters operate in the PQ control mode. However, it may be noted that if they are made to operate in the PV control mode, the corresponding elements of the correction and mismatch vectors have to be modified accordingly. Some typical elements of Eq. (3.23) are given in Appendix A.

Figure 3.5 depicts the flow chart of the unified AC-MVDC Newton-Raphson power-flow algorithm with the MVDC network configured in the PTP fashion.

3.3.1.2 Unified AC–DC Power-Flow Method for BTB Configuration

In Figure 3.3, again, without any loss of generality, if it is assumed that there are 'g' generators connected at the first 'g' buses of the 'n' bus AC system with bus 1 being the slack bus, then the Newton power-flow equation for the AC power system network incorporated with the 'q' terminal HVDC network connected in the BTB configuration can be written as

Solve: θ, **V**, **X**
Specified: **P**, **Q**, **R**

where

$$\theta = [\theta_2 \cdots \theta_n]^T, \; \mathbf{V} = [V_{g+1} \cdots V_n]^T, \; \theta_{sh} = [\theta_{sh1} \cdots \theta_{shq}]^T, \; \mathbf{m} = [m_1 \cdots m_q]^T$$
$$\text{and } \mathbf{X} = [\theta_{sh}^T \; \mathbf{m}^T]^T$$
$$\mathbf{P} = [P_2 \cdots P_n]^T, \; \mathbf{Q} = [Q_{g+1} \cdots Q_n]^T, \; \mathbf{P}_{sh} = [P_{sh2} \cdots P_{shq}], \; \mathbf{Q}_{sh} = [Q_{sh2} \cdots Q_{shq}],$$
$$\text{and } \mathbf{R} = [\; \mathbf{P}_{sh} \;\; \mathbf{Q}_{sh} \;\; V_{i+r-1} \;\; f_i \;]^T$$

Thus, the basic equation for the Newton power-flow analysis is

$$\mathbf{J} \begin{bmatrix} \Delta\theta \\ \Delta V \\ \Delta\theta_{sh} \\ \Delta m \end{bmatrix} = \begin{bmatrix} \Delta P \\ \Delta Q \\ \Delta R \end{bmatrix} \tag{3.24}$$

where $\mathbf{J} = \begin{bmatrix} & & \dfrac{\partial P}{\partial \theta_{sh}} & \dfrac{\partial P}{\partial m} \\ \mathbf{J}_{old} & & & \\ & & \dfrac{\partial Q}{\partial \theta_{sh}} & \dfrac{\partial Q}{\partial m} \\ \dfrac{\partial R}{\partial \theta} & \dfrac{\partial R}{\partial V} & \dfrac{\partial R}{\partial \theta_{sh}} & \dfrac{\partial R}{\partial m} \end{bmatrix}$ is the Jacobian matrix.

In Eq. (3.24), '$\mathbf{J}_{old}$' is the conventional power-flow Jacobian sub-block corresponding to the 'n' bus AC system. Also, in Eq. (3.24), '$\Delta \mathbf{R}$' is the vector comprising the mismatches of the control specifications of the VSC-HVDC. In the above formulation, it is assumed that all the slave converters operate in the PQ control mode. However, it may be noted that if they are made to operate in the PV control mode, the corresponding elements of the correction and mismatch vectors have to be modified accordingly.

Figure 3.6 depicts the flow chart of the unified AC-MVDC Newton-Raphson power-flow algorithm with the MVDC network configured in the BTB fashion.

3.3.2 Sequential AC–DC Power-Flow Method

In this method, the AC and DC equations are solved separately. The sequential method is used to solve the AC–DC power-flow equations with the MVDC system connected in the PTP and the BTB configurations. The PTP configuration is discussed below.

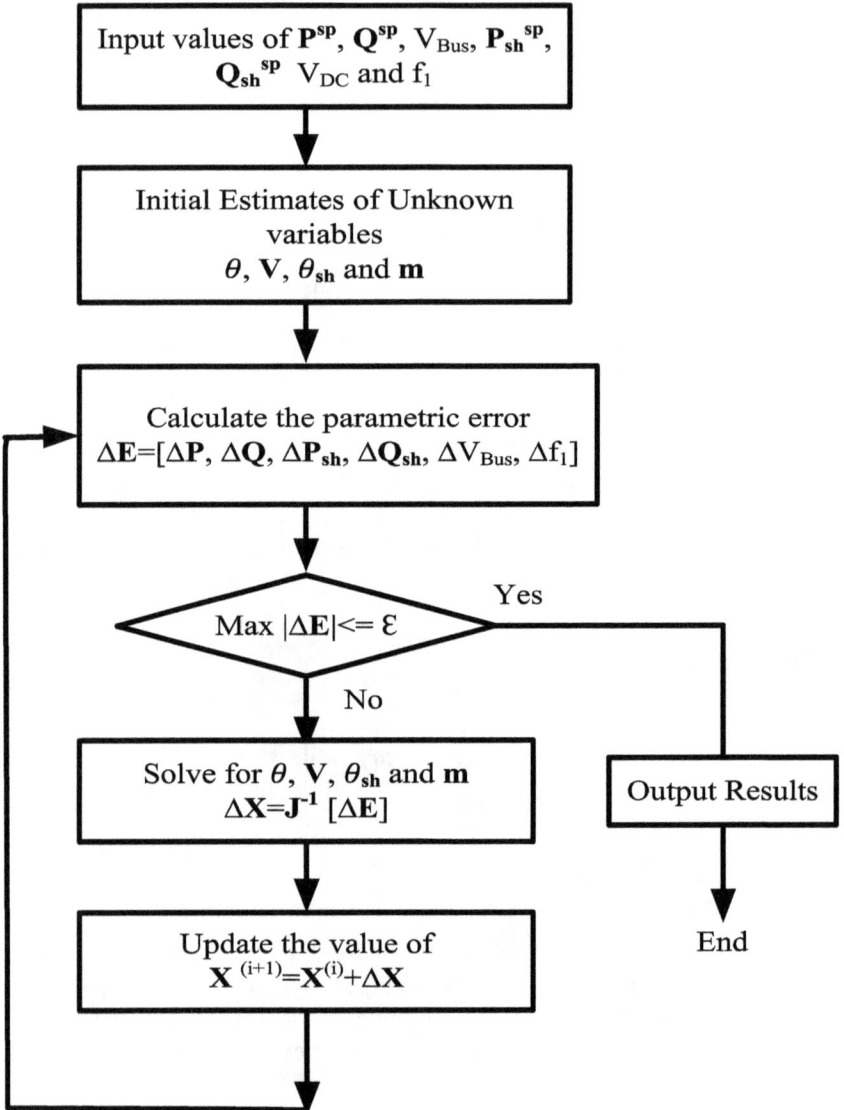

FIGURE 3.6 Flow chart of the unified NR AC-MVDC power-flow algorithm for the BTB connection.

3.3.2.1 Sequential AC–DC Power-Flow Method for PTP Configuration

In the sequential method, the effect of the 'q' VSCs represented as 'q' equivalent loads on the secondary sides of the converter transformers. This is shown in Figure 3.7. The secondaries of the 'q' converter transformers are shown connected

$$V_i \angle \Theta_i \qquad\qquad V_j \angle \Theta_j$$

Bus
'i'

$$\longleftarrow (P_{sh1} + j\, Q_{sh1})$$

$$(P_{sh1}{}^1 + j\, Q_{sh1}{}^1)$$

$$V_{i+1} \angle \Theta_{i+1} \qquad\qquad V_{j+1} \angle \Theta_{j+1}$$

Bus
'i+1'

$$\longleftarrow (P_{sh2} + j\, Q_{sh2})$$

$$(P_{sh2}{}^1 + j\, Q_{sh2}{}^1)$$

$$V_{i+q-1} \angle \Theta_{i+q-1} \qquad\qquad V_{j+q-1} \angle \Theta_{j+q-1}$$

Bus
'i+q-1'

$$\longleftarrow (P_{shq} + j\, Q_{shq})$$

$$(P_{shq}{}^1 + j\, Q_{shq}{}^1)$$

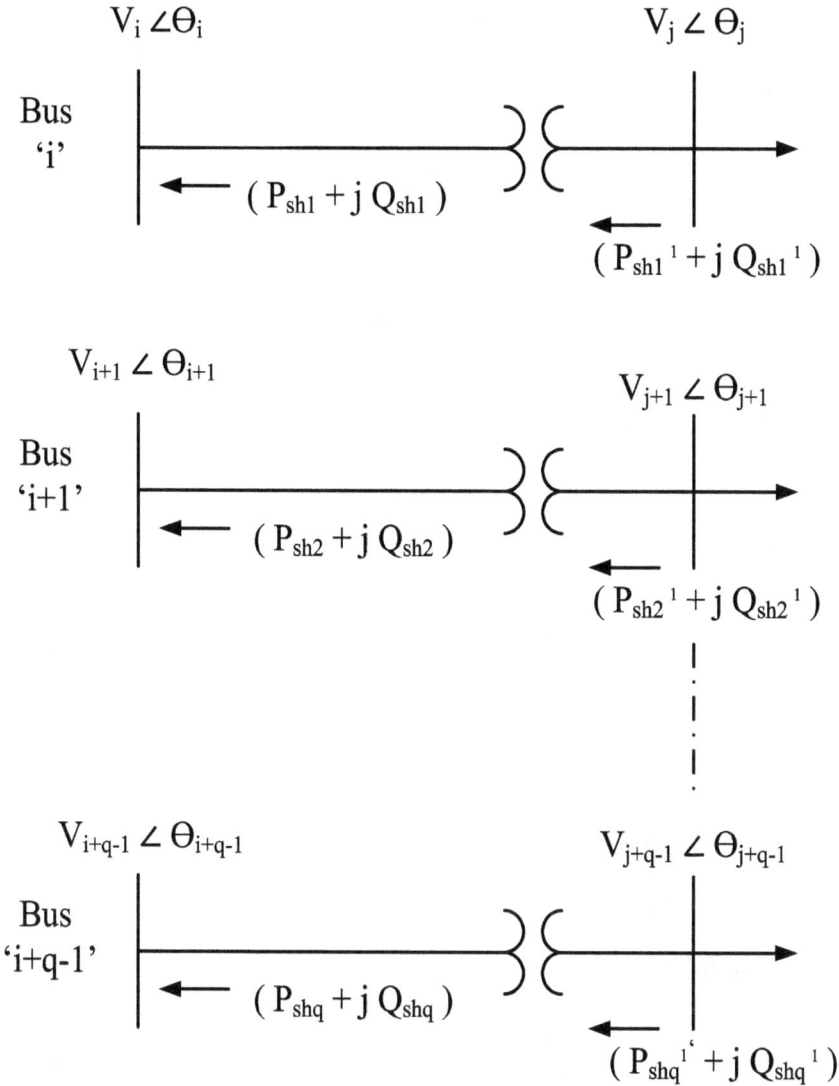

FIGURE 3.7 Representation of VSCs as equivalent complex load powers.

to 'q' fictitious AC buses 'j', '(j + 1)', and so on, up to bus '(j + q–1)', as shown in the figure. The effect of the 'q' VSCs are represented as 'q' equivalent complex loads at these fictitious AC buses 'j' to '(j + q–1)'. In this respect, it may be noted from Figure 3.7 that '(q–1)' complex powers ('S_{sh2}' to 'S_{shq}') are specified only in the terminal ends of the lines connected to their AC terminal buses {buses '(i + 1)' to '(i + q–1)'}. Let $S_{sha} = P_{sha} + jQ_{sha}$ ($2 \le a \le q$) be the complex power specified in the terminal end of the line connected to the AC bus '(i + a–1)' through the ath coupling transformer. Then, from Figure 3.7, the active component of the equivalent complex load power at the fictitious AC bus '(j + a–1)' {which represents the effect of the ath VSC} can be expressed as

$$P_{sha}{}' = P_{sha} + I_{sha}^2 \, R_{sha} = P_{sha}^{sp} + \left(\frac{P_{sha}^{sp\,2} + Q_{sha}^{sp\,2}}{V_{i+a-1}^2} \right) R_{sha} \qquad (3.25)$$

where P_{sha}^{sp} is the specified active power in the terminal end of the line connected to the AC bus '(i + a–1)'.

But from Figure 3.2, the active power delivered by the ath ($1 \le a \le q$) VSC at its AC terminal is

$$P_{sha}{}' = Re\left(V_{sha} \, I_{sha}^* \right)$$

$$= m_a^2 \, c^2 \, V_{DCa}^2 \, y_{sha} \, \cos\phi_{sha} - m_a c V_{DCa} V_{i+a-1} y_{sha} \cos\left(\theta_{sha} - \theta_{i+a-1} - \phi_{sha} \right) \quad (3.26)$$

Substituting Eq. (3.25) in (3.26), we get

$$m_a^2 \, c^2 \, V_{DCa}^2 \, y_{sha} \, \cos\phi_{sha} - m_a c \, V_{DCa} V_{i+a-1} y_{sha} \cos\left(\theta_{sha} - \theta_{i+a-1} - \phi_{sha} \right)$$

$$- P_{sha}^{sp} - \left(\frac{P_{sha}^{sp\,2} + Q_{sha}^{sp\,2}}{V_{i+a-1}^2} \right) R_{sha} = 0$$

$$\text{or,} \quad f_{2a} = 0 \qquad (3.27)$$

In a similar manner, the reactive component of the equivalent complex load power at the fictitious AC bus '(j + a–1)' {which represents the effect of the ath VSC} can be expressed as

$$-m_a^2 \, c^2 \, V_{DCa}^2 \, y_{sha} \, \sin\phi_{sha} - m_a c \, V_{DCa} V_{i+a-1} y_{sha} \sin\left(\theta_{sha} - \theta_{i+a-1} - \phi_{sha} \right)$$

$$- Q_{sha}^{sp} - \left(\frac{P_{sha}^{sp\,2} + Q_{sha}^{sp\,2}}{V_{i+a-1}^2} \right) X_{sha} = 0$$

$$\text{or,} \quad f_{3a} = 0 \qquad (3.28)$$

If the ath ($1 \le a \le q$) VSC operates in the PV control mode, 'Q_{sha}' is not specified, and hence, Eq. (3.26) is modified to {using Eq. (A.8) of Appendix A}

$$P_{sha}{}' - P_{sha}^{sp} - I_{sha}^2 R_{sha} = 0$$

$$\text{or,} \quad m_a^2 c^2 V_{DCa}^2 \, y_{sha} \, \cos\phi_{sha} - m_a c V_{DCa} V_{i+a-1} y_{sha} \cos\left(\theta_{sha} - \theta_{i+a-1} - \phi_{sha} \right) - P_{sha}^{sp}$$

$$- \left[V_{i+a-1}^2 y_{sha}^2 + m_a^2 \, c^2 V_{DCa}^2 \, y_{sha}^2 - 2m_{aghg} c V_{DCa} V_{i+a-1} \cos\left(\theta_{i+a-1} - \theta_{sha} \right) \right] R_{sha} = 0$$

$$\text{or,} \quad f_{4a} = 0 \qquad (3.29)$$

AC Network Solution

Corresponding to Figure 3.1, the Newton power-flow equation for the sequential solution of the AC power system variables can be written as

Solve: θ, **V**, **X**
Specified: **P**, **Q**, **R**

where

$$\theta = \left[\theta_2 \cdots \theta_n\right]^T, \mathbf{V} = \left[V_{g+1} \cdots V_n\right]^T, \theta_{sh} = \left[\theta_{sh1} \cdots \theta_q\right]^T, \mathbf{m} = \left[m_1 \cdots m_q\right]^T$$

and $\mathbf{X} = \left[\theta_{sh}^T, \mathbf{m}^T\right]^T$

$$\mathbf{P} = \left[P_2 \cdots P_n\right]^T, \mathbf{Q} = \left[Q_{g+1} \cdots Q_n\right]^T,$$

$\mathbf{f}_2 = \left[f_{22} \cdots f_{2q}\right], \mathbf{f}_3 = \left[f_{32} \cdots f_{3q}\right]$ and $\mathbf{R} = \left[\ V_{i+r-1}, \mathbf{f}_2, \mathbf{f}_3, f_{11}\right]^T$

Thus the basic Newton power-flow equation is given below:

$$
\begin{bmatrix}
& & \dfrac{\partial \mathbf{P}}{\partial \theta_{sh}} & \dfrac{\partial \mathbf{P}}{\partial \mathbf{m}} \\
\mathbf{J}_{old} & & & \\
& & \dfrac{\partial \mathbf{Q}}{\partial \theta_{sh}} & \dfrac{\partial \mathbf{Q}}{\partial \mathbf{m}} \\
\dfrac{\partial \mathbf{R}}{\partial \theta} & \dfrac{\partial \mathbf{R}}{\partial \mathbf{V}} & \dfrac{\partial \mathbf{R}}{\partial \theta_{sh}} & \dfrac{\partial \mathbf{R}}{\partial \mathbf{m}}
\end{bmatrix}
\begin{bmatrix}
\Delta\theta \\
\Delta\mathbf{V} \\
\Delta\theta_{sh} \\
\Delta\mathbf{m}
\end{bmatrix}
=
\begin{bmatrix}
\Delta\mathbf{P} \\
\Delta\mathbf{Q} \\
\Delta\mathbf{R}
\end{bmatrix}
\quad (3.30)
$$

In Eq. (3.30), $\mathbf{J}_{old}$ is the conventional load flow (without incorporating HVDC link) Jacobian sub-block given as follows:

$$
\mathbf{J}_{old} =
\begin{bmatrix}
\dfrac{\partial \mathbf{P}}{\partial \theta} & \dfrac{\partial \mathbf{P}}{\partial \mathbf{V}} \\
\dfrac{\partial \mathbf{Q}}{\partial \theta} & \dfrac{\partial \mathbf{Q}}{\partial \mathbf{V}}
\end{bmatrix}
$$

Also, in Eq. (3.30), '$\Delta\mathbf{P}$', '$\Delta\mathbf{Q}$' and '$\Delta\mathbf{R}$' represent the mismatch vectors. In addition, $\Delta\theta$, $\Delta\mathbf{V}$, $\Delta\theta_{sh}$ and $\Delta\mathbf{m}$ represent correction vectors. In the above formulation, it is assumed that all the slave converters operate in the PQ control mode. However, it may be noted that if they are made to operate in the PV control mode, the corresponding elements of the correction and mismatch vectors have to be modified accordingly. Some typical elements of Eq. (3.30) are given in Appendix A.

DC network solution

Again, corresponding to Figure 3.1, the Newton power-flow equation for the sequential solution of the DC variables of the 'q' bus MVDC system can be written as

Solve: $\mathbf{V}_{DC}$
Specified: $\mathbf{f}_5$

where

$\mathbf{V}_{DC} = \begin{bmatrix} V_{DC2} \cdots V_{DCq} \end{bmatrix}$ and $\mathbf{f}_5 = \begin{bmatrix} f_{12} \cdots f_{1q} \end{bmatrix}$

Thus the basic Newton power-flow equation is given below

$$\left[\frac{\partial \mathbf{f}_5}{\partial \mathbf{V}_{DC}} \right] [\Delta \mathbf{V}_{DC}] = [\Delta \mathbf{f}_5] \tag{3.31}$$

Also, in Eq. (3.31), '$\Delta \mathbf{f}_5$' is the mismatch vector whereas $\Delta \mathbf{V}_{DC}$ represents the correction vector comprising the DC voltages of all DC buses except the DC slack bus Figure 3.8 depicts the flow chart of the sequential AC-MVDC power flow algorithm with the MVDC network configured in the PTP fashion.

3.3.2.2 Sequential AC–DC Power-Flow Method for BTB Configuration

In the BTB configuration, the converters are at the same location and interconnected through a common DC link. Since this DC voltage is specified, the sequential method is not applicable for such AC–DC networks.

3.4 CASE STUDIES AND RESULTS

Several case studies were implemented to justify the results. For all successive studies, an M-VSC-HVDC network was integrated in the IEEE 300-bus [115] and European 1354-bus test systems [116]. The resistances and the leakage reactance of all the converter transformers were taken as 0.001 and 0.1 p.u., respectively, for the case studies. For all the case studies, the resistance of each DC link was chosen as 0.01 p.u. The converter loss constants 'a_1', 'b_1' and 'c_1' were selected as 0.011, 0.003 and 0.0043, respectively [18,105]. In addition, the value of 'c' for the VSC-based converters was uniformly adopted as $\frac{1}{2\sqrt{2}}$ [50]. The initial values of variables corresponding to VSC-based HVDC system were shown in Appendix A. A convergence tolerance of 10^{-8} p.u. was uniformly adopted for all the case studies. For all the case studies, 'NI' denotes the number of iterations for the algorithm to converge.

3.4.1 STUDIES WITH UNIFIED POWER-FLOW MODEL OF IEEE 300-BUS TEST SYSTEM INTEGRATED WITH VSC-BASED MULTI-TERMINAL DC (MVDC) GRIDS

Case I: Three-terminal MVDC network connected in the BTB configuration

In this study, a three-terminal BTB-connected VSC-HVDC network is incorporated in the IEEE 300-bus system between buses 266, 270 and 271. The VSC

Input values of $\mathbf{P^{sp}}$, $\mathbf{Q^{sp}}$, V_{Bus}, $\mathbf{P_{sh}}$, $\mathbf{Q_{sh}}$, V_{DC1}, $\mathbf{f_1}$, $\mathbf{f_2}$ and $\mathbf{f_3}$

Initial estimates of unknown variables θ, V, θ_{sh}, m and V_{DC}

Calculate the parametric Error for AC network
$\Delta E_{AC} = [\Delta \mathbf{P}, \Delta \mathbf{Q}, \Delta V_{Bus}, \Delta f_{11}, \Delta f_2 \text{ and } \Delta f_3]$

Solve for θ, V, θ_{sh}, and m
$\Delta \mathbf{X_{AC}} = \mathbf{J^{-1}} [\Delta \mathbf{E_{AC}}]$

Update the value of
$\mathbf{X_{AC}}^{(i+1)} = \mathbf{X_{AC}}^{(i)} + \Delta \mathbf{X_{AC}}$

Calculate the parametric Error for DC network
$\Delta \mathbf{E_{DC}} = [\Delta f_{12} .... \Delta f_{1q}]$

$\text{Max} [|\Delta \mathbf{E_{AC}}|, |\Delta \mathbf{E_{DC}}|] <= \varepsilon$

Output results

Solve for V_{DC}
$\Delta \mathbf{X_{DC}} = \mathbf{J^{-1}} [\Delta \mathbf{E_{DC}}]$

End

Update the value of
$\mathbf{X_{DC}}^{(i+1)} = \mathbf{X_{DC}}^{(i)} + \Delta \mathbf{X_{DC}}$

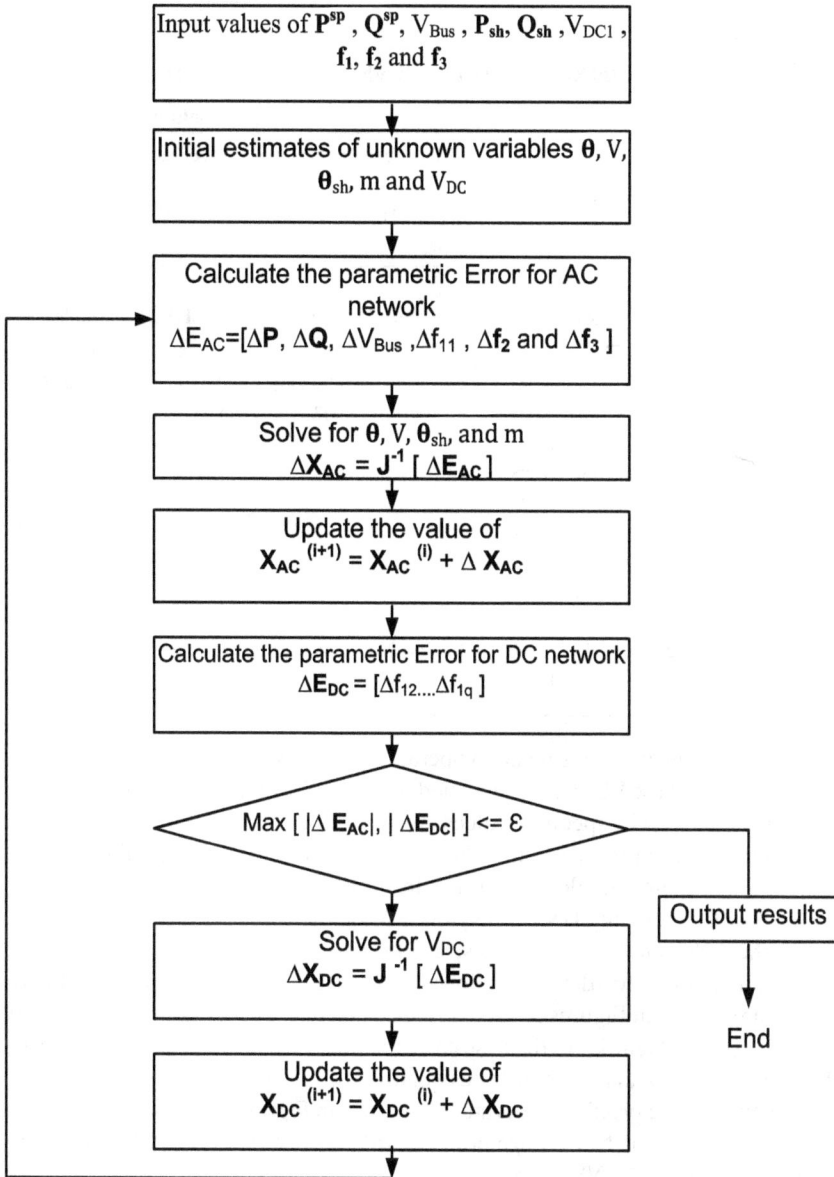

FIGURE 3.8   Flow chart of the sequential AC-MVDC power-flow algorithm for the PTP connection.

TABLE 3.1

Study of IEEE 300-Bus System with BTB VSC-HVDC Network

			Power-Flow Solution	
			Base Case Power Flow Converged in Six Iterations (NI = 6)	
HVDC Link Connection Details			$V_i = 1.011\angle -11.24;\ V_j = 1.011\angle -11.32;$ $V_k = 0.998\angle -17.67;$	
Master Converter	**Slave Converters**	**HVDC Link Specifications**	**AC Terminal Buses**	**HVDC Variables**
I	j, k	Master converter $V_{DC} = 3;$ $V_i = 1.02;$ Slave converters $P_{shj} = 0.5;$ $Q_{shj} = 0.2;$ $P_{shk} = 0.3;$ $Q_{shk} = 0.1;$	$V_j = 1.0207\angle -11.3819;$ $V_k = 1.0689\angle -10.3345;$ $\theta_i = -11.4007;$	Master converter $V_{shi} = 1.0410\angle -15.9687;$ $m_i = 0.9815; P_{lossi}\,(\%) = 1.66;$ Slave converters $V_{shj} = 1.0420\angle -8.6981;$ $V_{shk} = 1.0789\angle -8.8489;$ $m_j = 0.9824; m_k = 1.0172;$ $P_{lossj}\,(\%) = 1.38; P_{lossk}\,(\%) = 1.23;$ NI = 6

Note: For the above case study, i = 266, j = 270, k = 271. Values of voltage magnitudes, active and reactive powers are in p.u. Phase angles of voltages are in degrees.

connected to bus no. 266 is made to operate as the master converter. On the other hand, both the slave VSCs are connected to buses 270 and 271 and operate in the PQ control mode. The specified values are given in columns 1–3 of Table 3.1. The power-flow solution is shown in columns 4–5 of Table 3.1. From Table 3.1, it is observed that the power flow converges in six iterations, similar to the base case power flow (without any HVDC network). Also, the power-flow solution directly yields the VSC modulation indices, unlike existing models. The convergence characteristic plots for the power flows of the base case and BTB VSC-HVDC system are shown in Figures 3.9 and 3.10, respectively. From Figures 3.9 and 3.10, it is observed that similar to the base case, the proposed unified AC–DC Newton-Raphson algorithm also demonstrates quadratic convergence.

The bus voltage profile for this study is shown in Figure 3.11. From Figure 3.11, it is observed that the bus voltage profile hardly changes except for the AC terminal buses at which the VSCs are installed.

Case II: Three-terminal MVDC network connected in the PTP configuration
In this case study, two separate studies are conducted with a three-terminal, PTP-connected VSC-HVDC network incorporated in the IEEE-300 bus test system to demonstrate the versatility of the proposed model. In both the studies, the VSC-HVDC network is connected between AC buses 266, 270 and 271.

In the first study, the converter connected to AC bus no. '266' acts as the master converter while those connected to AC buses '270' and '271' act as slave

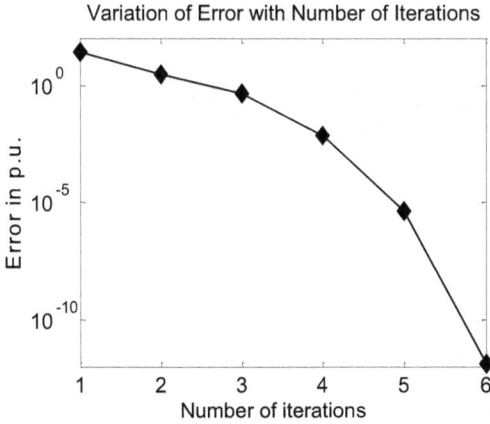

FIGURE 3.9 Convergence characteristic of the base case power flow in the IEEE-300 bus system.

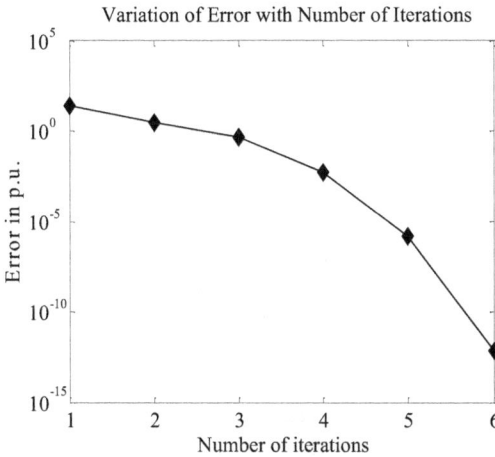

FIGURE 3.10 Convergence characteristic for the study of Table 3.1.

converters. Both the slave converters operate in the PQ control mode. The master converter maintains the voltage magnitude of AC bus no. 266 at a value of 1.02 p.u. The active powers at the terminal end of the lines connecting the converters to the AC buses '270' and '271' are specified as 0.7 and 0.5 p.u., respectively. In a similar manner, the corresponding line reactive powers are specified as 0.2 and 0.06 p.u., respectively. These specified values are shown in the third row and columns 1–3 of Table 3.2. The power-flow solution is shown in the third row and columns 4–5 of Table 3.2.

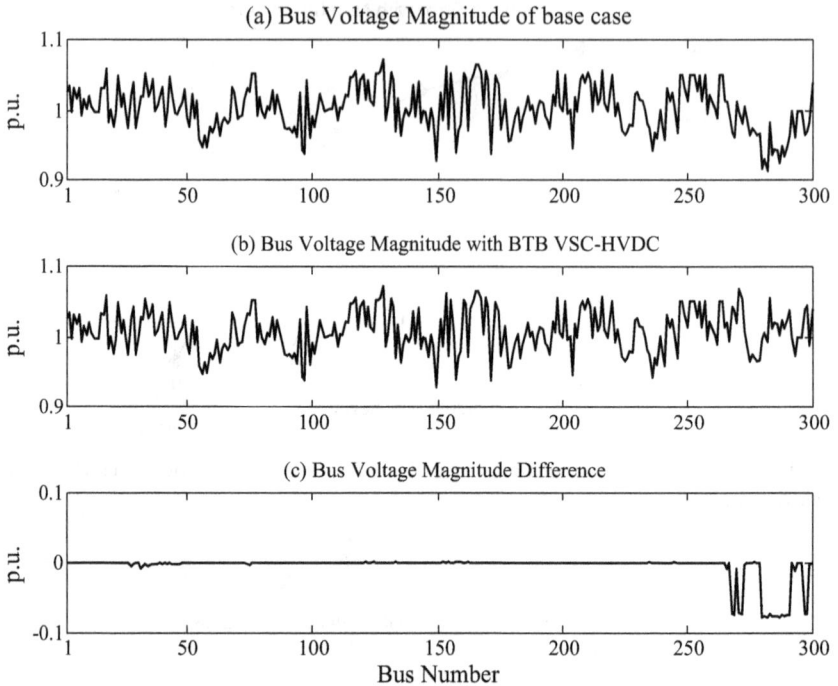

FIGURE 3.11   Bus voltage profile for the study of Table 3.1.

Subsequently a study is again conducted on the same AC–DC system but with both the slave converters (connected to AC buses '270' and '271') operated in the PV control mode. Their terminal end line active powers are specified as 0.4 and 0.5 p.u., respectively. The voltages of the AC buses '270' and '271' connected to the slave converters are specified as 1.02 and 1.0 p.u., respectively. The specified quantities are detailed in the fourth row and columns 1–3 of Table 3.2. The power-flow results are shown in the fourth row and columns 4–5 of Table 3.2.

From Table 3.2, it can be observed that for all the case studies, 'NI' is identical to that in the base case. It is observed that 'NI' is independent of the control strategies employed as well as the operating points specified. This demonstrates the robustness of the proposed algorithm.

The convergence characteristic plots for both the studies of Table 3.2 are shown in Figures 3.12 and 3.13, respectively. From Figures 3.12 and 3.13, again, it can be observed that the proposed model possesses quadratic convergence characteristics, similar to the base case power flow.

The bus voltage profiles for the first and second studies of Table 3.2 are shown in Figures 3.14 and 3.15, respectively. From Figures 3.14 and 3.15, it is observed that the bus voltage profile hardly changes except for the AC terminal buses at which the VSCs are installed.

TABLE 3.2

Study of IEEE 300-Bus System with Three-Terminal PTP VSC-HVDC Network

			Power-Flow Solution	
HVDC Link Connection Details			Base Case Power Flow Converged in Six Iterations (NI = 6) $V_i = 1.011\angle -11.24$; $V_j = 1.011\angle -11.32$; $V_k = 0.998\ \angle -17.67$;	
Master Converter	Slave Converters	HVDC Link Specifications	AC Terminal Buses	HVDC Variables
I	j, k	Master converter $V_{DCi} = 3$; $V_i = 1.02$; Slave converters $P_{shj} = 0.7$; $Q_{shj} = 0.2$; $P_{shk} = 0.5$; $Q_{shk} = 0.06$;	$\theta_i = -11.4398$; $\theta_j = -11.3827$; $\theta_k = -5.4606$; $V_j = 1.0209$; $V_k = 1.0503$;	Master converter $V_{shi} = 1.0516\angle -18.1648$; $m_i = 0.9914$; $P_{DCi} = 1.23$; $P_{lossi}(\%) = 2.15$; Slave converters $V_{shj} = 1.0434\angle -7.6254$; $V_{shk} = 1.0576\angle -2.8837$; $V_{DCj} = 2.9978$; $V_{DCk} = 2.9981$; $m_j = 0.9844$; $m_k = 0.9977$; $P_{DCj} = -0.7158$; $P_{DCk} = -0.5137$; $P_{lossj}(\%) = 1.53$; $P_{lossk}(\%) = 1.34$; NI = 6;
I	j, k	Master converter $V_{DCi} = 3$; $V_i = 1.02$; Master converter $P_{shj} = 0.4$; $V_j = 1.02$; $P_{shk} = 0.5$; $V_k = 1$;	$\theta_i = -11.4236$; $\theta_j = -11.4157$; $\theta_k = -5.0076$;	Master converter $V_{shi} = 1.0781\angle -16.3970$; $m_i = 1.0165$; $P_{DCi} = 0.9273$; $P_{lossi}(\%) = 1.93$; Slave converters $V_{shj} = 1.0213\angle -9.2151$; $V_{shk} = 0.9975\angle -2.1320$; $V_{DCj} = 2.9985$; $V_{DCk} = 2.9984$; $m_j = 0.9633$; $m_k = 0.9410$; $P_{DCj} = -0.4130$; $P_{DCk} = -0.5138$; $P_{lossj}(\%) = 1.28$; $P_{lossk}(\%) = 1.36$; NI = 6;

Note: For the above case study, i = 266, j = 270, k = 271. Values of voltage magnitudes, active and reactive powers are in p.u. Phase angles of voltages are in degrees.

Case III: Five-terminal MVDC network connected in the PTP configuration

In this case study, a five-terminal VSC-HVDC network is incorporated in the test system between buses 266, 270, 271, 272 and 273. The converter connected to bus no. 266 acts as the master converter while the converters connected to buses 270, 271, 272 and 273 act as slave converters. The master converter controls the voltage magnitude of AC bus no. 266 to a value of 1.02 p.u. The slave converters connected to buses 272 and 273 operate in the PV control mode while those

Variation of Error with Number of Iterations

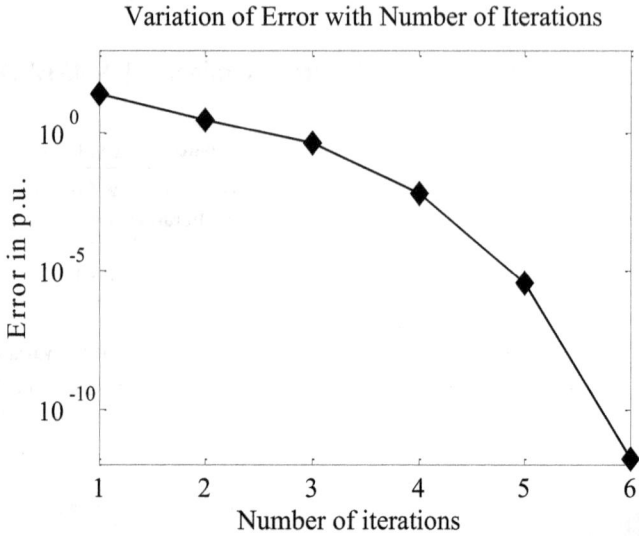

FIGURE 3.12   Convergence characteristic for the first study of Table 3.2.

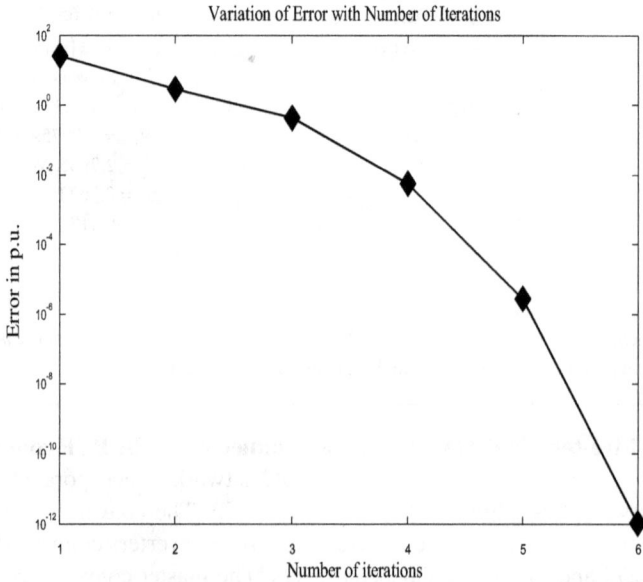

FIGURE 3.13   Convergence characteristic for the second study of Table 3.2.

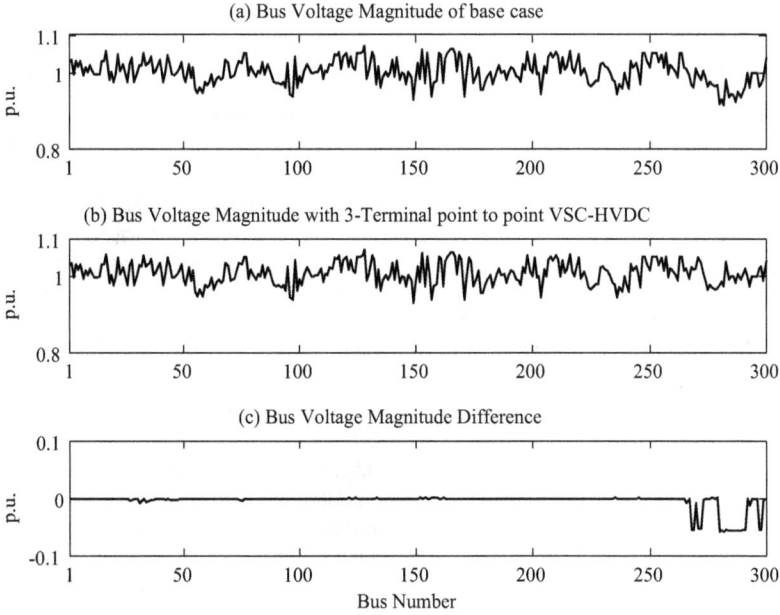

FIGURE 3.14   Bus voltage profile for the first study of Table 3.2.

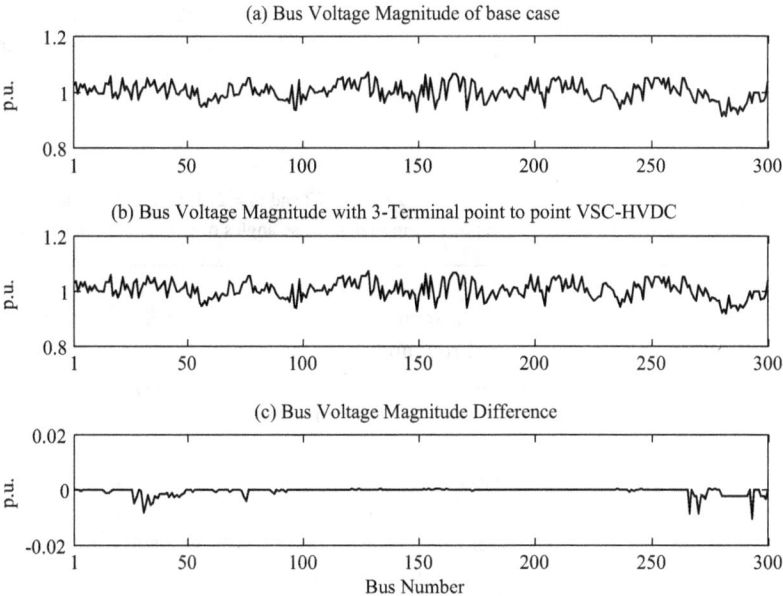

FIGURE 3.15   Bus voltage profile for the second study of Table 3.2.

TABLE 3.3

Study of IEEE 300-Bus System with Five-Terminal VSC-HVDC Network

			Power-Flow Solution	
			Base Case Power Flow Converged in Six Iterations (NI = 6) $V_i = 1.011\angle -11.24; \ V_j = 1.011\angle -11.32;$ $V_k = 0.998\angle -17.67;$	
HVDC Link			$V_l = 0.981\angle -19.46; \ V_m = 1.006\angle -17.47;$	
Master Converter	**Slave Converters**	**HVDC Link Specifications**	**AC Terminal Buses**	**HVDC Variables**
I	j, k, l, m	Master converter $V_{DCi} = 3;$ $V_i = 1.02;$ Slave converters $P_{shj} = 0.3;$ $V_l = 0.99;$ $P_{shk} = 0.2;$ $V_m = 1.01;$ $P_{shl} = 0.3;$ $P_{shm} = 0.3;$ $Q_{shj} = 0.2;$ $Q_{shk} = 0.05;$	$\theta_i = -11.5055;$ $\theta_j = -11.5252;$ $\theta_k = -5.1646;$ $\theta_l = -3.5273;$ $\theta_m = -6.5236;$ $V_j = 1.0206;$ $V_k = 1.0085;$	Master converter $V_{shi} = 1.0616\angle -17.7394;$ $m_i = 1.0158;$ $P_{DCi} = 1.1497; \ P_{lossi}(\%) = 2.09;$ Slave converters $V_{shj} = 1.0409\angle -9.9176;$ $V_{shk} = 1.0139\angle -4.0466;$ $V_{shl} = 0.9831\angle -1.7564;$ $V_{shm} = 1.0090\angle -4.8358;$ $V_{DCj} = 2.9989; \ V_{DCk} = 2.9988;$ $V_{DCl} = 2.9989; \ V_{DCm} = 2.9988;$ $m_j = 0.9722; \ m_k = 0.9545;$ $m_l = 0.9285; \ m_m = 0.9527;$ $P_{DCj} = -0.3127; \ P_{DCk} = -0.2118;$ $P_{DCl} = -0.3125;$ $P_{DCm} = -0.3124; \ P_{lossj}(\%) = 1.26;$ $P_{lossk}(\%) = 1.18; \ P_{lossl}(\%) = 1.24;$ $P_{lossm}(\%) = 1.23;$ NI = 6

Note: For the above study, i = 266, j = 270, k = 271, l = 272 and m = 273; Values of voltage magnitudes, active and reactive powers are in p.u. Phase angles of voltages are in degrees.

connected to buses 270 and 271 operate in the PQ control mode. The specified quantities are shown in the third row and columns 1–3 of Table 3.3. The power-flow solution is shown in columns 4–5 of Table 3.3. The convergence characteristic for this study is shown in Figure 3.16. From Figure 3.16, it is again observed that the proposed model retains the quadratic convergence characteristics of the unified AC–DC Newton-Raphson algorithm. It is also observed that 'NI' remains independent of the size of the MVDC network, the control strategies employed and the quantities specified (operating point). The bus voltage profile for the study of Table 3.3 is shown in Figure 3.17. From Figure 3.17, it is again observed that the bus voltage profile does not change much except the AC terminal buses at which the converters are connected.

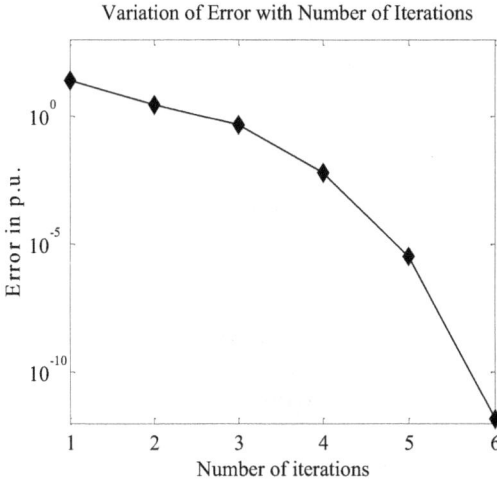

FIGURE 3.16   Convergence characteristic for the study of Table 3.3.

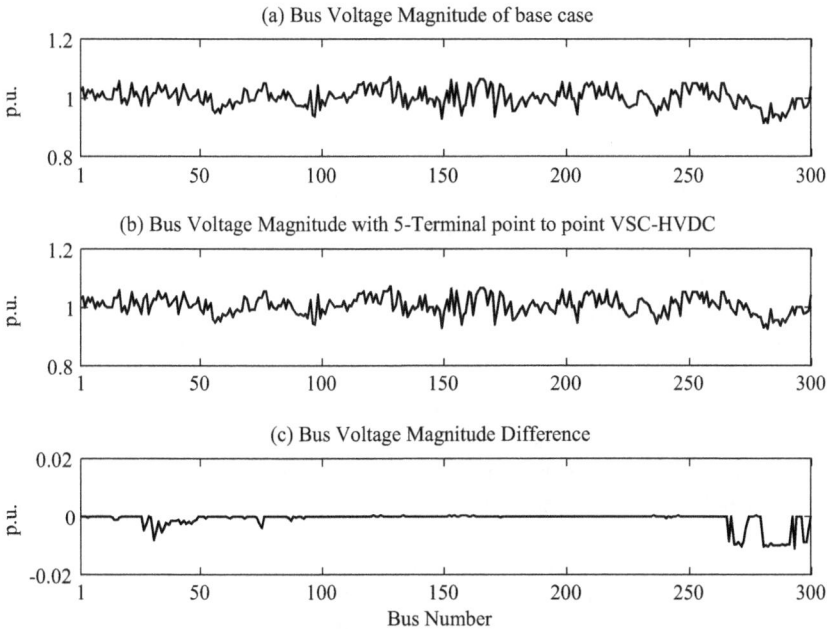

FIGURE 3.17   Bus voltage profile for the study of Table 3.3.

From Tables 3.1–3.3, it is observed that 'NI' remains identical to the base case, which shows the robustness of the proposed algorithm. The proposed algorithm is able to accommodate diverse VSC control strategies and operating point specifications. This shows the versatility of the model.

3.4.2 Studies with Unified Power-Flow Model of European 1354-Bus Test System Integrated with VSC-Based Multi-Terminal DC Grids

In this case study, a seven-terminal VSC-HVDC network is incorporated in the test system between buses 1280, 1281, 1286, 1292, 1313, 1324 and 1353. The converter connected to bus no. 1280 acts as the master converter while the converters connected to buses 1281, 1286, 1292, 1313, 1324 and 1353 act as slave converters. The master converter controls the voltage magnitude of AC bus no. 1280 to a value of 0.97 p.u. The slave converters connected to buses 1324 and 1353 operate in the PV control mode while those connected to buses 1281, 1286, 1292 and 1313 operate in the PQ control mode. The specified quantities are shown in the third row and columns 1–3 of Table 3.4. The power-flow solution is shown in columns 4–5 of Table 3.4. The convergence characteristic for this study is shown in Figure 3.18. From Figure 3.18, it is again observed that the proposed model retains the quadratic convergence characteristics of the unified AC–DC Newton-Raphson algorithm. The bus voltage profile for the study of Table 3.4 is shown in Figure 3.19. From Figure 3.19, it is again observed that the bus voltage profile does not change much except the AC terminal buses at which the converters are connected.

3.4.3 Studies with Sequential Power-Flow Model of IEEE 300-Bus Test System Integrated with MVDC Grids

Case I: Three-terminal MVDC network connected in the PTP configuration
In this case study, again two separate, sequential power-flow studies are conducted with a three-terminal, PTP-connected VSC-HVDC network incorporated in the IEEE 300-bus test system to demonstrate the versatility of the proposed model. In both the studies, the VSC-HVDC network is connected between AC buses 266, 270 and 271. The specified quantities for these studies are identical to those in the case study shown in Table 3.2 and are detailed in rows 3–4 and columns 1–3 of Table 3.5. Rows 3 and 4 correspond to the 'PQ' and 'PV' control modes of the slave VSCs, respectively. The corresponding power-flow solutions are shown in row 3 and row 4 (columns 4–5) of Table 3.5, respectively.

From Table 3.5, it can be observed that for both the studies, 'NI' is increased than that of the studies of Table 3.2 carried out using the unified AC–DC Newton-Raphson algorithm. The convergence characteristic plots (variation of mismatch error in p.u. with number of iterations) for first and second studies of Table 3.5 are shown in Figures 3.20 and 3.21, respectively. From Figures 3.20 and 3.21, it can be observed that the quadratic convergence characteristics of the Newton-Raphson method are lost due to the adoption of the sequential Newton-Raphson power-flow algorithm. The values of 'NI' have increased than those of Table 3.2. Also, the values of 'NI' are dependent on the VSC control strategy employed.

TABLE 3.4
Study of European 1354-Bus System with Seven-Terminal VSC-HVDC Network

			Power-Flow Solution	
			Base Case Power Flow Converged in Seven Iterations (NI = 7)	
			$V_i = 0.9554 \angle -35.2574$; $V_j = 1.0354 \angle -37.3082$; $V_k = 1.0156 \angle -12.3760$; $V_l = 1.0289 \angle -35.23066$; $V_m = 1.0351 \angle -37.4755$; $V_n = 0.9972 \angle -29.0208$;	
HVDC Link			$V_o = 0.9934 \angle -29.6628$;	

Master Converter	Slave Converters	HVDC Link Specifications	AC Terminal Buses	HVDC Variables
i	j, k, l, m, n, o	Master converter	$\theta_i = -43.2395$;	Master converter
		$V_{DCi} = 3$;	$\theta_j = -37.3070$;	$V_{shi} = 1.0746 \angle -53.8907$;
		$V_i = 0.97$;	$\theta_k = -12.0291$;	$m_i = 1.0132$;
		Slave	$\theta_l = -35.3598$;	$P_{DCi} = 1.8758$; $P_{lossi}(\%) = 3.7647$;
		converters	$\theta_m = -37.4745$;	Slave converters
		$P_{shj} = 0.3$;	$\theta_n = 28.6451$;	$V_{shj} = 1.0570 \angle -35.7493$;
		$P_{shk} = 0.4$;	$\theta_o = -29.5333$;	$V_{shk} = 1.0292 \angle -9.8473$;
		$P_{shl} = 0.3$;	$V_j = 1.0371$;	$V_{shl} = 1.0432 \angle -33.7695$;
		$P_{shm} = 0.3$;	$V_k = 1.0183$;	$V_{shm} = 1.0518 \angle -35.9057$;
		$P_{shn} = 0.2$;	$V_l = 1.0328$;	$V_{shn} = 1.0058 \angle -27.5094$;
		$P_{sho} = 0.3$;	$V_m = 1.0366$;	$V_{sho} = 1.0153 \angle -27.8484$;
		$Q_{shj} = 0.2$;		$V_{DCj} = 2.9990$; $V_{DCk} = 2.9989$;
		$Q_{shk} = 0.1$;		$V_{DCl} = 2.999$; $V_{DCm} = 2.999$;
		$Q_{shl} = 0.1$;		$V_{DCn} = 2.999$; $V_{DCo} = 2.999$;
		$Q_{shm} = 0.15$;		$m_j = 0.9969$; $m_k = 0.9707$;
		$V_n = 1.001$;		$m_l = 0.9838$; $m_m = 0.9920$;
		$V_o = 1$;		$m_n = 0.9486$; $m_o = 0.9576$;
				$P_{DCj} = -0.3127$; $P_{DCk} = -0.4131$;
				$P_{DCl} = -0.3124$; $P_{DCm} = -0.3125$;
				$P_{DCn} = -0.2118$; $P_{DCo} = -0.3126$;
				$P_{lossj}(\%) = 1.26$; $P_{lossk}(\%) = 1.29$;
				$P_{lossl}(\%) = 1.23$; $P_{lossm}(\%) = 1.24$;
				$P_{lossn}(\%) = 1.18$; $P_{losso}(\%) = 1.25$;
				NI = 7

Note: For the above study, i = 1280, j = 1281, k = 1286, l = 1292, m = 1313, n = 1324, o = 1353; Values of voltage magnitudes, active and reactive powers are in p.u. Phase angles of voltages are in degrees.

The bus voltage profiles for the first and second studies of Table 3.5 are shown in Figures 3.22 and 3.23, respectively. From Figures 3.22 and 3.23, it is observed that the bus voltage profile hardly changes except for the AC terminal buses at which the VSCs are connected.

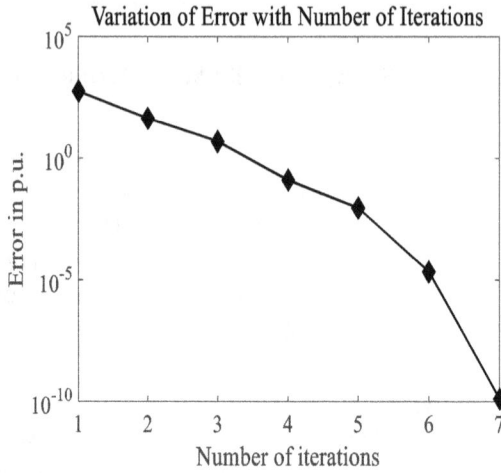

FIGURE 3.18   Convergence characteristic for the study of Table 3.4.

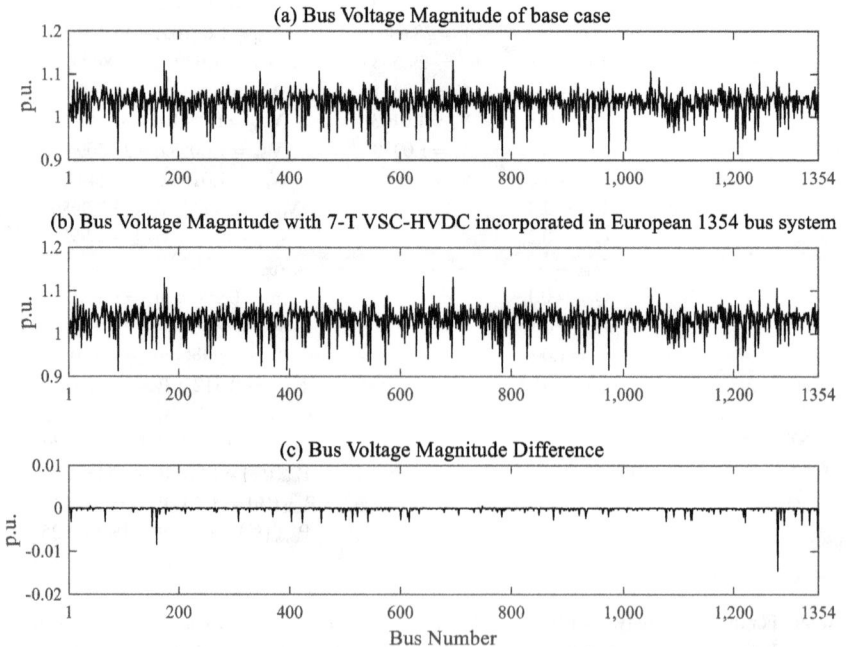

FIGURE 3.19   Bus voltage profile for the study of Table 3.4.

Case II: Five-terminal MVDC network connected in the PTP configuration

This case study is similar to that of the case study of the five-terminal MVDC network of Table 3.3. The specified quantities for the sequential power-flow are identical to those in columns 1–3 of Table 3.3 and are detailed again in columns 1–3

TABLE 3.5

Study of IEEE 300-Bus System with Three-Terminal VSC-HVDC Network

			Power-Flow Solution	
			Base Case Power Flow Converged in Six Iterations (NI = 6)	
HVDC Link Connection Details			$V_i = 1.011\angle -11.24$; $V_j = 1.011\angle -11.32$; $V_k = 0.998\angle -17.67$;	
Master Converter	**Slave Converters**	**HVDC Link Specifications**	**AC Terminal Buses**	**HVDC Variables**
I	j, k	Master converter $V_{DCi} = 3$; $V_i = 1.02$; Slave converters $P_{shj} = 0.7$; $Q_{shj} = 0.2$; $P_{shk} = 0.5$; $Q_{shk} = 0.06$;	$\theta_i = -11.4398$; $\theta_j = -11.3827$; $\theta_k = -5.4606$; $V_j = 1.0209$; $V_k = 1.0503$;	Master converter $V_{shi} = 1.0516\angle -18.1648$; $m_i = 0.9914$; $P_{DCi} = 1.2303$; $P_{lossi}(\%) = 2.15$; Slave converters $V_{shj} = 1.0434\angle -7.6254$; $V_{shk} = 1.0576\angle -2.8837$; $V_{DCj} = 2.9978$; $V_{DCk} = 2.9981$; $m_j = 0.9844$; $m_k = 0.9977$; $P_{DCj} = -0.7158$; $P_{DCk} = -0.5137$; $P_{lossj}(\%) = 1.53$; $P_{lossk}(\%) = 1.34$; NI = 10; CT = 2.05;
I	j, k	Master converter $V_{DCi} = 3$; $V_i = 1.02$; Slave converters $P_{shj} = 0.4$; $V_j = 1.02$; $P_{shk} = 0.5$; $V_k = 1$;	$\theta_i = -11.4236$; $\theta_j = -11.4157$; $\theta_k = -5.0076$;	Master converter $V_{shi} = 1.0781\angle -16.397$; $m_i = 1.0165$; $P_{DCi} = 0.9273$; $P_{lossi}(\%) = 1.9251$; Slave converters $V_{shj} = 1.0213\angle -9.2151$; $V_{shk} = 0.9975\angle -2.1320$; $V_{DCj} = 2.9985$; $V_{DCk} = 2.9984$; $m_j = 0.9633$; $m_k = 0.9410$; $P_{DCj} = -0.413$; $P_{DCk} = -0.5138$; $P_{lossj}(\%) = 1.28$; $P_{lossk}(\%) = 1.36$; NI = 9

Note: For the above case study, i = 266, j = 270, k = 271; Values of voltage magnitudes, active and reactive powers are in p.u. Phase angles of voltages are in degrees.

of Table 3.6. The power-flow solution is shown in row 3 and columns 4–5 of Table 3.6. The convergence characteristic for this case study is shown in Figure 3.24. From Figure 3.24, it can again be observed that the quadratic convergence of the unified Newton-Raphson algorithm is lost due to the adoption of the sequential power-flow method employed in this case. The bus voltage profile for the study of Table 3.6 is shown in Figure 3.25. From Figure 3.25, it is again observed that the bus voltage profile does not change much except at the AC terminal buses to which the converters are connected.

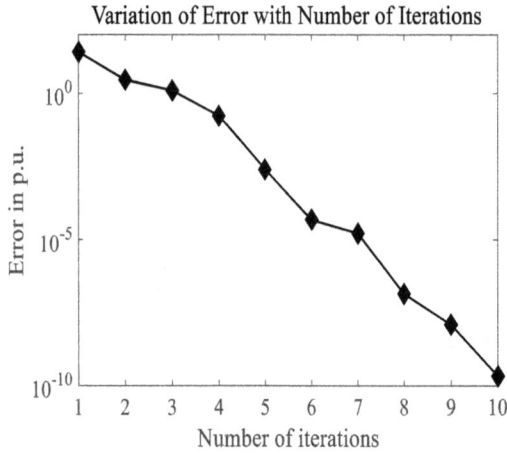

FIGURE 3.20 Convergence characteristic for the first study of Table 3.5.

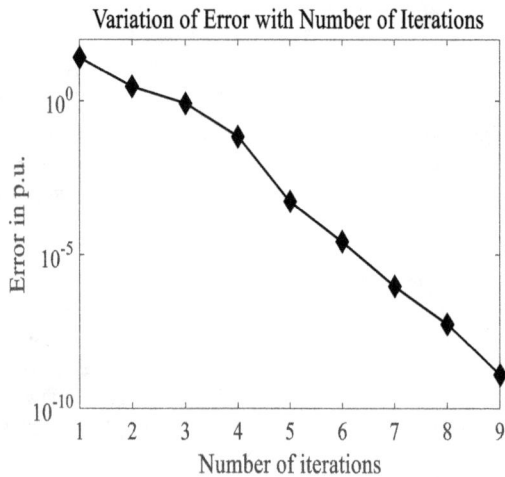

FIGURE 3.21 Convergence characteristic for the second study of Table 3.5.

3.4.4 Studies with Sequential Power-Flow Model of European 1354-Bus Test System Integrated with MVDC Grids

This case study is similar to that of the case study of the seven-terminal MVDC network of Table 3.4. The specified quantities for the sequential power-flow are identical to those in columns 1–3 of Table 3.4 and are detailed again in columns 1–3 of Table 3.7. The power-flow solution is shown in row 3 and columns 4–5 of Table 3.7. The convergence characteristic for this case study is shown in

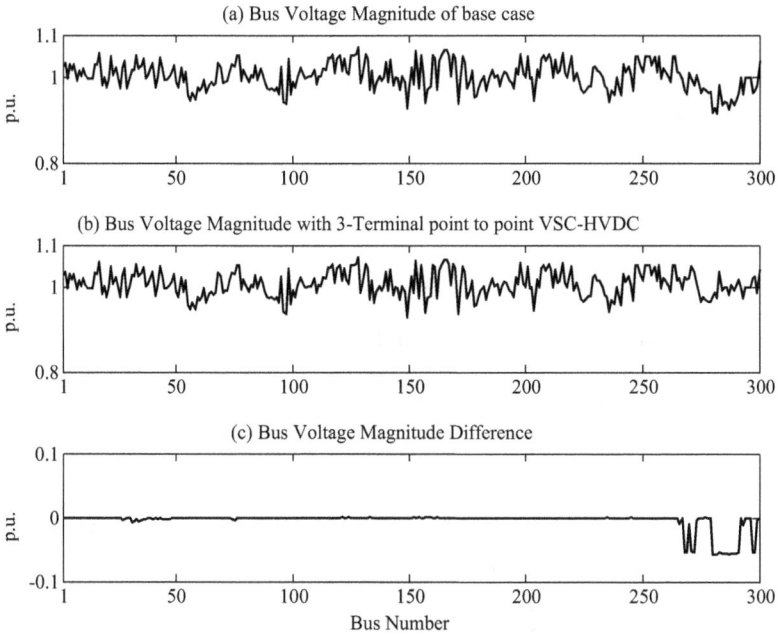

FIGURE 3.22   Bus voltage profile for the first study of Table 3.5.

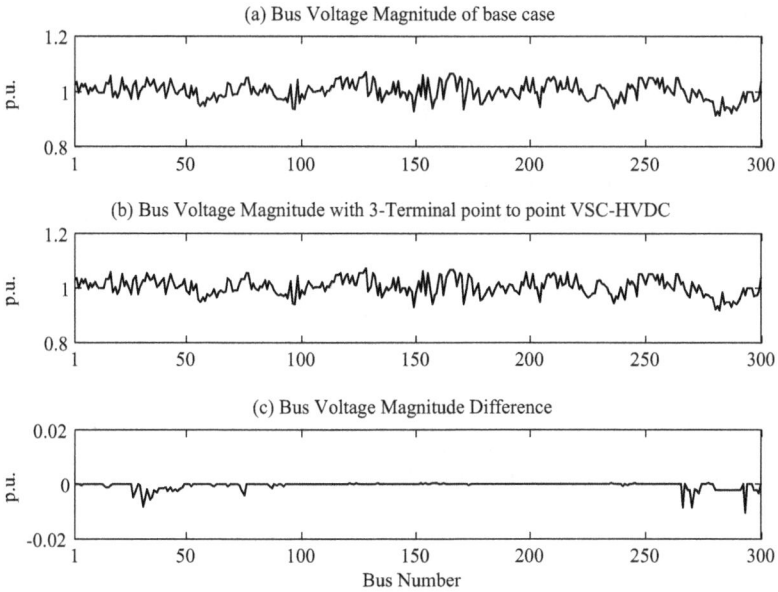

FIGURE 3.23   Bus voltage profile for the second study of Table 3.5.

TABLE 3.6
Study of IEEE 300-Bus System with Five-Terminal VSC-HVDC Network

				Power-Flow Solution
				Base Case Power Flow Converged in Six Iterations (NI = 6)
				$V_i = 1.011\angle -11.24$; $V_j = 1.011\angle -11.32$;
				$V_k = 0.998\angle -17.67$; $V_l = 0.981\angle -19.46$;
HVDC Link				$V_m = 1.006\angle -17.47$;
Master Converter	**Slave Converters**	**HVDC Link Specifications**	**AC Terminal Buses**	**HVDC Variables**
i	j, k, l, m	Master converter $V_{DCi} = 3$; $V_i = 1.02$;	$\theta_i = -11.5055$; $\theta_j = -11.5252$; $\theta_k = -5.1646$; $\theta_l = -3.5273$; $\theta_m = -6.5236$; $V_j = 1.0206$; $V_k = 1.0085$;	Master converter $V_{shi} = 1.0616\angle -17.7394$; $m_i = 1.008$; $P_{DCi} = 1.1497$; $P_{lossi}\,(\%) = 2.09$;
		Slave converters $P_{shj} = 0.3$; $V_l = 0.99$; $P_{shk} = 0.2$; $V_m = 1.01$; $P_{shl} = 0.3$; $P_{shm} = 0.3$; $Q_{shj} = 0.2$; $Q_{shk} = 0.05$;		Slave converters $V_{shj} = 1.0409\angle -9.9176$; $V_{shk} = 1.0139\angle -4.0466$; $V_{shl} = 0.9831\angle -1.7564$; $V_{shm} = 1.0090\angle -4.8358$; $V_{DCj} = 2.999$; $V_{DCk} = 2.9991$; $V_{DCl} = 2.999$; $V_{DCm} = 2.999$; $m_j = 0.9817$; $m_k = 0.9562$; $m_l = 0.9272$; $m_m = 0.9516$; $P_{DCj} = -0.3127$; $P_{DCk} = -0.2118$; $P_{DCl} = -0.3125$; $P_{DCm} = -0.3124$; $P_{lossj}\,(\%) = 1.26$; $P_{lossk}\,(\%) = 1.18$; $P_{lossl}\,(\%) = 1.24$; $P_{lossm}\,(\%) = 1.23$; NI = 12

Note: For the above study, i = 266, j = 270, k = 271, l = 272 and m = 273. Values of voltage magnitudes, active and reactive powers are in p.u. Phase angles of voltages are in degrees.

Figure 3.26. From Figure 3.26, it can again be observed that the quadratic convergence of the unified Newton-Raphson algorithm is lost due to the adoption of the sequential power-flow method employed in this case. The bus voltage profile for the study of Table 3.7 is shown in Figure 3.27. From Figure 3.27, it is again observed that the bus voltage profile does not change much except at the AC terminal buses to which the converters are connected.

A comparison of the convergence features of the proposed model vis-à-vis existing models in the literature is shown in Table 3.8.

Variation of Error with Number of Iterations

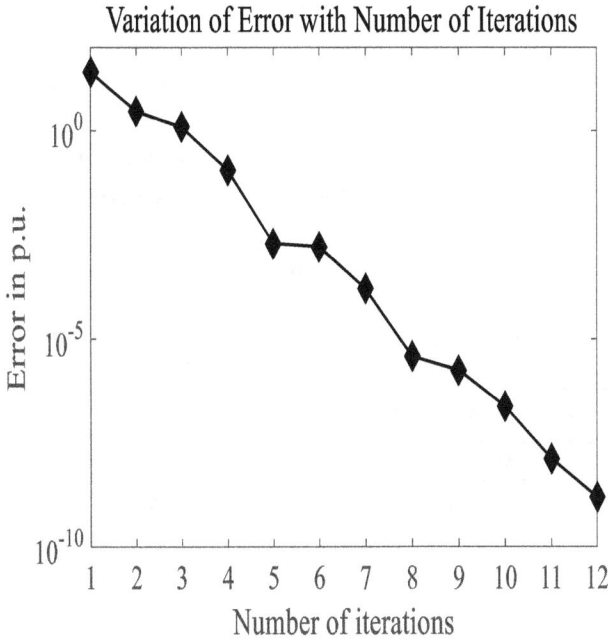

FIGURE 3.24 Convergence characteristic for the study of Table 3.6.

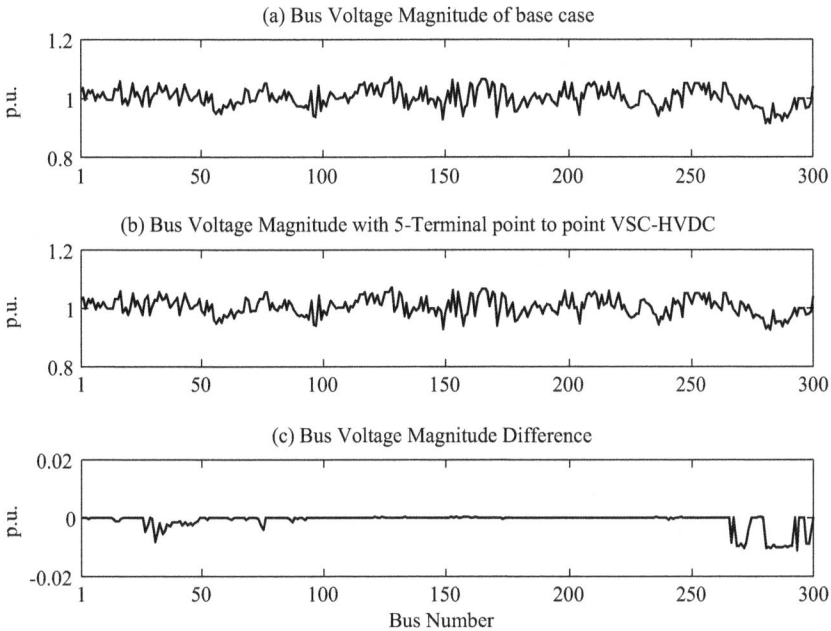

FIGURE 3.25 Bus voltage profile for the study of Table 3.6.

TABLE 3.7
Study of European 1354-Bus System with Seven-Terminal VSC-HVDC Network

				Power-Flow Solution
				Base Case Power Flow Converged in Six Iterations (NI = 7)
				$V_i = 0.9554 \angle -35.2574$; $V_j = 1.0354 \angle -37.3082$;
				$V_k = 1.0156 \angle -12.3760$;
				$V_l = 1.0289 \angle -35.23066$; $V_m = 1.0351 \angle -37.4755$;
HVDC Link				$V_n = 0.9972 \angle -29.0208$; $V_o = 0.9934 \angle -29.6628$;
Master Converter	**Slave Converters**	**HVDC Link Specifications**	**AC Terminal Buses**	**HVDC Variables**
I	j, k, l, m, n, o	Master converter	$\theta_i = -43.2395$;	Master converter
		$V_{DCi} = 3$;	$\theta_j = -37.3070$;	$V_{shi} = 1.0746 \angle -53.8907$;
		$V_i = 0.97$;	$\theta_k -12.0291$;	$m_i = 1.0132$;
		Slave	$\theta_l = -35.3598$;	$P_{DCi} = 1.8758$; $P_{lossi}(\%) = 3.7647$;
		converters	$\theta_m = -37.4745$;	Slave converters
		$P_{shj} = 0.3$;	$\theta_n = 28.6451$;	$V_{shj} = 1.0570 \angle -35.7493$;
		$P_{shk} = 0.4$;	$\theta_o = -29.5333$;	$V_{shk} = 1.0292 \angle -9.8473$;
		$P_{shl} = 0.3$;	$V_j = 1.0371$;	$V_{shl} = 1.0432 \angle -33.7695$;
		$P_{shm} = 0.3$;	$V_k = 1.0183$;	$V_{shm} = 1.0518 \angle -35.9057$;
		$P_{shn} = 0.2$;	$V_l = 1.0328$;	$V_{shn} = 1.0058 \angle -27.5094$;
		$P_{sho} = 0.3$;	$V_m = 1.0366$;	$V_{sho} = 1.0153 \angle -27.8484$;
		$Q_{shj} = 0.2$;		$V_{DCj} = 2.9990$; $V_{DCk} = 2.9989$;
		$Q_{shk} = 0.1$;		$V_{DCl} = 2.999$; $V_{DCm} = 2.999$;
		$Q_{shl} = 0.1$;		$V_{DCn} = 2.999$; $V_{DCo} = 2.999$;
		$Q_{shm} = 0.15$;		$m_j = 0.9969$; $m_k = 0.9707$;
		$V_n = 1.001$;		$m_l = 0.9838$; $m_m = 0.9920$;
		$V_o = 1$;		$m_n = 0.9486$; $m_o = 0.9576$;
				$P_{DCj} = -0.3127$; $P_{DCk} = -0.4131$;
				$P_{DCl} = -0.3124$; $P_{DCm} = -0.3125$;
				$P_{DCn} = -0.2118$; $P_{DCo} = -0.3126$;
				$P_{lossj}(\%) = 1.26$; $P_{lossk}(\%) = 1.29$;
				$P_{lossl}(\%) = 1.23$; $P_{lossm}(\%) = 1.24$;
				$P_{lossn}(\%) = 1.18$; $P_{losso}(\%) = 1.25$;
				NI = 10

Note: For the above study, i = 1280, j = 1281, k = 1286, l = 1292, m = 1313, n = 1324, o = 1353; Values of voltage magnitudes, active and reactive powers are in p.u. Phase angles of voltages are in degrees.

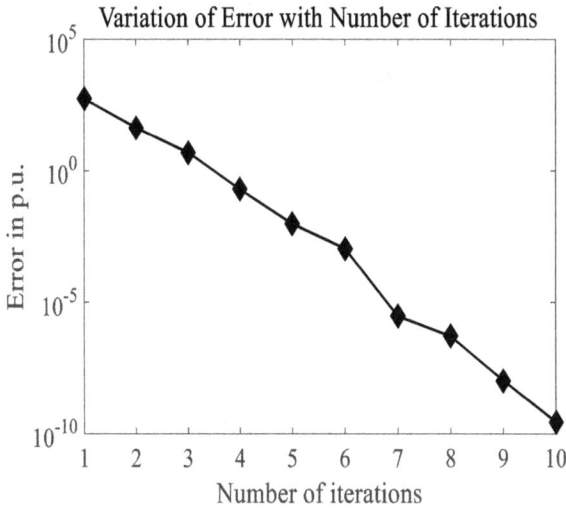

FIGURE 3.26   Convergence characteristic for the study of Table 3.7.

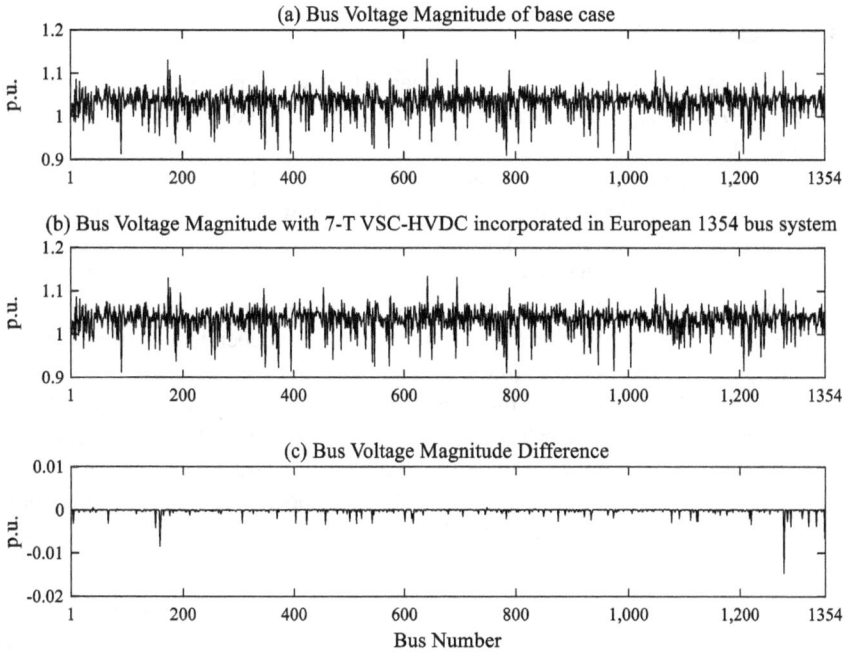

FIGURE 3.27   Bus voltage profile for the study of Table 3.7.

TABLE 3.8
A Comparison of Convergence Features with Existing Models

Reference No.	Tolerance (p.u.)	No. of Buses in the System AC	No. of Buses in the System DC	NI	
[82]	10^{-8}	29	5	Min: 3; max: 15 (depending on wind power generation)	
[78]	10^{-6}	9	4	6	
		32	4	7	
Proposed model	10^{-8}	300	NIL (base)	6	
Proposed model	10^{-8}	1354	NIL (base)	7	
Unified	10^{-8}	300	BTB	6	
			3	6	
			5	6	
		1354	7	7	
Sequential	10^{-8}	300	3	PQ control 10	PV control 9
			5		12
		1354	7		10

3.5 SUMMARY

In this chapter, both unified and sequential Newton power-flow models of integrated AC–DC systems have been developed. Unlike existing models, the modulation indices of the VSCs can be expressed as unknowns in both the models. Both the algorithms are implemented by employing diverse control strategies in different topologies of multi-terminal DC networks incorporated in the IEEE 300-bus and European 1354-bus test systems. It is observed that the unified method possesses the quadratic convergence characteristics as in the base case power flow. Further, the number of iterations taken by the unified Newton-Raphson power-flow algorithm for convergence is independent of the MTDC topology, the MTDC control strategies employed and the operating point specifications. On the other hand, the quadratic convergence characteristics of the Newton-Raphson method are lost if the sequential AC–DC power-flow algorithm is adopted. Also, the number of iterations taken by the unified Newton power-flow algorithm for convergence is dependent on both the operating point specifications and the MTDC control strategies employed.

4 Power-Flow Modelling of AC Power Systems Integrated with VSC-Based Multi-Terminal DC (AC-MVDC) Grids Employing DC Voltage Droop Control

4.1 INTRODUCTION

The power-flow modelling of voltage-sourced converter-based integrated AC–DC systems employing DC slack bus control (also known as master-slave control) is detailed in Chapter 3. As already discussed in Chapter 3, the main disadvantage of this control scheme is the DC grid instability following a failure of the master converter. This problem can be tackled by ensuring that individual converters take part in the DC voltage regulation scheme by adjusting their active power flow in response to changes in the DC voltage with the operating point, known as DC voltage droop control [12,20]. For MTDC systems, different types of DC voltage droop control have been envisaged to ensure proper sharing based on the converter ratings. These include Voltage-Power (V-P) droop, Voltage-Current (V-I) droop, Voltage Margin (VM) control and Voltage-Power (V-P) droop with Dead Band (DB) [12,25,32,38].

For planning, operation and control of integrated AC-MVDC systems employing DC voltage droop control, their power-flow solution is an essential requirement. In this respect, [91–94] present some comprehensive research works on the development of efficient power-flow algorithms of integrated AC-MVDC systems employing DC voltage droop control.

However, the above research works do not address the following issues:

1. Development of a unified Newton power-flow model of integrated AC-MVDC systems employing DC voltage droop control. In this respect, [91,92] have presented sequential Newton power-flow models of integrated AC-MVDC systems incorporating DC voltage droop control.

DOI: 10.1201/9781003252078-4

However, the quadratic convergence characteristic of the unified AC/DC Newton-Raphson algorithm is lost in Refs. [91,92] due to the adoption of the sequential AC–DC power-flow algorithm.

2. The VSC modulation index 'm' has not been considered as an unknown in any of the above models. For VSC applications, 'm' is a crucial parameter. Typically, $0 < m \leq 1$. Some of the factors that put a cap on the lower and upper bounds of 'm' have been reported in Ref. [11].

This chapter presents a generalized approach to both the unified and the sequential power-flow modelling of integrated AC-MVDC grids employing DC voltage droop control. Unlike most of the research works published in this area, in the proposed model, 'm' is considered as an unknown and can be obtained directly from the power-flow solution. The proposed work also includes converter losses.

4.2 MODELLING OF INTEGRATED AC-MVDC SYSTEMS EMPLOYING DC VOLTAGE DROOP CONTROL

For modelling of hybrid AC-MVDC systems employing DC voltage droop control, a generalized AC–DC network already considered in Chapter 3 is shown in Figure 4.1. It comprises an 'n' bus AC power system network integrated with

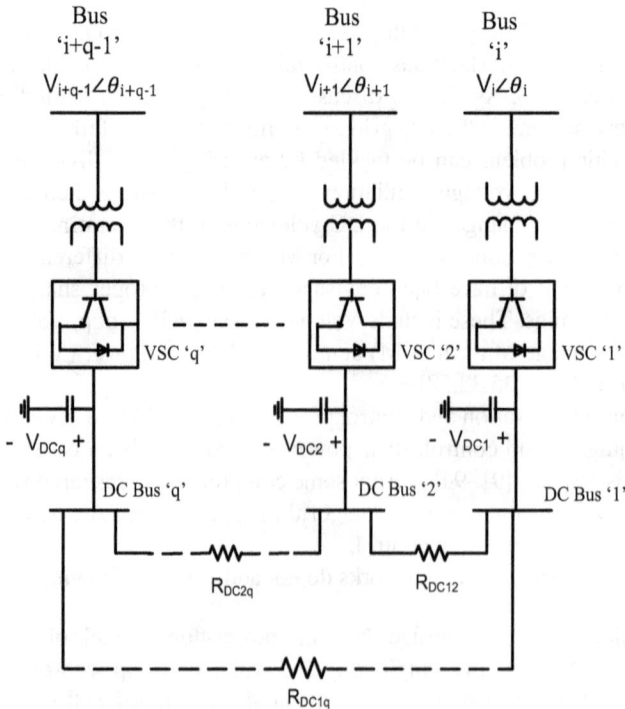

FIGURE 4.1 Schematic diagram of an integrated AC-MTDC system.

a MVDC grid using 'q' VSCs and their respective converter transformers. The assumptions adopted for the modelling are also identical to those in Chapter 3. As shown in Figure 4.1, the AC buses connected to the 'q' VSCs are numbered as 'i', '(i+1)', and so on, up to '(i+q−1)', while the 'q' VSCs are connected in the PTP configuration on their DC sides. The PTP connection is considered, being more prevalent in practical HVDC installations over the world. Figure 4.2 shows the equivalent circuit of the network shown in Figure 4.1 with 'q' fundamental frequency, positive sequence voltage sources pertaining to the 'q' VSCs. V_{sha} represents the voltage phasor pertaining to the ath ($1 \le a \le q$) VSC. The ath ($1 \le a \le q$) VSC is connected to AC terminal bus '(i+a−1)' whose voltage is represented by the phasor $V_{i+a-1} = V_{i+a-1} \angle \theta_{i+a-1}$.

From Figure 4.2, the current in the link (not shown) connecting the ath VSC and its AC terminal bus is

$$I_{sha} = y_{sha}\left(V_{sha} - V_{i+a-1}\right) \tag{4.1}$$

where $V_{sha} = V_{sha} \angle \theta_{sha} = m_a c V_{DCa} \angle \theta_{sha}$, $y_{sha} = 1/Z_{sha}$, $Z_{sha} = R_{sha} + jX_{sha}$, R_{sha} and X_{sha} are the resistance and the leakage reactance of the ath converter

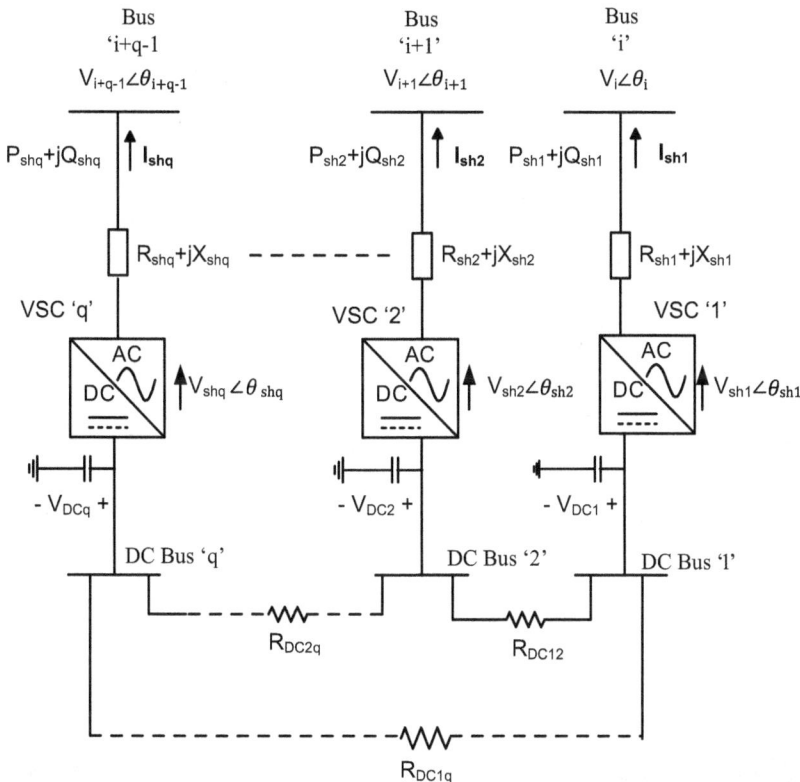

FIGURE 4.2 Equivalent circuit of Figure 4.1.

transformer, respectively, 'm$_a$' is the VSC modulation index and the constant 'c' is representative of the VSC architecture [11].

As already detailed in Chapter 3 and from Figures 4.1 and 4.2, the net current injection at the AC bus '(i+a–1)' connected to the ath ($1 \le a \le p$) converter can be written as

$$I_{i+a-1} = \sum_{k=1}^{n} Y_{(i+a-1)k} V_k - y_{sha} V_{sha} \tag{4.2}$$

In the above equation, $Y_{(i+a-1)(i+a-1)} = Y_{(i+a-1)(i+a-1)}^{old} + y_{sha}$ and $Y_{(i+a-1)(i+a-1)}^{old} = y_{(i+a-1)0} + \sum_{k=1,\, k \ne i+a-1}^{n} y_{(i+a-1)k}$ are the values of the self-admittances of bus '(i+a–1)' with the ath VSC connected and in the original 'n' bus AC system without any VSC, respectively. Similarly, '$y_{(i+a-1)0}$' accounts for the shunt capacitances of all the transmission lines connected to bus '(i+a–1)'.

4.3 POWER-FLOW EQUATIONS OF INTEGRATED AC-MVDC SYSTEMS EMPLOYING DC VOLTAGE DROOP CONTROL

From Figure 4.2, the net active and reactive power injections at the AC bus '(i+a–1)' are

$$P_{i+a-1} = \sum_{k=1}^{n} V_{i+a-1} V_k Y_{(i+a-1)k} \cos\left[\theta_{i+a-1} - \theta_k - \phi_{(i+a-1)k} \right]$$

$$- m_a c V_{DCa} V_{i+a-1} y_{sha} \cos\left(\theta_{i+a-1} - \theta_{sha} - \phi_{sha} \right) \tag{4.3}$$

$$Q_{i+a-1} = \sum_{k=1}^{n} V_{i+a-1} V_k Y_{(i+a-1)k} \sin\left[\theta_{i+a-1} - \theta_k - \phi_{(i+a-1)k} \right]$$

$$- m_a c V_{DCa} V_{i+a-1} y_{sha} \sin\left(\theta_{i+a-1} - \theta_{sha} - \phi_{sha} \right) \tag{4.4}$$

Also, from Figure 4.2, the active and reactive power flows at the terminal end of the line connecting the ath VSC to the AC bus '(i+a–1)' are

$$P_{sha} = \mathrm{Re}\left[V_{i+a-1} I_{sha}^* \right] = m_a c\, V_{DCa} V_{i+a-1} y_{sha} \cos\left(\theta_{i+a-1} - \theta_{sha} - \phi_{sha} \right) - V_{i+a-1}^2 y_{sha} \cos\phi_{sha} \tag{4.5}$$

and $$Q_{sha} = m_a c\, V_{DCa} V_{i+a-1} y_{sha} \sin\left(\theta_{i+a-1} - \theta_{sha} - \phi_{sha} \right) + V_{i+a-1}^2 y_{sha} \sin\phi_{sha} \tag{4.6}$$

Also, from Figure 4.2, by virtue of the power balance on the AC and DC sides of the ath VSC,

$$\text{Re}\left(\mathbf{V_{sha}}\ \mathbf{I_{sha}^*}\right) + \sum_{v=1}^{q} V_{DCa} V_{DCv} Y_{DCav} = -P_{lossa} \tag{4.7}$$

Substitution of Eq. (4.1) in Eq. (4.7) gives

$$\left(m_a c\ V_{DCa}\right)^2 y_{sha} \cos\phi_{sha} - m_a c\ V_{DCa} V_{i+a-1} y_{sha} \cos\left(\theta_{sha} - \theta_{i+a-1} - \phi_{sha}\right)$$

$$+ \sum_{v=1}^{q} V_{DCa} V_{DCv} Y_{DCav} + P_{lossa} = 0$$

$$\text{or,} \quad f_{1a} = 0 \qquad \forall\ 1 \le a \le q \tag{4.8}$$

Thus, for 'q' VSCs, 'q' independent equations are obtained.

In Eqs. (4.7) and (4.8), $Y_{DCav} = -\dfrac{1}{R_{DCav}}$, where '$R_{DCav}$' is the resistance of the DC link between DC buses 'a' and 'v'. Also, 'P_{lossa}' represents the losses [18,102] of the ath VSC as already detailed in Chapter 3 and is again given below:

$$P_{lossa} = a_1 + b_1 I_{sha} + c_1 I_{sha}^2 \tag{4.9}$$

where 'a_1', 'b_1' and 'c_1' are loss factors [18,92] and

$$I_{sha} = y_{sha}\left[V_{i+a-1}^2 + \left(m_a c\ V_{DCa}\right)^2 - 2V_{i+a-1} m_a c V_{DCa}\ \text{os}\left(\theta_{i+a-1} - \theta_{sha}\right)\right]^{1/2} \tag{4.10}$$

The derivation of Eq. (4.10) is given in Appendix A.

Now, in the AC-MTDC system (Figure 4.2) with 'q' VSCs, if it is assumed that the rth $(1 \le r \le q)$ VSC is used for voltage control of its corresponding AC bus, we have

$$V_{i+a-1}^{sp} - V_{i+a-1}^{cal} = 0 \qquad \forall a, 1 \le a \le q,\ a = r \tag{4.11}$$

Also, not more than '$(q-1)$' line active and reactive power flows {Eqs. (4.5) and (4.6)} can be specified, which give us '$(2q-2)$' independent equations given as

$$P_{sha}^{sp} - P_{sha}^{cal} = 0 \tag{4.12}$$

$$Q_{sha}^{sp} - Q_{sha}^{cal} = 0 \tag{4.13}$$

$$\forall a, 1 \le a \le q,\ a \ne r.$$

Instead of PQ control mode, if a VSC operates in the PV one, Eq. (4.13) changes to

$$V_{i+a-1}^{sp} - V_{i+a-1}^{cal} = 0 \qquad \forall a,\ 1 \le a \le q,\ a \ne r \tag{4.14}$$

Further, the net reactive power injection at AC bus '(i+r−1)' can be specified as its voltage is controlled by the rth VSC. Thus, we get

$$Q^{sp}_{i+a-1} - Q^{cal}_{i+a-1} = 0 \quad \forall a, \ 1 \le a \le q, \ a = r \qquad (4.15)$$

In Eqs. (4.11)–(4.15), V^{sp}_{i+a-1}, Q^{sp}_{i+a-1}, P^{sp}_{sha} and Q^{sp}_{sha} are specified values while V^{cal}_{i+a-1}, Q^{cal}_{i+a-1}, P^{cal}_{sha} and Q^{cal}_{sha} are calculated values {using Eqs. (4.4)–(4.6)}.

4.4   DC VOLTAGE DROOP CONTROL IN MVDC SYSTEMS

In DC voltage droop control [12,20,25–32,38], multiple converters participate in the DC voltage control scheme. Droop control comprises both linear and nonlinear voltage droop characteristics. Among the linear ones, Voltage-Power (V-P) and Voltage-Current (V-I) droops have been the two most popular and widely used strategies for DC voltage droop control. Nonlinear voltage droop control characteristics include dead-bands and limits. Among the nonlinear ones, Voltage Margin, V-P droop with power dead band and V-P droop with voltage limits are some of the more widely used characteristics. Some of these are elaborated below.

1. Voltage-Power (V-P) droop
   If the ath VSC follows a linear V-P droop characteristic, its rectifying power can be expressed as

$$P_{DCa} = R_a \left(V^{*}_{DCa} - V_{DCa} \right) + P^{*}_{DCa} \qquad (4.16)$$

   where 'V^{*}_{DCa}' and 'P^{*}_{DCa}' represent the DC voltage and power references of its droop characteristics and 'R_a' is the droop control gain.
2. Voltage-Current (V-I) droop
   If the ath VSC follows a linear V-I droop characteristic, the net DC current injection at its terminal can be expressed as

$$I_{DCa} = R_a \left(V^{*}_{DCa} - V_{DCa} \right) + I^{*}_{DCa} \qquad (4.17)$$

   where 'I^{*}_{DCa}' and 'V^{*}_{DCa}' are the DC current and voltage references of its droop line and 'R_a' is the droop control gain.
   Thus, the rectifying power of the VSC can be expressed as

$$P_{DCa} = V_{DCa} \left[R_a \left(V^{*}_{DCa} - V_{DCa} \right) + I^{*}_{DCa} \right] \qquad (4.18)$$

   Figure 4.3 depicts the linear V-P and V-I droop characteristics for any arbitrary VSC 'a'.
   Computation of 'V^{*}_{DCa}' and 'P^{*}_{DCa}'
   The values of the DC voltage and power references 'V^{*}_{DCa}', 'P^{*}_{DCa}' and 'I^{*}_{DCa}' in Eqs. (4.16)–(4.18) for all the 'q' converters are either pre-specified or obtained from a DC power flow. While carrying out the

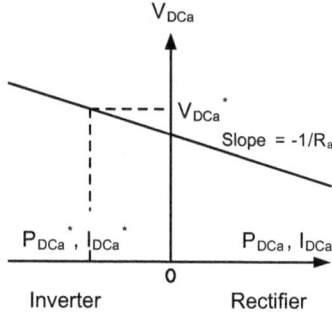

FIGURE 4.3 Linear voltage droop characteristic of the ath VSC.

DC power flow, if DC slack bus control is assumed, then, by convention, the voltage 'V_{DC1}^*' (of DC terminal 1) is specified and it does not operate on droop control. This problem can be circumvented if the average voltage of all the 'q' DC terminals 'V_{DCav}^*' is specified instead of 'V_{DC1}^*' [92]. In that case, the DC power-flow equations are

Solve $\mathbf{V_{DC}^*}$ specified $\mathbf{P_{DC}^*}$ and V_{DCav}^*

where $\mathbf{V_{DC}^*} = \begin{bmatrix} V_{DC1}^* \cdots V_{DCq}^* \end{bmatrix}^T$, $\mathbf{P_{DC}^*} = \begin{bmatrix} P_{DC2}^* \cdots P_{DCq}^* \end{bmatrix}^T$.

Subsequently, P_{DC1}^* and 'I_{DCa}^*' ($1 \le a \le q$) can be computed.

3. Voltage-Power (V-P) droop with dead band

In practical DC grids, the droop characteristics can be a combination of multiple linear or nonlinear functions of the DC voltage. Figure 4.4 shows the voltage droop characteristics with dead-band and voltage

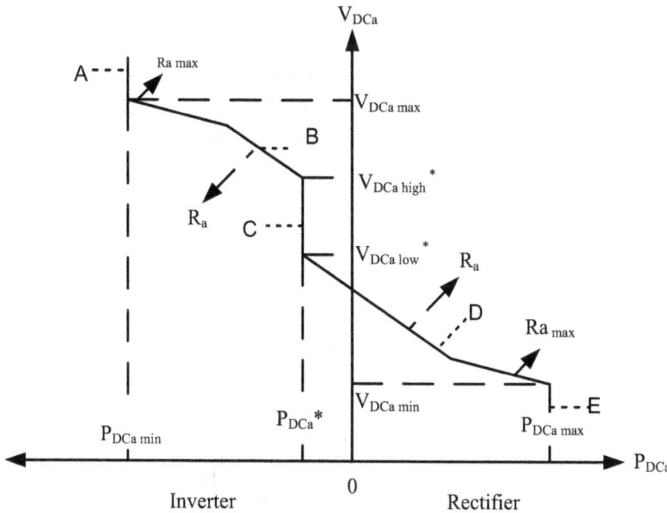

FIGURE 4.4 Nonlinear voltage droop characteristic of the ath VSC.

limits. If the ath VSC follows a nonlinear droop characteristic as shown in Figure 4.4, the converters operate in constant power control mode when DC voltage is maintained between '$V^*_{DCa\,high}$' and '$V^*_{DCa\,low}$'. When the DC voltage lies outside the dead-band zone, the DC terminal follows the linear V-P characteristic. Again, beyond a threshold DC voltage, the droop control gain increases ($R_{a\,max}$) to maintain the DC voltage within an acceptable limit for the stability of the DC grid.

The composite droop characteristic is shown in Figure 4.4 and can be expressed as

$$P_{DCa} = R_{a\,max}\left(V_{DCa\,max} - V_{DCa}\right) + \left[R_a\left(V^*_{DCa\,high} - V_{DCa\,max}\right) + P^*_{DCa}\right]$$

$$\text{for } V_{DCa} \geq V_{DCa\,max} \tag{4.19}$$

$$= R_a\left(V^*_{DCa\,high} - V_{DCa}\right) + P^*_{DCa} \qquad \text{for } V^*_{DCa\,high} < V_{DCa} < V_{DCa\,max} \tag{4.20}$$

$$= 0.\left(V^*_{DCa\,high} - V_{DCa}\right) + P^*_{DCa} \qquad \text{for } V^*_{DCa\,low} \leq V_{DCa} \leq V^*_{DCa\,high} \tag{4.21}$$

$$= R_a\left(V^*_{DCa\,low} - V_{DCa}\right) + P^*_{DCa} \qquad \text{for } V^*_{DCa\,min} < V_{DCa} < V^*_{DCa\,low} \tag{4.22}$$

$$= R_{a\,max}\left(V_{DCa\,min} - V_{DCa}\right) + \left[R_a\left(V^*_{DCa\,low} - V_{DCa\,min}\right) + P^*_{DCa}\right] \text{for } V_{DCa} \leq V_{DCa\,min} \tag{4.23}$$

In this scheme, the droop control gains can be specified according to the rating of the converters to share the additional power.
4. DC voltage margin control

In this mode of DC voltage control, each VSC regulates the DC voltage as long as its DC power is within the minimum and maximum power limits and the reference DC voltages of the different VSCs are offset from one another by a voltage margin [92]. The V-P characteristic corresponding to this scheme is shown in Figure 4.5.

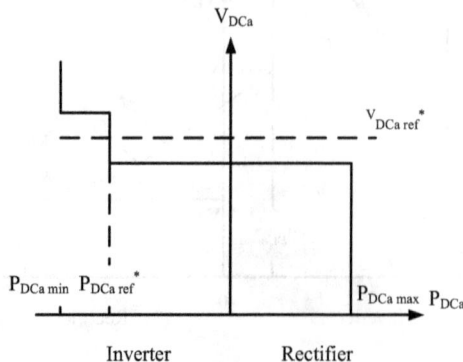

FIGURE 4.5   Voltage margin characteristic of the ath VSC.

4.5   MODELLING OF AC-MVDC SYSTEMS WITH DC VOLTAGE DROOP CONTROL

Let us assume now that all the 'q' VSCs in the AC-MTDC system shown in Figure 4.1 operate on droop control. To simplify matters, let all the 'q' VSCs follow linear V-P droops. Then, for the ath VSC ($1 \le a \le q$), from Eq. (4.16), we have

$$P_{DCa} = V_{DCa}I_{DCa} = \sum_{v=1}^{q} V_{DCa}V_{DCv}Y_{DCav} = R_a\left(V_{DCa}^* - V_{DCa}\right) + P_{DCa}^*$$

$$\text{or, } \sum_{v=1}^{q} V_{DCa}V_{DCv}Y_{DCav} + R_aV_{DCa} - R_aV_{DCa}^* - P_{DCa}^* = 0 \qquad (4.24)$$

$$\text{or, } f_{2a} = 0 \quad \forall a, \ 1 \le a \le q \qquad (4.25)$$

Equation (4.25) represents 'q' independent equations.

As already mentioned earlier in Section 4.4, in Eq. (4.24), the values of the DC voltage and power references 'V_{DCa}^*' and 'P_{DCa}^*' for all the 'q' converters are either pre-specified or obtained from a DC power flow. Now, two distinctly different models can be realized depending on whether the values of 'V_{DCa}^*' and 'P_{DCa}^*' are specified or not. These are elaborated below.

Model 'A': Values of 'V_{DCa}^*' and 'P_{DCa}^*' are known a priori

In some models [92], the values of 'V_{DCa}^*' and 'P_{DCa}^*' for all the 'q' converters are pre-specified or obtained by carrying out a DC power flow. In such cases, the 'q' independent droop equations represented by Eq. (4.24) are sufficient to compute the values of the DC bus voltages 'V_{DCa}' directly and, subsequently, the DC bus power injections 'P_{DCa}' ($1 \le a \le q$). Once 'P_{DCa}' are known, for the AC-MTDC power flow, the active powers 'P_{sha}' {Eq. (4.5)} in the lines joining the '(q–1)' VSCs to their corresponding AC buses cannot be specified, as this would be tantamount to knowing the losses in the converter transformers and the VSC, prior to the power flow. This is detailed in Figure 4.6. This is not in line with practical considerations which are targeted to maintain a specified 'P_{sha}'. This is a major drawback of the model.

Now, under the assumption that there are 'g' generators connected at the first 'g' buses of the 'n' bus AC system with bus 1 being the slack bus, the unified AC-MTDC power-flow problem corresponding to model 'A' is of the form

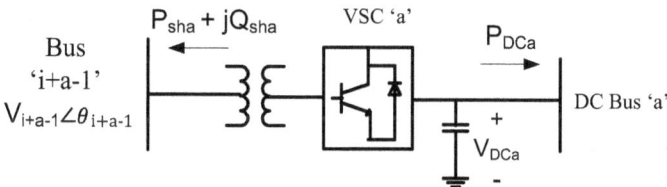

FIGURE 4.6   Power flows for the ath VSC connected to its AC bus '(i+a−1)'.

Compute: θ, $\mathbf{V}$, $\mathbf{X}$
Given: $\mathbf{P}$, $\mathbf{Q}$, $\mathbf{R}$
with

$$\theta = \left[\theta_2 \cdots \theta_n\right]^T, \; \mathbf{V} = \left[V_{g+1} \cdots V_n\right]^T, \; \theta_{sh} = \left[\theta_{sh1} \cdots \theta_{shq}\right]^T, \; \mathbf{m} = \left[m_1 \cdots m_q\right]^T,$$

$$\mathbf{X} = \left[\theta_{sh}^T \quad \mathbf{m}^T\right]^T$$

$$\mathbf{P} = \left[P_2 \cdots P_n\right]^T, \; \mathbf{Q} = \left[Q_{g+1} \cdots Q_n\right]^T, \; \mathbf{Q}_{sh} = \left[Q_{sh2} \cdots Q_{shq}\right],$$

$$\mathbf{f}_1 = \left[f_{11} \cdots f_{1q}\right], \; \mathbf{R} = \left[\mathbf{Q}_{sh} \; V_{i+r-1} \; \mathbf{f}_1\right]^T$$

For this model, it is presumed that VSC 'r' is employed for the voltage control of the AC bus '(i+r–1)' unlike the other '(q–1)' VSCs, which control the line reactive power flows.

The Newton power-flow equation is

$$\mathbf{J}\left[\Delta\theta^T \; \Delta\mathbf{V}^T \; \Delta\theta_{sh}^T \; \Delta\mathbf{m}^T\right]^T = \left[\Delta\mathbf{P}^T \; \Delta\mathbf{Q}^T \; \Delta\mathbf{R}^T\right]^T \tag{4.26}$$

where $\mathbf{J}$ is the power-flow Jacobian.

In Eq. (4.26), '$\Delta\mathbf{P}$', '$\Delta\mathbf{Q}$' and '$\Delta\mathbf{R}$' represent the mismatch vectors, while $\Delta\theta$, $\Delta\mathbf{V}$, $\Delta\theta_{sh}$ and $\Delta\mathbf{m}$ represent the correction vectors. The elements of '$\mathbf{J}$' can be obtained very easily from Eq. (4.26).

It can be observed that 'V_{DCa}' ($1 \le a \le q$) can be solved using Eq. (4.24), independent of the AC-MTDC power flow {Eq. (4.26)}, if 'V_{DCa}^*' and 'P_{DCa}^*' are known.

After all the unknowns are solved, the line active power flows 'P_{sha}' can be computed {using Eq. (4.5)}.

Thus, to summarize, Model 'A' addresses the problem 'given the DC voltage and power (or current) references of the VSC droop lines and the target line reactive power flows, what should be the line active power flow values?'

Figure 4.7 depicts the flow chart of the Newton-Raphson (NR) power-flow algorithm for droop control model 'A'.

Model 'B': Values of 'V_{DCa}^*' and 'P_{DCa}^*' are not known a priori

If the DC voltage ('V_{DCa}^*') and power ('P_{DCa}^*') reference values of the 'q' VSCs are not known, the DC bus voltages 'V_{DCa}' and hence the DC bus power injections 'P_{DCa}' ($1 \le a \le q$) cannot be computed independently {using Eq. (4.24)}. This enables the line active power-flow values 'P_{sha}' to be specified control objectives {Figure 4.6}, which is in line with practical MTDC control. This is an advantage over model 'A'.

Unified method of Model B

For the above modelling strategy, the unified AC-MTDC power-flow problem is of the form

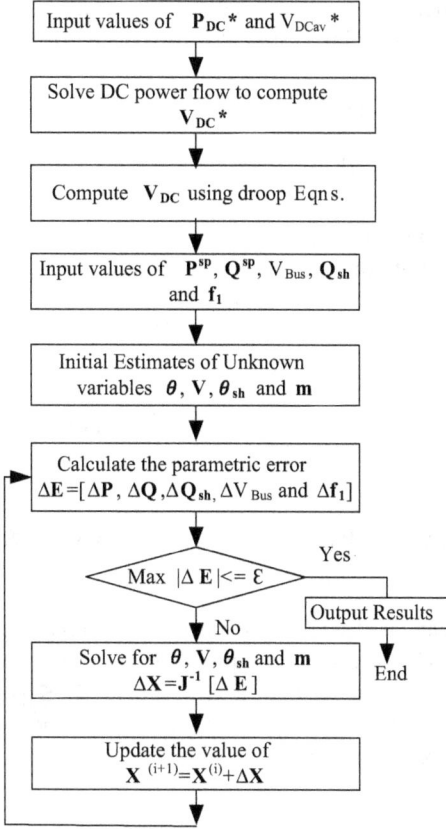

FIGURE 4.7 Flow chart of NR power-flow algorithm (Model A).

Compute: θ, $\mathbf{V}$, $\mathbf{X}$
Given: $\mathbf{P}$, $\mathbf{Q}$, $\mathbf{R}$
with

$$\theta = \left[\theta_2 \cdots \theta_n\right]^T, \ \mathbf{V} = \left[V_{g+1} \cdots V_n\right]^T, \ \theta_{sh} = \left[\theta_{sh1} \cdots \theta_{shq}\right]^T, \ \mathbf{m} = \left[m_1 \cdots m_q\right]^T,$$

$$\mathbf{V_{DC}} = \left[V_{DC1} \cdots V_{DCq}\right]^T$$

$$\mathbf{X} = \left[\theta_{sh}^T \quad \mathbf{m}^T \quad \mathbf{V}_{DC}^T\right]^T$$

$$\mathbf{P} = \left[P_2 \cdots P_n\right]^T, \ \mathbf{Q} = \left[Q_{g+1} \cdots Q_n\right]^T, \ \mathbf{P}_{sh} = \left[P_{sh2} \cdots P_{shq}\right], \ \mathbf{Q}_{sh} = \left[Q_{sh2} \cdots Q_{shq}\right],$$

$$\mathbf{f}_1 = \left[f_{11} \cdots f_{1q}\right]$$

and $\mathbf{R} = \left[\mathbf{P}_{sh} \quad \mathbf{Q}_{sh} \quad V_{i+r-1} \quad V_{DCav} \quad \mathbf{f}_1\right]^T$

For this model too, it is presumed that VSC 'r' is employed for the control of voltage magnitude of the AC bus '(i+r−1)' unlike the other '(q−1)' VSCs, which control the line active as well as reactive power flows.

The unified AC-MTDC power-flow equation is

$$\mathbf{J}\left[\Delta\boldsymbol{\theta}^T \ \Delta\mathbf{V}^T \ \Delta\boldsymbol{\theta}_{sh}^T \ \Delta\mathbf{m}^T \ \Delta\mathbf{V}_{DC}^T\right]^T = \left[\Delta\mathbf{P}^T \ \Delta\mathbf{Q}^T \ \Delta\mathbf{R}^T\right]^T \tag{4.27}$$

where $\mathbf{J}$ is the power-flow Jacobian.

It is important to note that for Model 'B', after 'V_{DCa}' $(1 \le a \le q)$ is obtained from the AC-MTDC power flow {Eq. (4.27)}, 'P_{DCa}' $(1 \le a \le q)$ is computed.

Subsequently, to compute the DC voltage ('V_{DCa}^*'), power ('P_{DCa}^*') or the current ('I_{DCa}^*') references of the 'q' VSC droop lines, we proceed as follows.

Substituting $P_{DCa}^* = V_{DCa}^* I_{DCa}^* = \sum\limits_{v=1}^{q} V_{DCa}^* V_{DCv}^* Y_{DCav}$ in Eq. (4.16), we get

$$\sum\limits_{v=1}^{q} V_{DCa}^* V_{DCv}^* Y_{DCav} + R_a V_{DCa}^* - R_a V_{DCa} - P_{DCa} = 0 \qquad \forall a, \ 1 \le a \le q \tag{4.28}$$

From Eq. (4.28), it can be observed that since 'V_{DCa}' $(1 \le a \le q)$ is already known from the AC-MTDC power-flow solution, 'P_{DCa}' can be computed as well and, hence, ('V_{DCa}^*') can be solved. After 'V_{DCa}^*' $(1 \le a \le q)$ is obtained {from Eq. (4.28)}, the power ('P_{DCa}^*') or the current ('I_{DCa}^*') references are also computed.

Sequential method of model B

AC network solution

Compute: $\boldsymbol{\theta}$, $\mathbf{V}$, $\mathbf{X}$

Given: $\mathbf{P}$, $\mathbf{Q}$, $\mathbf{R}$

with

$$\boldsymbol{\theta} = \left[\theta_2 \cdots \theta_n\right]^T, \ \mathbf{V} = \left[V_{g+1} \cdots V_n\right]^T, \ \boldsymbol{\theta}_{sh} = \left[\theta_{sh1} \cdots \theta_{shq}\right]^T, \ \mathbf{m} = \left[m_1 \cdots m_q\right]^T,$$

$$\mathbf{X} = \left[\boldsymbol{\theta}_{sh}^T \ \mathbf{m}^T\right]^T$$

$$\mathbf{P} = \left[P_2 \cdots P_n\right]^T, \ \mathbf{Q} = \left[Q_{g+1} \cdots Q_n\right]^T, \ \mathbf{P}_{sh} = \left[P_{sh2} \cdots P_{shq}\right], \ \mathbf{Q}_{sh} = \left[Q_{sh2} \cdots Q_{shq}\right],$$

and $\mathbf{R} = \left[\mathbf{P}_{sh} \ \mathbf{Q}_{sh} \ V_{i+r-1} \ f_{11}\right]^T$

For this model too, it is presumed that VSC 'r' is employed for the control of voltage magnitude of the AC bus '(i+r−1)' unlike the other '(q−1)' VSCs, which control the line active as well as reactive power flows.

The sequential AC-MTDC power-flow equation is

$$\mathbf{J}\left[\Delta\boldsymbol{\theta}^T \ \Delta\mathbf{V}^T \ \Delta\boldsymbol{\theta}_{sh}^T \ \Delta\mathbf{m}^T\right]^T = \left[\Delta\mathbf{P}^T \ \Delta\mathbf{Q}^T \ \Delta\mathbf{R}^T\right]^T \tag{4.29}$$

where $\mathbf{J}$ is the power-flow Jacobian.

DC network solution

The power-flow solution pertaining to DC network solution can be formed as
Solve: $\mathbf{V_{DC}}$
Specified: V_{DCavg}, $\mathbf{f_3}$
where

$$\mathbf{V_{DC}} = \left[V_{DC1} \cdots V_{DCq} \right],\ \mathbf{f_3} = P_{DC2} \cdots P_{DCq}$$

and $\mathbf{R_{DC}} = \left[V_{DCavg}\ \mathbf{f_3} \right]^{T}$.

The Newton power-flow equation for the above is

$$\mathbf{J_{DC}}\ \Delta \mathbf{V_{DC}^T} = \Delta \mathbf{R_{DC}} \tag{4.30}$$

In Eq. (4.30), '$\Delta \mathbf{R_{DC}}$' and $\Delta \mathbf{V_{DC}}$ are the mismatch and correction vectors, respectively.

Similar to the unified method, the DC reference values are calculated using droop equations.

Figures 4.8 and 4.9 depict the flow charts of the proposed unified and sequential NR algorithms for droop control model 'B'.

Thus, to summarize, Model 'B' addresses the problem 'given the target line active and reactive power flows, what should be the voltage and power references of the VSC droop lines?'

It may be noted that the above analysis was done considering that the 'q' VSCs follow linear V-P voltage droop characteristics. In a similar manner, if all the 'q' VSCs follow linear V-I droop characteristics, substituting $I_{DCa}^{*} = \sum_{v=1}^{q} V_{DCv}^{*} Y_{DCav}$ in

Eq. (4.17), we get

$$I_{DCa} = R_a \left(V_{DCa}^{*} - V_{DCa} \right) + \sum_{v=1}^{q} V_{DCv}^{*} Y_{DCav}$$

$$\text{or,}\quad \sum_{v=1}^{q} V_{DCv}^{*} Y_{DCav} + R_a V_{DCa}^{*} - R_a V_{DCa} - I_{DCa} = 0 \tag{4.31}$$

Thus, for Model 'B', subsequent to 'V_{DCa}' ($1 \le a \le q$) being obtained from the AC-MTDC power flow, 'I_{DCa}' ($1 \le a \le q$) is computed. Then, the DC voltage ('V_{DCa}^{*}') references of the 'q' VSC droop lines can be solved from the 'q' droop equations. After 'V_{DCa}^{*}' ($1 \le a \le q$) are computed, the current ('I_{DCa}^{*}') or the power ('P_{DCa}^{*}') references can also be computed.

Similarly, if the ath VSC follows a nonlinear voltage droop characteristic as shown in Figure 4.4 and operates at point 'A' of the characteristic, we have

$$P_{DCa} = R_{a\,max} \left(V_{DCa\,max} - V_{DCa} \right) + \left[R_a \left(V_{DCa\,high}^{*} - V_{DCa\,max} \right) + P_{DCa}^{*} \right]$$

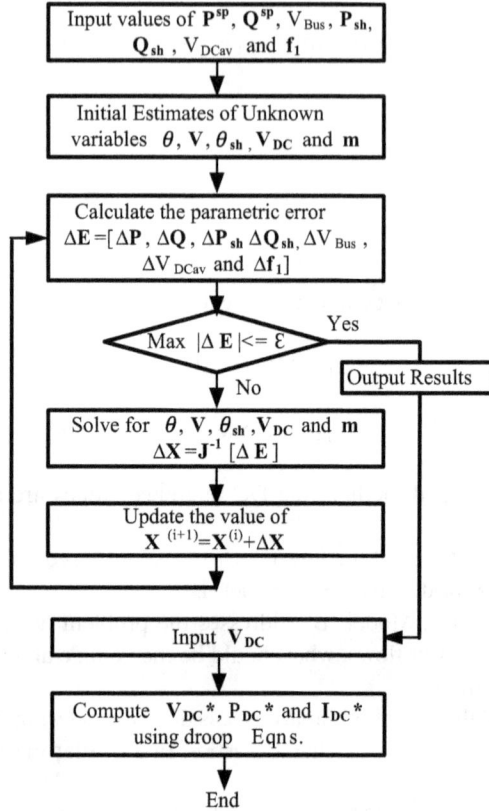

FIGURE 4.8   Flow chart of the unified NR power-flow algorithm (Model B).

$$\text{or,} \quad \sum_{v=1}^{q} V_{DCa}^{*} V_{DCv}^{*} Y_{DCav} + R_a \left(V_{DCa\,high}^{*} - V_{DCa\,max} \right)$$

$$+ R_{a\,max} \left(V_{DCa\,max} - V_{DCa} \right) - P_{DCa} = 0 \qquad (4.32)$$

In a similar manner, the equations corresponding to other operating points ('B', 'C', 'D' and 'E') of the nonlinear voltage droop characteristics shown in Figure 4.4 can also be derived very easily.

4.6   CASE STUDIES AND RESULTS

For validation of the above models, a large number of studies were carried out by employing diverse DC voltage droop control strategies on MTDC grids embedded within the IEEE 300-bus and European 1354-bus networks [115,116].

In all the occurrences, the VSC constant was selected as $c = \dfrac{1}{2\sqrt{2}}$ [11]. Also,

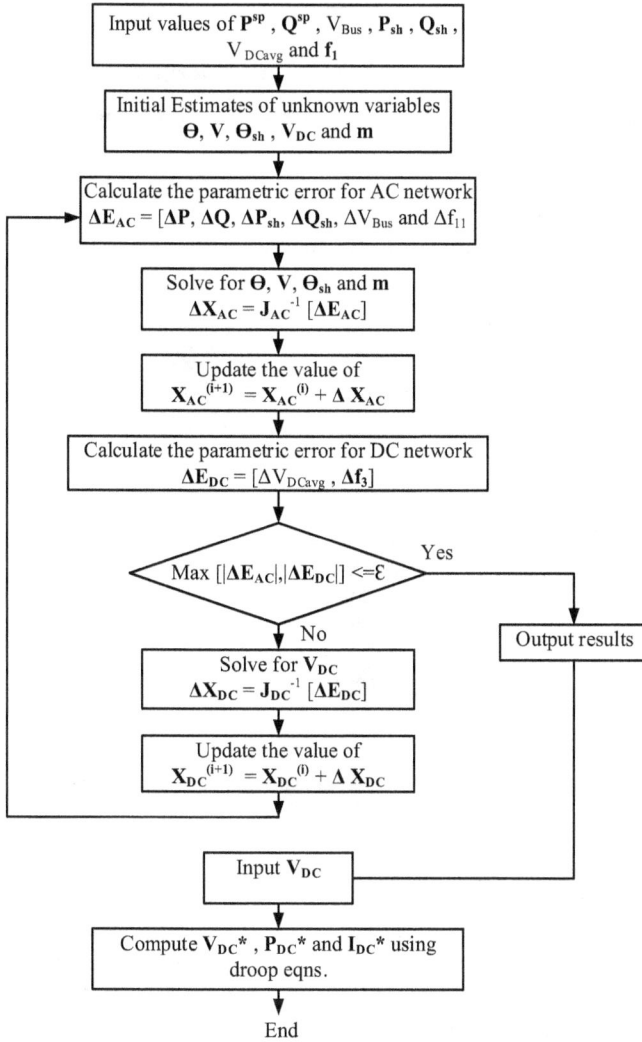

FIGURE 4.9 Flow chart of the sequential NR power-flow algorithm (Model B).

for all the VSC coupling transformers, $R_{sha} = 0.001$ p.u. and $X_{sha} = 0.1$ p.u. ($\forall a, 1 \leq a \leq q$). The converter loss constants 'a_l', 'b_l' and 'c_l' were chosen as 0.011, 0.003 and 0.0043, respectively [18,105]. For interconnections between DC terminals, $R_{DCuv} = 0.01$ p.u. ($\forall u, v, 1 \leq u \leq q, 1 \leq v \leq q, u \neq v$), throughout the chapter [83]. In all occurrences, a termination error tolerance of 10^{-8} p.u. was selected. 'NI' denotes the number of iterations. In all the results given in Tables 4.1–4.8, values of bus voltage magnitudes, current magnitudes, active and reactive powers and droop control gains are denoted in p.u., while phase angles of voltage phasors are denoted in degrees.

TABLE 4.1

Study of IEEE 300-Bus System with 5-Terminal VSC-HVDC Network Employing Linear DC Voltage Droop Characteristics (Model A)

Base case power flow (NI=6)

$V_{266} = 1.011 \angle -11.24$; $V_{270} = 1.011 \angle -11.32$; $V_{271} = 0.998 \angle -17.67$;

$V_{272} = 0.981 \angle -19.46$; $V_{273} = 1.006 \angle -17.47$;

DC (Five-Terminal) Power Flow

Given Quantities	Solution
$V^*_{DCav} = 3$;	DC power flow converged in three iterations
$P^*_{DC2} = 0.35$;	$V^*_{DC1} = 2.9991$; $V^*_{DC2} = 3.0002$; $V^*_{DC3} = 3.0002$; $V^*_{DC4} = 3.0002$;
$P^*_{DC3} = 0.3$;	$V^*_{DC5} = 3.0003$; $P^*_{DC1} = -1.3495$; $I^*_{DC3} = 0.1$; $I^*_{DC4} = 0.1$;
$P^*_{DC4} = 0.3$;	$I^*_{DC5} = 0.1333$;
$P^*_{DC5} = 0.4$;	NI=3;

Computation of V_{DC} from Droop Equations

Given Quantities	Solution
$V^*_{DC1} = 2.9991$; $V^*_{DC2} = 3.0002$;	$V_{DC1} = 2.9991$; $V_{DC2} = 3.0002$; $V_{DC3} = 3.0002$;
$V^*_{DC3} = 3.0002$; $V^*_{DC4} = 3.0002$;	$V_{DC4} = 3.0002$; $V_{DC5} = 3.0003$;
$V^*_{DC5} = 3.0003$; $P^*_{DC1} = -1.3495$;	$P_{DC1} = -1.3495$; $P_{DC2} = 0.3500$;
$P^*_{DC2} = 0.35$; $I^*_{DC3} = 0.1$	$P_{DC3} = 0.3000$; $P_{DC4} = 0.3000$;
$I^*_{DC4} = 0.1$; $I^*_{DC5} = 0.1333$;	$P_{DC5} = 0.4000$;
$R_1 = 20$; $R_2 = 15$; $R_3 = 10$;	NI = 4;
$R_4 = 10$; $R_5 = 10$;	

AC-MTDC Power Flow with Linear Voltage Droop Control

Given Quantities	Solution	
	AC Buses	VSC
$V_{DC1} = 2.9991$;	$\theta_{266} = -11.6616$;	$\theta_{sh1} = -4.7995$; $m_1 = 1.0203$;
$V_{DC2} = 3.0002$;	$V_{270} = 1.0201 \angle -11.8087$;	$\theta_{sh2} = -13.7810$; $m_2 = 0.9804$;
$V_{DC3} = 3.0002$;	$V_{271} = 1.022 \angle -34.5310$;	$\theta_{sh3} = -36.2298$; $m_3 = 0.9775$;
$V_{DC4} = 3.0002$;	$V_{272} = 1.0118 \angle -39.7253$;	$\theta_{sh4} = -41.4609$; $\theta_{sh5} = -36.3384$;
$V_{DC5} = 3.0003$;	$V_{273} = 1.0014 \angle -33.9952$;	$m_4 = 0.9652$; $m_5 = 0.9538$;
$V_{266} = 1.02$;	$P_{sh1} = 1.3239$; $P_{sh2} = -0.3631$;	
$Q_{sh2} = 0.2$;	$P_{sh3} = -0.3126$; $P_{sh4} = -0.3126$;	
$Q_{sh3} = 0.15$;	$P_{sh5} = -0.4132$;	
$Q_{sh4} = 0.12$;	Converter loss (%)	
$Q_{sh5} = 0.1$;	$P_{loss1} = 2.37$; $P_{loss2} = 1.29$; $P_{loss3} = 1.25$; $P_{loss4} = 1.25$; $P_{loss5} = 1.31$;	
	NI = 6	

TABLE 4.2

Study of IEEE 300-Bus System with 5-Terminal VSC-HVDC Network Incorporating Nonlinear Voltage Droop Characteristics with Dead Band

Base case power flow (NI=6)

$V_{266} = 1.011 \angle -11.24$; $V_{270} = 1.011 \angle -11.32$; $V_{271} = 0.998 \angle -17.67$;

$V_{272} = 0.981 \angle -19.46$; $V_{273} = 1.006 \angle -17.47$;

DC Power Flow to Calculate DC Reference Values

Specified Quantities	Power-Flow Solution
$V^*_{DCav} = 3.02$; $P^*_{DC2} = 0.5$;	$V^*_{DC1} = 3.0189$; $V^*_{DC2} = 3.0203$; $V^*_{DC3} = 3.0203$; $V^*_{DC4} = 3.0202$;
$P^*_{DC3} = 0.4$; $P^*_{DC4} = 0.3$;	$V^*_{DC5} = 3.0203$; $P^*_{DC1} = -1.5993$; $I^*_{DC3} = 0.1325$; $I^*_{DC4} = 0.0993$;
$P^*_{DC5} = 0.4$;	$I^*_{DC5} = 0.1325$;
	NI = 3;

Computation of V_{DC} from Droop Equations

Specified Quantities	Power-Flow Solution	
Converter	Control mode	
1,2	V-P droop	
3	V-P droop with dead band	
4,5	V-I droop	
$V_{DCmax} = 3.014$;	$V_{DC1} = 3.0174$; $V_{DC2} = 3.01883$; $V_{DC3} = 3.0186$;	
$V^*_{DChigh} = 3.013$;	$V_{DC4} = 3.0187$; $V_{DC5} = 3.0187$; $P_{DC1} = -1.5681$;	
$V^*_{DClow} = 3.012$; $V_{DCmin} = 3.009$;	$P_{DC2} = 0.5234$; $P_{DC3} = 0.2523$; $P_{DC4} = 0.3466$;	
$R_1 = 20$; $R_2 = 15$; $R_3 = 10$;	$P_{DC5} = 0.4465$;	
$R_4 = 10$; $R_5 = 10$;	NI = 4;	
$R_{max} = 30$;		

AC-MTDC power flow with linear voltage droop and nonlinear droop with dead band (VSCs connected to AC buses 266, 270, 271, 272 and 273)

Specified Quantities	Power-Flow Solution	
$V_{266} = 1.02$;	AC terminal buses	VSCs
$Q_{sh2} = 0.1$;	$\theta_{266} = -11.7581$; $V_{270} = 1.0196 \angle -11.9345$;	$\theta_{sh1} = -4.1614$;
$Q_{sh3} = 0.05$;	$V_{271} = 0.8849 \angle -38.2872$;	$\theta_{sh2} = -14.8734$;
$Q_{sh4} = 0.04$;	$V_{272} = 0.8494 \angle -45.8518$;	$\theta_{sh3} = -40.2153$;
$Q_{sh5} = 0.1$;	$V_{273} = 0.9948 \angle -36.1943$; $P_{sh1} = 1.5345$;	$\theta_{sh4} = -48.6957$;
	$P_{sh2} = -0.5376$; $P_{sh3} = -0.2647$;	$\theta_{sh5} = -38.8364$;
	$P_{sh4} = -0.3598$; $P_{sh5} = -0.4601$	$m_1 = 1.0597$;
	$Q_{sh1} = 1.0107$;	$m_2 = 0.9653$;
		$m_3 = 0.8347$;
		$m_4 = 0.8009$;
		$m_5 = 0.9421$;
	Converter loss (%)	
	$P_{loss1} = 0.0358$; $P_{loss2} = 1.38$; $P_{loss3} = 1.23$; $P_{loss4} = 1.31$; $P_{loss5} = 1.34$;	
	NI = 6	

TABLE 4.3

Study of IEEE 300-Bus System with 5-Terminal VSC-HVDC Network Incorporating Nonlinear DC Voltage Droop Characteristics (Voltage Margin)

Base case power flow (NI = 6)

$V_{266} = 1.011 \angle -11.24$; $V_{270} = 1.011 \angle -11.32$; $V_{271} = 0.998 \angle -17.67$;

$V_{272} = 0.981 \angle -19.46$; $V_{273} = 1.006 \angle -17.47$;

DC Power Flow to Calculate DC Reference Values

Specified Quantities	Power-Flow Solution
$V^*_{DCav} = 3.03$; $P^*_{DC2} = 0.5$; $P^*_{DC3} = 0.4$;	$V^*_{DC1} = 3.0289$; $V^*_{DC2} = 3.0303$; $V^*_{DC3} = 3.0303$;
$P^*_{DC4} = 0.3$; $P^*_{DC5} = 0.4$;	$V^*_{DC4} = 3.0302$; $V^*_{DC5} = 3.0303$; $P^*_{DC1} = -1.5993$;
	$I^*_{DC3} = 0.1320$; $I^*_{DC4} = 0.0990$; $I^*_{DC5} = 0.1320$;
	NI = 3;

Computation of V_{DC} from Droop Equations

Specified Quantities	Power-Flow Solution
Converter	Control mode
1,2	V-P droop
3	V-P droop with dead band
4,5	V-I droop
$V_{DCmax} = 3.01$; $V^*_{DChigh} = 3.00$;	$V_{DC1} = 3.0142$; $V_{DC2} = 3.0155$; $V_{DC3} = 3.0144$;
$V^*_{DClow} = 2.99$; $V_{DCmin} = 2.98$;	$V_{DC4} = 3.0155$; $V_{DC5} = 3.0156$; $P_{DC1} = -1.3036$;
$R_1 = 20$; $R_2 = 15$; $R_3 = 1000$;	$P_{DC2} = 0.7224$; $P_{DC3} = -1$; $P_{DC4} = 0.7413$;
$R_4 = 10$; $R_5 = 10$	$P_{DC5} = 0.8408$;
	NI = 5;

AC-MTDC power flow with linear voltage droop and nonlinear droop with dead band (VSCs connected to AC buses 266, 270, 271, 272 and 273)

Specified Quantities	Power-Flow Solution	
$V_{266} = 1.02$;	AC terminal buses	VSCs
$Q_{sh2} = 0.1$;	$\theta_{266} = -12.0060$;	$\theta_{sh1} = -5.8424$;
$Q_{sh3} = 0.05$;	$V_{270} = 1.0195\angle -12.2210$;	$\theta_{sh2} = -16.2548$;
$Q_{sh4} = 0.04$;	$V_{271} = 0.9646 \angle -13.7538$;	$\theta_{sh3} = -7.7795$;
$Q_{sh5} = 0.1$;	$V_{272} = 0.9022 \angle -25.2104$;	$\theta_{sh4} = -30.5161$;
	$V_{273} = 0.8695 \angle -64.0396$;	$\theta_{sh5} = -70.4611$;
	$P_{sh1} = 1.2713$; $P_{sh2} = -0.7384$;	$m_1 = 1.0787$; $m_2 = 0.9671$;
	$P_{sh3} = 0.9805$; $P_{sh4} = -0.7586$;	$m_3 = 0.9159$; $m_4 = 0.8533$;
	$P_{sh5} = -0.8601$; $Q_{sh1} = 1.2406$	$m_5 = 0.8306$;
	Converter loss (%)	
	$P_{loss1} = 2.93$; $P_{loss2} = 1.55$; $P_{loss3} = 1.85$; $P_{loss4} = 1.66$; $P_{loss5} = 1.83$;	
	NI = 6	

TABLE 4.4
Study of European 1354-Bus System with 7-Terminal VSC-HVDC Network Employing Linear DC Voltage Droop Characteristics (Model A)

Base case power flow converged in seven iterations (NI = 7)

$V_{1280} = 0.9554 \angle -35.2574$; $V_{1281} = 1.0354 \angle -37.3082$; $V_{1286} = 1.0156 \angle -12.3760$;

$V_{1292} = 1.0289 \angle -35.23066$; $V_{1313} = 1.0351 \angle -37.4755$; $V_{1324} = 0.9972 \angle -29.0208$;

$V_{1353} = 0.9934 \angle -29.6628$;

DC (Seven-Terminal) Power Flow

Given Quantities	Solution
$V^*_{DCav} = 3.02$; $P^*_{DC2} = 0.5$;	DC power flow converged in three iterations
$P^*_{DC3} = 0.4$; $P^*_{DC4} = 0.3$;	$V^*_{DC1} = 3.0189$; $V^*_{DC2} = 3.0202$; $V^*_{DC3} = 3.0202$; $V^*_{DC4} = 3.0201$;
$P^*_{DC5} = 0.4$; $P^*_{DC6} = 0.4$;	$V^*_{DC5} = 3.0202$; $V^*_{DC6} = 3.0202$; $V^*_{DC7} = 3.0201$; $P^*_{DC1} = -2.299$;
$P^*_{DC7} = 0.3$;	$I^*_{DC5} = 0.1324$; $I^*_{DC6} = 0.1324$; $I^*_{DC7} = 0.0993$;
	NI = 3;

Computation of V_{DC} from Droop Equations

Given Quantities	Solution
$V^*_{DC1} = 3.0189$; $V^*_{DC2} = 3.0202$;	$V_{DC1} = 3.0189$; $V_{DC2} = 3.0202$;
$V^*_{DC3} = 3.0202$; $V^*_{DC4} = 3.0201$;	$V_{DC3} = 3.0202$; $V_{DC4} = 3.0201$;
$V^*_{DC5} = 3.0202$; $V^*_{DC6} = 3.0202$;	$V_{DC5} = 3.0202$; $V_{DC6} = 3.0202$;
$V^*_{DC7} = 3.0201$; $P^*_{DC1} = -2.299$;	$V_{DC7} = 3.0201$;
$I^*_{DC5} = 0.1324$; $I^*_{DC6} = 0.1324$;	$P_{DC1} = -2.2990$; $P_{DC2} = 0.5$;
$I^*_{DC7} = 0.0993$; $R_1 = 20$; $R_2 = 15$;	$P_{DC3} = 0.4$; $P_{DC4} = 0.3$;
$R_3 = 10$; $R_4 = 10$; $R_5 = 10$;	$P_{DC5} = 0.4$; $P_{DC6} = 0.4$;
$R_6 = 15$; $R_7 = 20$;	$P_{DC7} = 0.3$;
	NI = 4;

AC-MTDC Power Flow with Linear Voltage Droop Control

Given Quantities	Solution	
	AC Buses	VSC
$V_{DC1} = 3.0189$;	$\theta_{1280} = -26.6472$;	$\theta_{sh1} = -13.4234$;
$V_{DC2} = 3.0202$;	$V_{1281} = 1.0359 \angle -37.8067$;	$\theta_{sh2} = -40.5112$;
$V_{DC3} = 3.0202$;	$V_{1292} = 1.016 \angle -12.7774$;	$\theta_{sh3} = -15.0538$;
$V_{DC4} = 3.0201$;	$V_{1313} = 1.0305 \angle -35.2997$;	$\theta_{sh4} = -36.9753$;
$V_{DC5} = 3.0202$;	$V_{1324} = 1.0356 \angle -37.9622$;	$\theta_{sh5} = -40.1472$;
$V_{DC6} = 3.0202$;	$V_{1324} = 0.9971 \angle -30.2104$;	$\theta_{sh6} = -32.5752$;
$V_{DC7} = 3.0201$;	$V_{1353} = 0.9965 \angle -29.9515$;	$\theta_{sh7} = -31.7373$;
$V_{1280} = 0.98$;	$P_{sh1} = 2.2531$; $P_{sh2} = -0.5141$;	$m_1 = 0.9418$; $m_2 = 0.9888$;
$Q_{sh2} = 0.2$;	$P_{sh3} = -0.4132$; $P_{sh4} = -0.3125$;	$m_3 = 0.9611$; $m_4 = 0.9744$;
$Q_{sh3} = 0.1$;	$P_{sh5} = -0.4132$; $P_{sh6} = -0.4132$;	$m_5 = 0.9837$; $m_6 = 0.9427$;
$Q_{sh4} = 0.1$;	$P_{sh7} = -0.3127$;	$m_7 = 0.9475$;
$Q_{sh5} = 0.15$;	Converter loss (%)	
$Q_{sh6} = 0.09$	$P_{loss1} = 4.06$; $P_{loss2} = 1.38$; $P_{loss3} = 1.30$; $P_{loss4} = 1.23$;	
$Q_{sh7} = 0.15$;	$P_{loss5} = 1.30$; $P_{loss6} = 1.30$; $P_{loss7} = 1.26$;	
	NI = 7	

TABLE 4.5
Study of European 1354-Bus System with 7-Terminal VSC-HVDC Network Incorporating Nonlinear DC Voltage Droop Characteristics with Dead Band (Model A)

Base case power flow converged in seven iterations (NI=7)

$V_{1280} = 0.9554\ \angle -35.2574$; $V_{1281} = 1.0354\ \angle -37.3082$; $V_{1286} = 1.0156\ \angle -12.3760$;

$V_{1292} = 1.0289\ \angle -35.23066$; $V_{1313} = 1.0351\ \angle -37.4755$; $V_{1324} = 0.9972\ \angle -29.0208$;

$V_{1353} = 0.9934\ \angle -29.6628$;

DC Power Flow to Calculate DC Reference Values

Specified Quantities	Power-Flow Solution
$V^*_{DCav} = 3.02$; $P^*_{DC2} = 0.5$;	$V^*_{DC1} = 3.0189$; $V^*_{DC2} = 3.0202$; $V^*_{DC3} = 3.0202$;
$P^*_{DC3} = 0.4$; $P^*_{DC4} = 0.3$;	$V^*_{DC4} = 3.0201$; $V^*_{DC5} = 3.0202$; $V^*_{DC6} = 3.0202$;
$P^*_{DC5} = 0.4$; $P^*_{DC6} = 0.4$;	$V^*_{DC7} = 3.0201$; $P^*_{DC1} = -2.299$; $I^*_{DC5} = 0.1324$;
$P^*_{DC7} = 0.3$;	$I^*_{DC6} = 0.1324$; $I^*_{DC7} = 0.0993$;
	NI = 3;

Computation of V_{DC} from Droop Equations

Converter	Control Mode
1,2,3	V-P droop
4	V-P droop with dead band
5,6,7	V-I droop

Given Quantities	Solution
$V^*_{DC1} = 3.0189$; $V^*_{DC2} = 3.0202$;	$V_{DC1} = 3.0180$; $V_{DC2} = 3.0193$;
$V^*_{DC3} = 3.0202$; $V^*_{DC4} = 3.0201$;	$V_{DC3} = 3.0193$; $V_{DC4} = 3.0191$;
$V^*_{DC5} = 3.0202$; $V^*_{DC6} = 3.0202$;	$V_{DC5} = 3.0193$; $V_{DC6} = 3.0193$;
$V^*_{DC7} = 3.0201$; $P^*_{DC1} = -2.299$;	$V_{DC7} = 3.0192$;
$I^*_{DC5} = 0.1324$; $I^*_{DC6} = 0.1324$;	$P_{DC1} = -2.2807$; $P_{DC2} = 0.5138$;
$I^*_{DC7} = 0.0993$; $V_{DCmax} = 3.014$;	$P_{DC3} = 0.4092$; $P_{DC4} = 0.1359$;
$V^*_{DChigh} = 3.013$; $V^*_{DClow} = 3.012$;	$P_{DC5} = 0.4275$; $P_{DC6} = 0.4410$;
$V_{DCmin} = 3.009$;	$P_{DC7} = 0.3543$;
$R_1 = 20$; $R_2 = 15$; $R_3 = 10$; $R_4 = 10$;	NI = 4;
$R_5 = 10$; $R_6 = 15$; $R_7 = 20$; $R_{max} = 30$;	

AC-MTDC power flow with linear voltage droop and nonlinear droop with dead band (VSCs connected to AC buses 1280, 1281, 1286, 1292, 1313, 1324 and 1353)

(Continued)

TABLE 4.5 (*Continued*)
Study of European 1354-Bus System with 7-Terminal VSC-HVDC Network Incorporating Nonlinear DC Voltage Droop Characteristics with Dead Band (Model A)

Specified Quantities	Power-Flow Solution	
	AC Buses	**VSC**
$V_{DC1} = 3.0180$;	$\theta_{1280} = -26.6732$;	$\theta_{sh1} = -13.5517$;
$V_{DC2} = 3.0193$;	$V_{1281} = 1.0359 \angle -37.8186$;	$\theta_{sh2} = -40.5961$;
$V_{DC3} = 3.0193$;	$V_{1292} = 1.016 \angle -12.7862$;	$\theta_{sh3} = -15.1137$;
$V_{DC4} = 3.0191$;	$V_{1313} = 1.0313 \angle -34.9713$;	$\theta_{sh4} = -35.7642$;
$V_{DC5} = 3.0193$;	$V_{1324} = 1.0355 \angle -37.9750$;	$\theta_{sh5} = -40.3057$;
$V_{DC6} = 3.0193$;	$V_{1324} = 0.9966 \angle -30.3349$;	$\theta_{sh6} = -32.9376$;
$V_{DC7} = 3.0192$;	$V_{1353} = 0.9961 \angle -30.0326$;	$\theta_{sh7} = -32.1304$;
$V_{1280} = 0.98$; $Q_{sh2} = 0.2$;	$P_{sh1} = 2.2352$; $P_{sh2} = -0.5280$;	$m_1 = 0.9418$; $m_2 = 0.9891$;
$Q_{sh3} = 0.1$;	$P_{sh3} = -0.4225$; $P_{sh4} = -0.1475$;	$m_3 = 0.9614$; $m_4 = 0.9752$;
$Q_{sh4} = 0.1$;	$P_{sh5} = -0.4409$; $P_{sh6} = -0.4545$;	$m_5 = 0.9841$; $m_6 = 0.9426$;
$Q_{sh5} = 0.15$;	$P_{sh7} = -0.3673$;	$m_7 = 0.9476$;
$Q_{sh6} = 0.09$	Converter loss (%)	
$Q_{sh7} = 0.15$;	$P_{loss1} = 4.02$; $P_{loss2} = 1.39$; $P_{loss3} = 1.30$; $P_{loss4} = 1.16$;	
	$P_{loss5} = 1.32$; $P_{loss6} = 1.33$; $P_{loss7} = 1.28$;	
	NI = 7	

TABLE 4.6
Study of European 1354-Bus System with 7-Terminal VSC-HVDC Network Incorporating Nonlinear DC Voltage Droop Characteristics with Voltage Margin (Model A)

Base case power flow converged in seven iterations (NI = 7)
$V_{1280} = 0.9554 \angle -35.2574$; $V_{1281} = 1.0354 \angle -37.3082$; $V_{1286} = 1.0156 \angle -12.3760$;
$V_{1292} = 1.0289 \angle -35.23066$; $V_{1313} = 1.0351 \angle -37.4755$; $V_{1324} = 0.9972 \angle -29.0208$;
$V_{1353} = 0.9934 \angle -29.6628$;

DC Power Flow to Calculate DC Reference Values

Specified Quantities	Power-Flow Solution
$V_{DCav}^* = 3.03$; $P_{DC2}^* = 0.5$;	$V_{DC1}^* = 3.0289$; $V_{DC2}^* = 3.0302$; $V_{DC3}^* = 3.0302$;
$P_{DC3}^* = 0.4$; $P_{DC4}^* = 0.3$;	$V_{DC4}^* = 3.0301$; $V_{DC5}^* = 3.0302$; $V_{DC6}^* = 3.0302$;
$P_{DC5}^* = 0.4$; $P_{DC6}^* = 0.4$;	$V_{DC7}^* = 3.0301$; $P_{DC1}^* = -2.299$; $I_{DC5}^* = 0.1320$;
$P_{DC7}^* = 0.3$;	$I_{DC6}^* = 0.1320$; $I_{DC7}^* = 0.0990$;
	NI = 3;

(*Continued*)

TABLE 4.6 (*Continued*)

Study of European 1354-Bus System with 7-Terminal VSC-HVDC Network Incorporating Nonlinear DC Voltage Droop Characteristics with Voltage Margin (Model A)

Computation of V_{DC} from Droop Equations

Converter	Control Mode
1,2,3	V-P droop
4	V-P droop with voltage margin
5,6,7	V-I droop

Given Quantities	Solution
$V_{DC1}^* = 3.0289$; $V_{DC2}^* = 3.0302$;	$V_{DC1} = 3.0216$;
$V_{DC3}^* = 3.0302$; $V_{DC4}^* = 3.0301$;	$V_{DC2} = 3.0230$;
$V_{DC5}^* = 3.0302$; $V_{DC6}^* = 3.0302$;	$V_{DC3} = 3.0229$;
$V_{DC7}^* = 3.0301$; $P_{DC1}^* = -2.299$;	$V_{DC4} = 3.0222$;
$I_{DC5}^* = 0.1320$; $I_{DC6}^* = 0.1320$;	$V_{DC5} = 3.0230$;
$I_{DC7}^* = 0.0990$; $V_{DCmax}^* = 3.01$;	$V_{DC6} = 3.0230$;
$V_{DChigh}^* = 3.0$; $V_{DClow}^* = 2.99$;	$V_{DC7} = 3.0230$;
$V_{DCmin} = 2.98$;	$P_{DC1} = -2.1537$; $P_{DC2} = 0.6092$;
$R_1 = 20$; $R_2 = 15$; $R_3 = 10$;	$P_{DC3} = 0.4730$; $P_{DC4} = -1$;
$R_4 = 1000$; $R_5 = 10$; $R_6 = 15$;	$P_{DC5} = 0.6176$; $P_{DC6} = 07246$;
$R_7 = 20$;	$P_{DC7} = 0.7304$;
$P_{DCmin} = -1$; $P_{DCmax} = 1$;	NI = 5;

AC-MTDC power flow with linear voltage droop and nonlinear droop with voltage margin (VSCs connected to AC buses 1280, 1281, 1286, 1292, 1313, 1324 and 1353)

Specified Quantities	Power-Flow Solution	
$V_{DC1} = 3.0216$;	**AC Buses**	**VSC**
$V_{DC2} = 3.0230$;		
$V_{DC3} = 3.0229$;	$\theta_{1280} = -26.8738$;	$\theta_{sh1} = -13.5517$;
$V_{DC4} = 3.0222$;	$V_{1281} = 1.0357 \angle -37.9214$;	$\theta_{sh2} = -40.5961$;
$V_{DC5} = 3.0230$;	$V_{1292} = 1.0158 \angle -12.8500$;	$\theta_{sh3} = -15.1137$;
$V_{DC6} = 3.0230$;	$V_{1313} = 1.0356 \angle -32.7538$;	$\theta_{sh4} = -35.7642$;
$V_{DC7} = 3.0230$;	$V_{1324} = 1.0353 \angle -38.0837$;	$\theta_{sh5} = -40.3057$;
$V_{1280} = 0.98$; $Q_{sh2} = 0.2$;	$V_{1324} = 0.9934 \angle -1.2140$;	$\theta_{sh6} = -32.9376$;
$Q_{sh3} = 0.1$;	$V_{1353} = 0.9936 \angle -30.6158$;	$\theta_{sh7} = -32.1304$;
$Q_{sh4} = 0.1$;	$P_{sh1} = 2.1116$; $P_{sh2} = -0.6242$;	$m_1 = 0.9386$; $m_2 = 0.9881$;
$Q_{sh5} = 0.15$;	$P_{sh3} = -0.4867$; $P_{sh4} = 0.9813$;	$m_3 = 0.9603$; $m_4 = 0.9831$;
$Q_{sh6} = 0.09$	$P_{sh5} = -0.6326$; $P_{sh6} = -0.7409$;	$m_5 = 0.9834$; $m_6 = 0.9398$;
$Q_{sh7} = 0.15$;	$P_{sh7} = -0.7468$; $Q_{sh1} = 0.0279$;	$m_7 = 0.9457$;
	Converter loss (%)	
	$P_{loss1} = 3.74$; $P_{loss2} = 1.46$; $P_{loss3} = 1.35$; $P_{loss4} = 1.77$;	
	$P_{loss5} = 1.46$; $P_{loss6} = 1.57$; $P_{loss7} = 1.58$;	
	NI = 7	

TABLE 4.7

Study of IEEE 300-Bus System with 5-Terminal VSC-HVDC Network Incorporating Linear DC Voltage Droop Characteristics (Model B)

Base case power flow (NI = 6)

$V_{266} = 1.011\ \angle-11.24;\ V_{270} = 1.011\ \angle-11.32;\ V_{271} = 0.998\ \angle-17.67;$

$V_{272} = 0.981\ \angle-19.46;\ V_{273} = 1.006\ \angle-17.47;$

AC-MTDC power flow with linear voltage droop control (VSCs connected to AC buses 266, 270, 271, 272 and 273)

Given Quantities		Solution	
		AC Buses	**VSC**
V_{DCav}	3.0	$\theta_{266} = -11.5503;$	$\theta_{sh1} = -18.4481;\ \theta_{sh2} = -9.4048;\ \theta_{sh3} = -2.1709;$
V_{266}	1.03	$\theta_{270} = -11.5408;$	$\theta_{sh4} = -1.0782;\ \theta_{sh5} = -5.2494;\ m_1 = 1.0579;$
P_{sh2}	0.4	$\theta_{271} = -3.6012;$	$m_2 = 0.9803;\ m_3 = 1.0361;\ m_4 = 1.0384;$
Q_{sh2}	0.09	$\theta_{272} = -2.4997;$	$m_5 = 0.9876;\ V_{DC1} = 3.0009;\ V_{DC2} = 2.9997;$
P_{sh3}	0.3	$\theta_{273} = -6.8247;$	$V_{DC3} = 2.9998;\ V_{DC4} = 2.9998;\ V_{DC5} = 2.9998;$
Q_{sh3}	0.08	$V_{270} = 1.0299;$	$P_{DC1} = 1.3504;\ P_{DC2} = -0.4130;\ P_{DC3} = -0.3123;$
P_{sh4}	0.3	$V_{271} = 1.0909;$	$P_{DC4} = -0.3122;\ P_{DC5} = -0.3124;$
Q_{sh4}	0.05	$V_{272} = 1.0962;$	Converter loss (%)
P_{sh5}	0.3	$V_{273} = 1.0391;$	$P_{loss1} = 2.66;\ P_{loss2} = 1.29;\ P_{loss3} = 1.22;\ P_{loss4} = 1.23;$
Q_{sh5}	0.08	NI = 6;	$P_{loss5} = 1.23;$

Computation of references V_{DC}^* from droop equations

$V_{DC1} = 3.0009;\ V_{DC2} = 2.9997;$ $V_{DC1}^* = 3.0031;\ V_{DC2}^* = 3.002;\ V_{DC3}^* = 3.0022;\ V_{DC4}^* = 3.0020;$

$V_{DC3} = 2.9998;\ V_{DC4} = 2.9998;$ $V_{DC5}^* = 3.002;\ I_{DC3}^* = -0.0223;\ I_{DC4}^* = -0.1264;$

$V_{DC5} = 2.9998;\ R_1 = 20;$ $I_{DC5}^* = -0.1372;\ P_{DC1}^* = 1.3056;\ P_{DC2}^* = -0.4468;$

$R_2 = 15;\ R_3 = 10;$ $P_{DC3}^* = -0.0669;\ P_{DC4}^* = -0.3794;\ P_{DC5}^* = -0.4120;$

$R_4 = 10;\ R_5 = 15;$ NI = 3;

4.6.1 STUDIES OF 5-TERMINAL VSC-HVDC NETWORK INCORPORATED IN THE IEEE 300 BUS SYSTEM (MODEL A)

Case I: Model A employing linear V-P and V-I droop characteristics

For this study, at the outset, the base case power flow (in the absence of any MTDC grid) is carried out. The results are given in row 1 of Table 4.1. Then a five-terminal MTDC grid is integrated with the IEEE 300-bus test system at AC buses '266', '270', '271', '272' and '273'. While the VSCs connected to AC buses '266' and '270' follow linear V-P droop characteristics, the VSCs connected to AC buses '271', '272' and '273' operate on linear V-I droop characteristics. The droop control gains of VSCs 1 and 2 are set to 20 and 15, respectively [24,82]. The droop control gains of VSCs 3, 4 and 5 are all set to a value of 10.

Next, a DC power flow is carried out to obtain the voltage ('V_{DCa}^*'), power ('P_{DCa}^*') and current ('I_{DCa}^*') references ($1 \le a \le 5$) for the droop lines of the VSCs. The results are given in row 4 of Table 4.1.

TABLE 4.8

Study of IEEE 300-Bus System with 5-Terminal VSC-HVDC Network Incorporating Nonlinear Voltage Droop Characteristics with Dead Band (Model B)

Base case power flow (NI = 6)

$V_{266} = 1.011 \angle -11.24$; $V_{270} = 1.011 \angle -11.32$; $V_{271} = 0.998 \angle -17.67$;

$V_{272} = 0.981 \angle -19.46$; $V_{273} = 1.006 \angle -17.47$;

AC-MTDC power flow with linear voltage droop control (VSCs connected to AC buses 266, 270, 271, 272 and 273)

Given Quantities		AC Buses	Solution VSC
V_{DCav}	3.02	$\theta_{266} = -11.4982$;	$\theta_{sh1} = -18.0960$; $\theta_{sh2} = -9.6028$; $\theta_{sh3} = -4.5960$;
V_{266}	1.02	$\theta_{270} = -11.5041$;	$\theta_{sh4} = -4.9446$; $\theta_{sh5} = -5.1988$; $m_1 = 0.9779$;
P_{sh2}	0.35	$\theta_{271} = -5.9761$;	$m_2 = 0.9657$; $m_3 = 1.0481$; $m_4 = 1.0549$; $m_5 = 0.9839$;
Q_{sh2}	0.1	$\theta_{272} = -5.8507$;	$V_{DC1} = 3.0208$; $V_{DC2} = 3.0198$; $V_{DC3} = 3.0198$;
P_{sh3}	0.3	$\theta_{273} = -6.760$;	$V_{DC4} = 3.0199$; $V_{DC5} = 3.0198$; $I_{DC3} = -1.034$;
Q_{sh3}	0.1	$V_{270} = 1.0203$;	$I_{DC4} = -0.0701$; $I_{DC5} = -0.1035$;
P_{sh4}	0.2	$V_{271} = 1.1094$;	DC power
Q_{sh4}	0.09	$V_{272} = 1.1179$;	$P_{DC1} = 1.1996$; $P_{DC2} = -0.3627$; $P_{DC3} = -0.3123$;
P_{sh5}	0.3	$V_{273} = 1.0401$;	$P_{DC4} = -0.2118$; $P_{DC5} = -0.3124$;
			Converter loss (%)
			$P_{loss1} = 2.0959$; $P_{loss2} = 1.2618$; $P_{loss3} = 1.22$;
			$P_{loss4} = 1.1754$; $P_{loss5} = 1.2310$;
Q_{sh5}	0.1	NI = 6	

Computation of references V_{DC}^* from droop equations

$V_{DC1} = 3.0208$; $V_{DC2} = 3.0198$;	$V_{DC1}^* = 3.0214$; $V_{DC2}^* = 3.0204$; $V_{DC3}^* = 3.0204$; $V_{DC4}^* = 3.0204$;
$V_{DC3} = 3.0198$; $V_{DC4} = 3.0199$;	$V_{DC5}^* = 3.0204$; $P_{DC1} = 1.1996$; $P_{DC2} = -0.3627$; $P_{DC3} = -0.3123$;
$V_{DC5} = 3.0198$; $I_{DC3} = -1.034$;	$P_{DC4} = -0.2118$; $P_{DC5} = -0.3124$; $P_{DC1}^* = 1.1882$; $P_{DC2}^* = -0.2913$;
$I_{DC4} = -0.0701$; $I_{DC5} = -0.1035$;	$P_{DC3}^* = -0.3295$; $P_{DC4}^* = -0.2290$; $P_{DC5}^* - 0.3380$; $I_{DC1}^* = 0.3933$;
$V_{DCmax} = 3.02$; $V_{DChigh}^* = 3.015$;	$I_{DC2}^* = -0.0965$; $I_{DC3}^* = -0.1091$; $I_{DC4}^* - 0.0758$; $I_{DC5}^* = -0.1119$;
$V_{DClow}^* = 3.01$; $V_{DCmin} = 3.009$;	NI = 4
$R_1 = 20$; $R_2 = 15$; $R_3 = 10$;	
$R_4 = 10$; $R_5 = 15$;	

Thereafter, the DC voltages {'V_{DCa}' ($1 \leq a \leq 5$)} are computed from the voltage ('V_{DCa}^*'), power ('P_{DCa}^*') and current ('I_{DCa}^*') references using droop Eqn. (4.24). The results are given in row 7 of Table 4.1.

After obtaining 'V_{DCa}', the AC-MTDC power flow is carried out. The results are given in row 10 of Table 4.1. The computed values of the active power flows in lines ('P_{sh2}', 'P_{sh3}', 'P_{sh4}' and 'P_{sh5}') are shown in bold in row 10 of Table 4.1.

Table 4.1 shows that for both the base case and the AC-MTDC power flow with droop model 'A', 'NI' remains the same.

Case II: Model A employing nonlinear V-P droop with dead band

This case study is similar to the previous study of Table 4.1 except VSC 3 which employs a nonlinear V-P droop with dead band. The droop control gains are identical to that of the study of Table 4.1. First, a separate DC power flow is carried out to calculate the reference values for the droop control lines. The results are shown in row 4 of Table 4.2. Thereafter, the DC voltages $\{`V_{DCa}` \ (1 \le a \le 5)\}$ are computed from the voltage ($`V^*_{DCa}`$), power ($`P^*_{DCa}`$) and current ($`I^*_{DCa}`$) references using droop Eqn. (4.24). The results are given in row 11 of Table 4.2.

Subsequently, the AC-MTDC power flow is carried out. The results are shown in rows 12–14 of Table 4.2.

From the power-flow solution, it is observed that the converter connected at AC bus 271 operates at the point 'A' as shown in Figure 4.4.

Case III: Model A employing voltage margin control

This case study is conducted on the same AC-MTDC system of Table 4.2, but the VSC connected to AC bus 271 employs voltage margin control. The droop control gains of the VSCs are identical to that of Table 4.2, except VSC 3, which operates in voltage margin control. The droop gain for voltage margin control is set to a value of 1000. For voltage margin control characteristic, the maximum and minimum DC powers are set to 1.0 and −1.0 p.u., respectively.

First, a separate DC power flow is carried out to calculate the reference values for the droop control lines. The results are shown in row 4 of Table 4.3. Thereafter, the DC voltages $\{`V_{DCa}` \ (1 \le a \le 5)\}$ are computed from the voltage ($`V^*_{DCa}`$), power ($`P^*_{DCa}`$) and current ($`I^*_{DCa}`$) references using droop Eqn. (4.24). The results are given in row 11 of Table 4.3.

Subsequently, the AC-MTDC power flow is carried out. The results are shown in rows 12–14 of Table 4.3.

The convergence characteristics of base case, listed in Tables 4.1, 4.2 and 4.3 are shown in Figures 4.10, 4.11, 4.12 and 4.13, respectively. From Figures 4.10 to 4.13, it is observed that the proposed algorithm demonstrates excellent convergence characteristics, converging in six iterations. The bus voltage profiles for the studies of Tables 4.1, 4.2 and 4.3 are shown in Figures 4.14, 4.15 and 4.16, respectively. From Figures 4.14–4.16, it can be observed that the bus voltage profiles do not change much except at the AC terminal buses to which the converters are connected.

4.6.2 STUDIES OF 7-TERMINAL VSC-HVDC NETWORK INCORPORATED IN THE EUROPEAN 1354 BUS SYSTEM (MODEL A)

Case I: Model A employing linear V-P and V-I droop characteristics

For this study, at the outset, the base case power flow (in the absence of any MTDC grid) is carried out. The results are given in row 1 of Table 4.4. Then a seven-terminal MTDC grid is integrated with the European 1354-bus test system at AC buses '1280', '1281', '1286', '1292', '1313', '1324' and '1353'. While the VSCs connected to AC buses '1280', '1281', '1286' and '1292' follow linear V-P droop characteristics, the VSCs connected to AC buses '1313', '1324' and '1353'

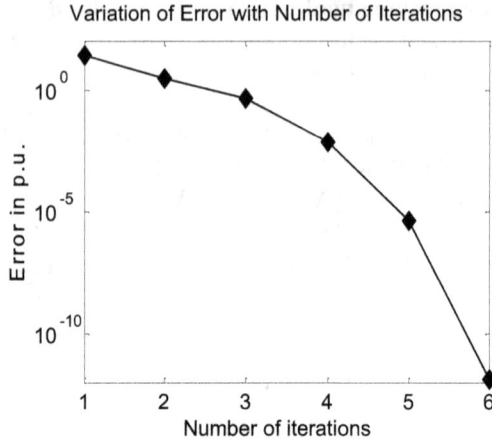

FIGURE 4.10 Convergence characteristic for the base case power flow in IEEE 300-bus system.

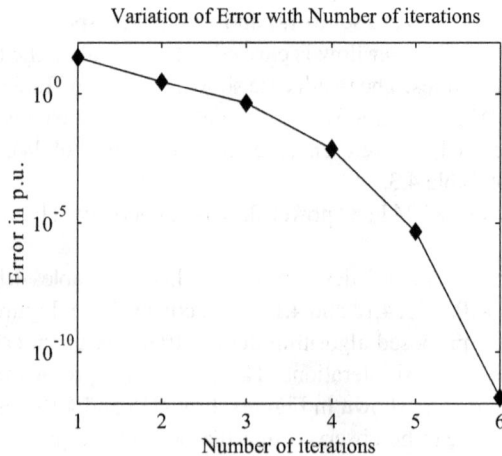

FIGURE 4.11 Convergence characteristic for the case study of Table 4.1.

operate on linear V-I droop characteristics. The droop control gains of VSCs 1, 2, 3, 4, 5, 6 and 7 are set to 20, 15, 10, 10,10, 15 and 20, respectively [24,82].

Next, a DC power flow is carried out to obtain the voltage ('V^*_{DCa}'), power ('P^*_{DCa}') and current ('I^*_{DCa}') references ($1 \leq a \leq 7$) for the droop lines of the VSCs. The results are given in row 4 of Table 4.4.

Thereafter, the DC voltages {'V_{DCa}' ($1 \leq a \leq 5$)} are computed from the voltage ('V^*_{DCa}'), power ('P^*_{DCa}') and current ('I^*_{DCa}') references using droop Eqn. (4.24). The results are given in row 7 of Table 4.4.

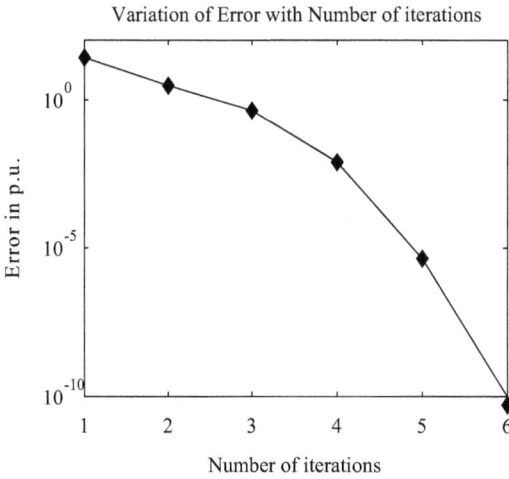

FIGURE 4.12   Convergence characteristic for the study of Table 4.2.

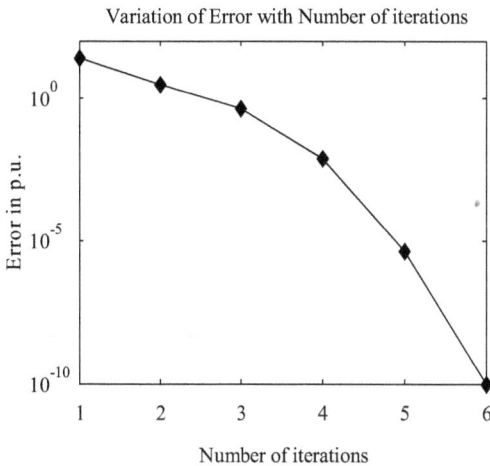

FIGURE 4.13   Convergence characteristic for the study of Table 4.3.

After obtaining 'V_{DCa}', the AC-MTDC power flow is carried out. The results are given in row 10 of Table 4.4. The computed values of the active power flows in lines ('P_{sh2}', 'P_{sh3}', 'P_{sh4}' and 'P_{sh5}') are shown in bold in row 10 of Table 4.4.

Table 4.4 shows that for both the base case and the AC-MTDC power flow with droop model 'A', 'NI' remains the same.

Case II: Model A employing nonlinear V-P droop with dead band

This case study is similar to the previous study of Table 4.4 except VSC 4 which employs a nonlinear V-P droop with dead band. The droop control gains

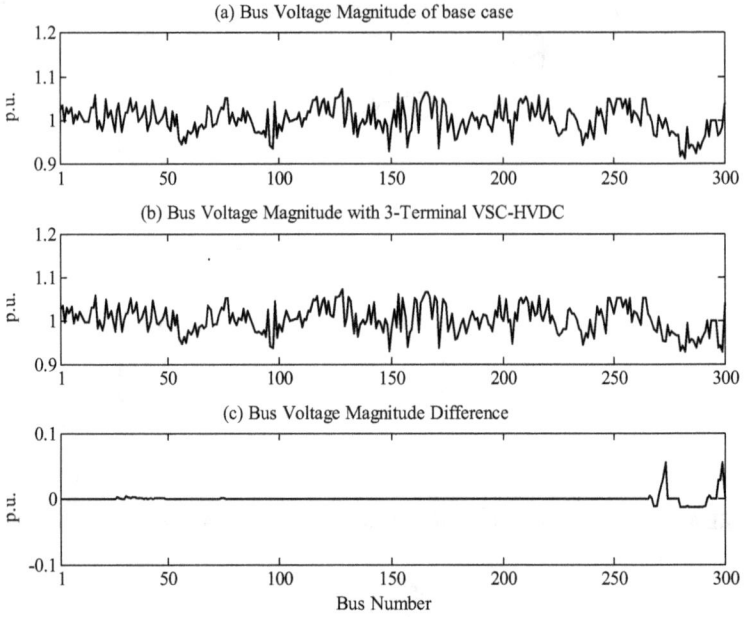

FIGURE 4.14   Bus voltage profile for the study of Table 4.1.

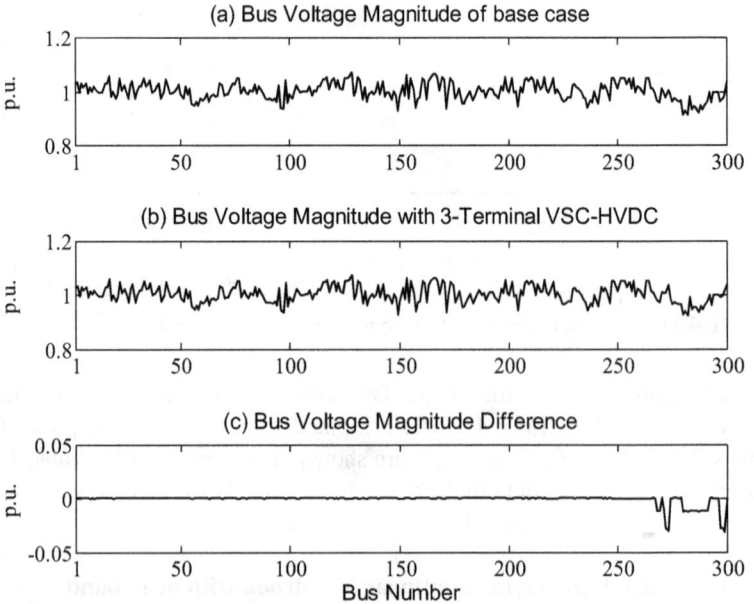

FIGURE 4.15   Bus voltage profile for the study of Table 4.2.

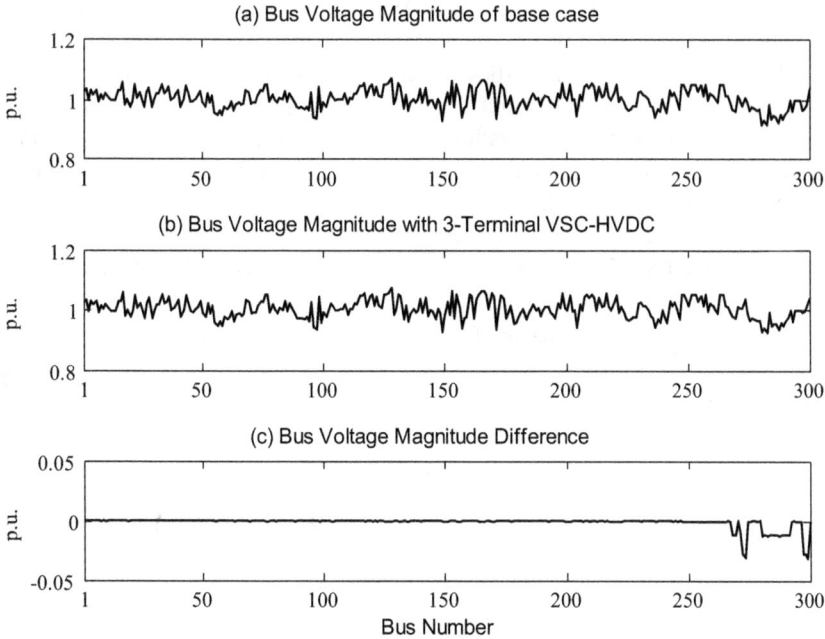

FIGURE 4.16   Bus voltage profile for the study of Table 4.3.

are identical to that of the study of Table 4.4. First, a separate DC power flow is carried out to calculate the reference values for the droop control lines. The results are shown in row 4 of Table 4.5. Thereafter, the DC voltages {'V_{DCa}' ($1 \le a \le 5$)} are computed from the voltage ('V_{DCa}^*'), power ('P_{DCa}^*') and current ('I_{DCa}^*') references using droop Eqn. (4.24). The results are given in row 11 of Table 4.5.

Subsequently, the AC-MTDC power flow is carried out. The results are shown in rows 12–14 of Table 4.5.

From the power-flow solution, it is observed that the converter connected at AC bus 1292 operates at the point 'A' as shown in Figure 4.4.

Case III: Model A employing voltage margin control

This case study is conducted on the same AC-MTDC system of Table 4.5 but the VSC connected to AC bus-1292 employs voltage margin control. The droop control gains of the VSCs are identical to that of Table 4.5, except VSC 4, which operates in voltage margin control. The droop gain for voltage margin control is set to a value of 1000. For voltage margin control characteristic, the maximum and minimum DC powers are set to 1.0 and −1.0 p.u., respectively.

First, a separate DC power flow is carried out to calculate the reference values for the droop control lines. The results are shown in row 4 of Table 4.6. Thereafter, the DC voltages {'V_{DCa}' ($1 \le a \le 5$)} are computed from the voltage ('V_{DCa}^*'), power ('P_{DCa}^*') and current ('I_{DCa}^*') references using droop Eqn. (4.24). The results are given in row 11 of Table 4.6.

Subsequently, the AC-MTDC power flow is carried out. The results are shown in rows 12–14 of Table 4.6.

The convergence characteristics of base case, listed in Tables 4.4, 4.5 and 4.6 are shown in Figures 4.17, 4.18 and 4.19, respectively. From Figures 4.17 to 4.19, it is observed that the proposed algorithm demonstrates excellent convergence characteristics, converging in seven iterations. The bus voltage profiles for the studies of Tables 4.4, 4.5 and 4.6 are shown in Figures 4.20, 4.21 and 4.22, respectively. From Figures 4.20–4.22, it can be observed that the bus voltage profiles do not change much except at the AC terminal buses to which the converters are connected.

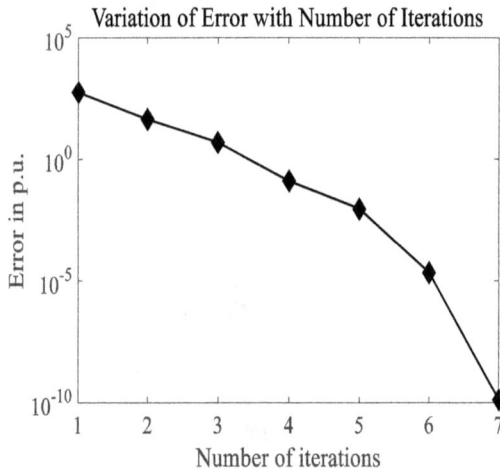

FIGURE 4.17 Convergence characteristic of the base case power flow in the European 1354-bus system.

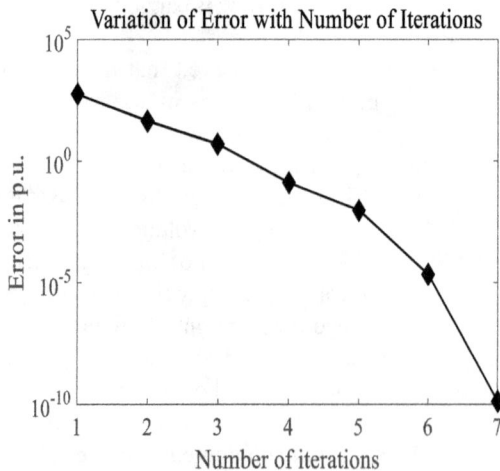

FIGURE 4.18 Convergence characteristic for the study of Table 4.4.

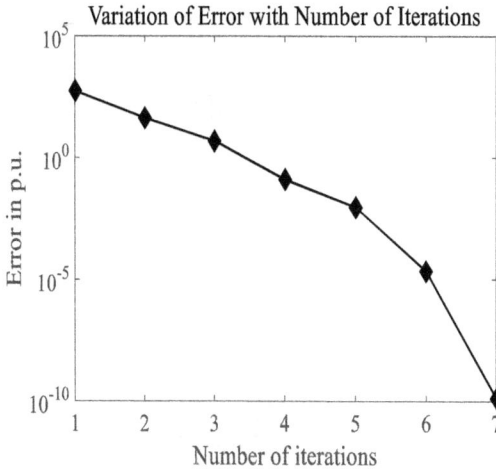

FIGURE 4.19   Convergence characteristic for the study of Table 4.5.

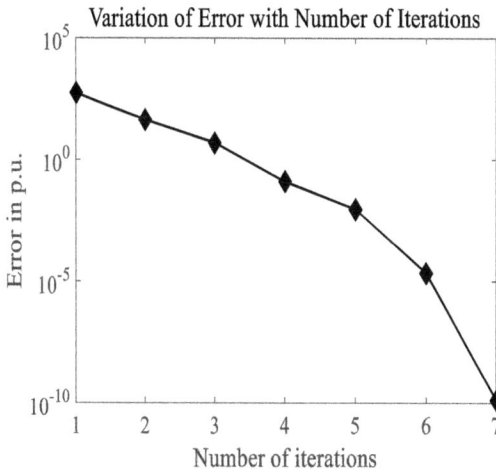

FIGURE 4.20   Convergence characteristic for the study of Table 4.6.

4.6.3   STUDIES WITH UNIFIED POWER-FLOW MODEL OF IEEE 300-BUS TEST SYSTEM INTEGRATED WITH 5-TERMINAL MVDC GRID (MODEL B)

Case I: Model B employing linear V-P and V-I droop characteristics
    In this case study, the five-terminal MTDC grid is again integrated with the AC system at AC buses '266', '270', '271', '272' and '273' and the unified method of Model B is used to solve the network. In this study both the line active and reactive powers are specified. First, the AC-MTDC power flow is carried out. The specified values are given in rows 3–13 and column 1 of Table 4.7.

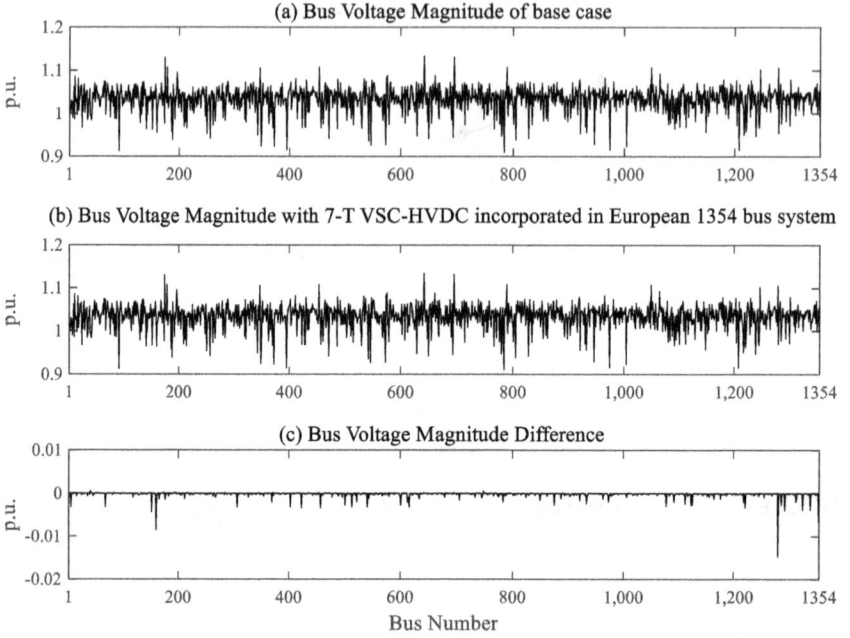

FIGURE 4.21 Bus voltage profile for the study of Table 4.4.

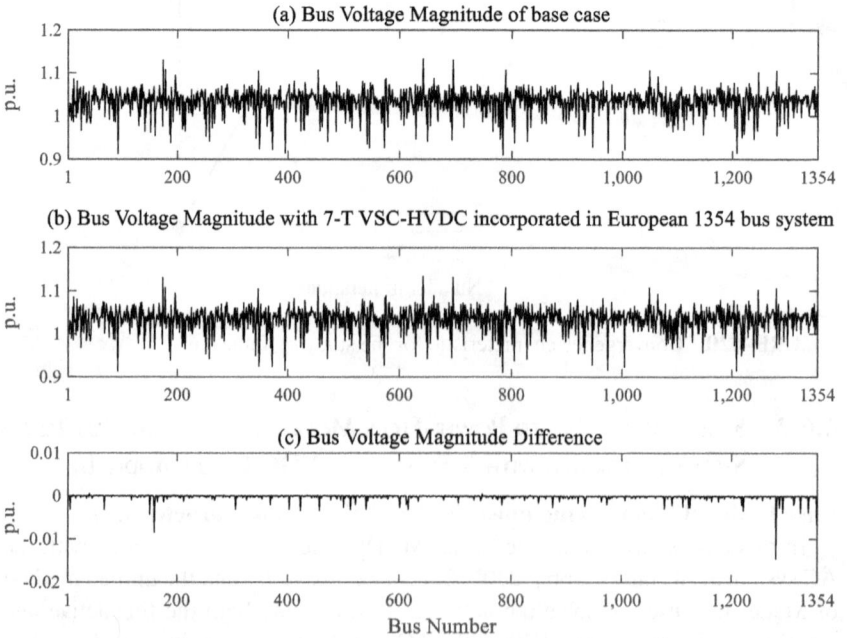

FIGURE 4.22 Bus voltage profile for the study of Table 4.5.

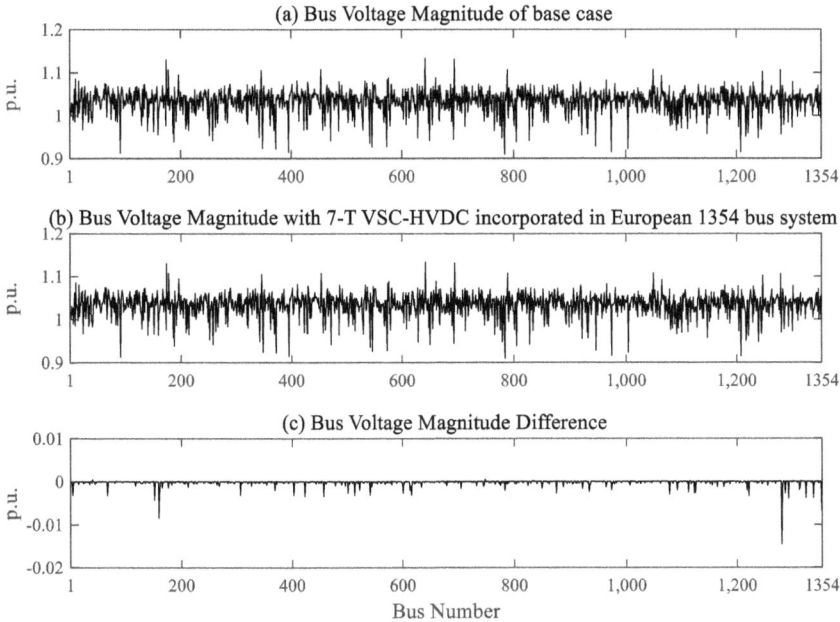

FIGURE 4.23 Bus voltage profile for the study of Table 4.6.

The results of the AC-MTDC power flow are given in rows 3–13 and columns 2–3 of Table 4.7. The AC-MTDC power flow converges in six number of iterations.

Subsequent to the AC-MTDC power flow, using the values of 'V_{DCa}' obtained (and hence, 'P_{DCa}' and 'I_{DCa}'), the DC voltage ('V_{DCa}^*') and thereafter, the power ('P_{DCa}^*') and current ('I_{DCa}^*') references for the droop lines of the VSCs are computed using the droop equations {Eqs. (4.28) and (4.29)}. The results are given in row 15 of Table 4.7.

Case II: Model B employing nonlinear V-P droop with dead band

This case study is similar to the previous study of Table 4.7 except VSC 2 which employs a nonlinear V-P droop with dead band. The droop control gains are identical to that of the study of Table 4.7. First, the AC-MTDC power flow is carried out. The results are shown in rows 3–13 of Table 4.8. Subsequently, a separate DC power flow is carried out to calculate the reference values for the droop control lines. The results are shown in row 15 of Table 4.8.

From the power-flow solution, it is observed that the converter connected at AC bus 272 operates at the point 'A' as shown in Figure 4.4.

The convergence characteristics of Tables 4.7 and 4.8 are shown in Figures 4.24 and 4.25, respectively. From Figures 4.24 and 4.25, it is observed that the proposed algorithm demonstrates excellent convergence characteristics, converging in six iterations. The bus voltage profiles for the studies of Tables 4.7 and 4.8 are shown in Figures 4.26 and 4.27, respectively. From Figures 4.26 and 4.27, it can

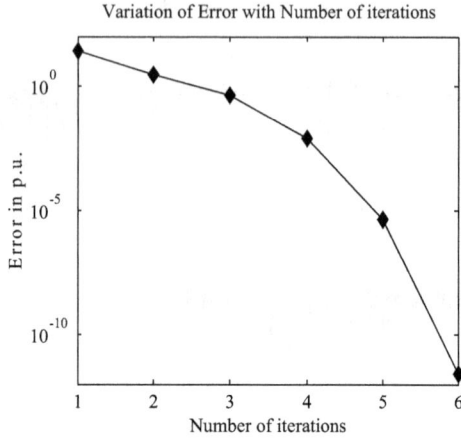

FIGURE 4.24 Convergence characteristic for the study of Table 4.7.

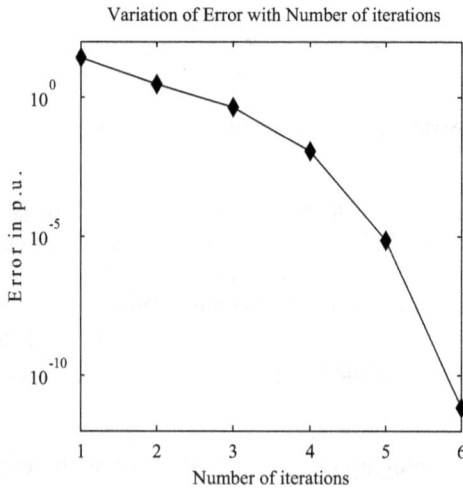

FIGURE 4.25 Convergence characteristic for the study of Table 4.8.

be observed that the bus voltage profiles do not change much except at the AC terminal buses to which the converters are connected.

4.6.4 STUDIES WITH UNIFIED POWER-FLOW MODEL OF EUROPEAN 1354-BUS TEST SYSTEM INTEGRATED WITH 7-TERMINAL MVDC GRID (MODEL B)

Case I: Model B employing linear V-P and V-I droop characteristics

In this case study, the seven-terminal MTDC grid is again integrated with the AC system at AC buses '1280', '1281', '1286', '1292' '1313', '1324' and '1353' and

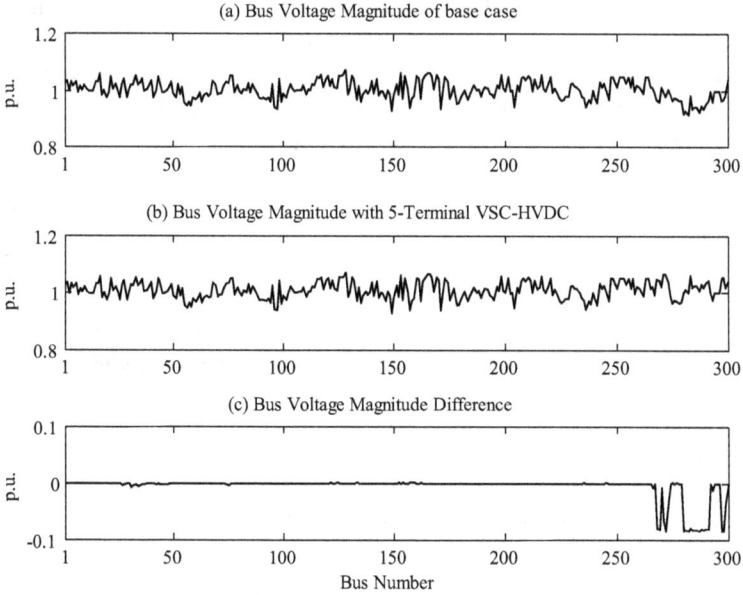

FIGURE 4.26   Bus voltage profile for the study of Table 4.7.

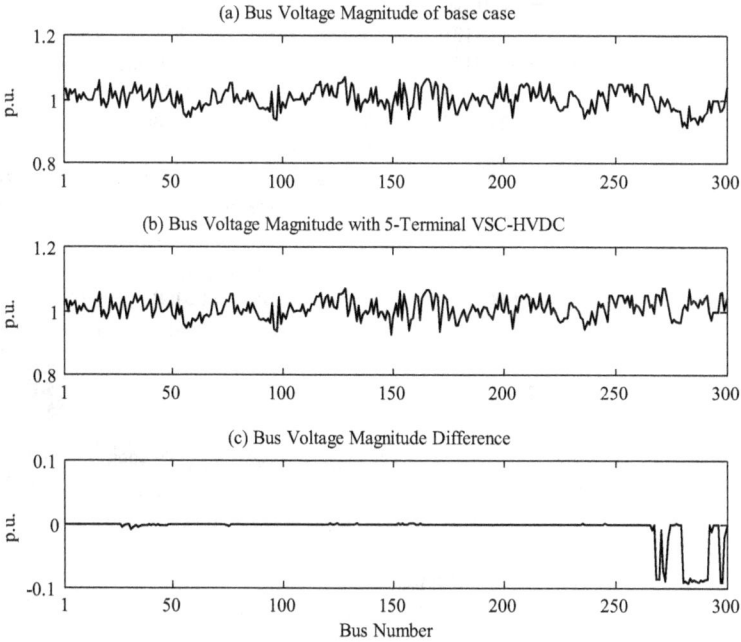

FIGURE 4.27   Bus voltage profile for the study of Table 4.8.

the unified method of Model B is used to solve the network. In this study both the line active and reactive powers are specified. First, the AC-MTDC power flow is carried out. The specified values are given in rows 3–17 and columns 1–2 of Table 4.9. The results of the AC-MTDC power flow are given in rows 3–17 and columns 3–4 of Table 4.9. The AC-MTDC power flow converges in seven number of iterations.

TABLE 4.9

Study of European 1354-Bus System with 7-Terminal VSC-HVDC Network Incorporating Linear DC Voltage Droop Characteristics (Model B)

Base case power flow converged in seven iterations (NI = 7)

$V_{1280} = 0.9554 \angle -35.2574$; $V_{1281} = 1.0354 \angle -37.3082$;

$V_{1286} = 1.0156 \angle -12.3760$; $V_{1292} = 1.0289 \angle -35.23066$;

$V_{1353} = 0.9934 \angle -29.6628$;

AC-MTDC power flow with linear voltage droop control (VSCs connected to AC buses 1280, 1281, 1286, 1292, 1313, 1324 and 1353)

Given Quantities		Solution	
		AC Buses	**VSC**
V_{DCav}	3.0	$\theta_{1280} = -42.2332$;	$\theta_{sh1} = -51.5357$; $\theta_{sh2} = -34.9973$; $\theta_{sh3} = 10.4817$;
V_{1280}	0.98	$\theta_{1281} = -37.0767$;	$\theta_{sh4} = -34.3566$; $\theta_{sh5} = -35.6830$; $\theta_{sh6} = -27.4798$;
P_{sh2}	0.4	$\theta_{1286} = -12.1181$;	$\theta_{sh7} = -28.5106$; $m_1 = 1.0244$; $m_2 = 0.9971$;
Q_{sh2}	0.2	$\theta_{1292} = -35.4158$;	$m_3 = 0.9698$; $m_4 = 0.9830$; $m_5 = 0.9917$;
P_{sh3}	0.3	$\theta_{1313} = -37.2516$;	$m_6 = 0.9546$; $m_7 = 0.9570$;
Q_{sh3}	0.1	$\theta_{1324} = -28.6030$;	$V_{DC1} = 3.008$; $V_{DC2} = 2.9998$; $V_{DC3} = 2.9999$;
P_{sh4}	0.2	$\theta_{1353} = 29.6316$;	$V_{DC4} = 2.9999$; $V_{DC5} = 2.9999$; $V_{DC6} = 2.9999$;
Q_{sh4}	0.1	$V_{1281} = 1.0372$;	$V_{DC7} = 2.9999$; $P_{DC1} = 1.6747$; $P_{DC2} = -0.4133$;
P_{sh5}	0.3	$V_{1286} = 1.0180$;	$P_{DC3} = -0.3124$; $P_{DC4} = -0.2119$; $P_{DC5} = -0.3125$;
Q_{sh5}	0.15	$V_{1292} = 1.0325$;	$P_{DC6} = -0.2119$; $P_{DC7} = -0.2121$;
P_{sh6}	0.2	$V_{1313} = 1.0367$;	Converter loss (%)
Q_{sh6}	0.09	$V_{1324} = 1.0031$;	$P_{loss1} = 3.39$; $P_{loss2} = 1.31$; $P_{loss3} = 1.23$;
P_{sh7}	0.2	$V_{1353} = 0.9996$;	$P_{loss4} = 1.19$; $P_{loss5} = 1.24$; $P_{loss6} = 1.19$;
Q_{sh7}	0.15	NI = 7	$P_{loss7} = 1.20$;

Computation of references V_{DC}^* from droop equations

$V_{DC1} = 3.008$; $V_{DC2} = 2.9998$;

$V_{DC3} = 2.9999$; $V_{DC4} = 2.9999$;

$V_{DC5} = 2.9999$; $V_{DC6} = 2.9999$;

$V_{DC7} = 2.9999$;

$R_1 = 20$; $R_2 = 15$; $R_3 = 10$;

$R_4 = 10$; $R_5 = 10$; $R_6 = 15$;

$R_7 = 20$;

$V_{DC1}^* = 3.0009$; $V_{DC2}^* = 2.9997$; $V_{DC3}^* = 2.9998$;

$V_{DC4}^* = 2.9998$; $V_{DC5}^* = 2.9997$; $V_{DC6}^* = 2.9997$;

$V_{DC7}^* = 2.9997$; $I_{DC4}^* = -0.0101$; $I_{DC5}^* = -0.1143$;

$I_{DC6}^* = -0.0857$; $I_{DC7}^* = -0.0907$;

DC power

$P_{DC1}^* = 1.6543$; $P_{DC2}^* = -0.4286$;

$P_{DC3}^* = -0.3227$; $P_{DC4}^* = -0.0304$;

$P_{DC5}^* = -0.3430$; $P_{DC6}^* = -0.2572$; $P_{DC7}^* = -0.2720$;

NI = 3

Subsequent to the AC-MTDC power flow, using the values of 'V_{DCa}' obtained (and hence, 'P_{DCa}' and 'I_{DCa}'), the DC voltage ('V^*_{DCa}') and thereafter, the power ('P^*_{DCa}') and current ('I^*_{DCa}') references for the droop lines of the VSCs are computed using the droop equations {Eqs. (4.28) and (4.29)}. The results are given in row 19 of Table 4.9.

Case II: Model B employing nonlinear droop characteristics

This case study is similar to the previous study of Table 4.9 except VSC 2 which employs a nonlinear V-P droop with dead band. The droop control gains are identical to that of the study of Table 4.9. First, the AC-MTDC power flow is carried out. The results are shown in rows 3–17 of Table 4.10. Subsequently, a separate DC power flow is carried out to calculate the reference values for the droop control lines. The results are shown in row 19 of Table 4.10.

From the power-flow solution, it is observed that the converter connected at AC bus 1281 operates at the point 'A' as shown in Figure 4.4.

The convergence characteristics of Tables 4.9 and 4.10 are shown in Figures 4.28 and 4.29, respectively. From Figures 4.28 and 4.29, it is observed that the proposed algorithm demonstrates excellent convergence characteristics, converging in seven iterations. The bus voltage profiles for the studies of Tables 4.9 and 4.10 are shown in Figures 4.30 and 4.31, respectively. From Figures 4.30 and 4.31, it can be observed that the bus voltage profiles do not change much except at the AC terminal buses to which the converters are connected.

4.6.5 STUDIES WITH SEQUENTIAL POWER-FLOW MODEL OF IEEE 300-BUS TEST SYSTEM INTEGRATED WITH 5-TERMINAL MVDC GRID (MODEL B)

Case I: Model B employing linear V-P and V-I droop characteristics

In this case study, the five-terminal MTDC grid is again integrated with the AC system at AC buses '266', '270', '271', '272' and '273' and the sequential method of Model B is used to solve the network. In this study both the line active and reactive powers are specified. First, the AC-MTDC power flow is carried out. The specified values are given in rows 3–13 and column 1 of Table 4.11.

The results of the AC-MTDC power flow are given in rows 3–17 and columns 2–3 of Table 4.11. The AC-MTDC power flow converges in six number of iterations.

Subsequent to the AC-MTDC power flow, using the values of 'V_{DCa}' obtained (and hence, 'P_{DCa}' and 'I_{DCa}'), the DC voltage ('V^*_{DCa}') and thereafter, the power ('P^*_{DCa}') and current ('I^*_{DCa}') references for the droop lines of the VSCs are computed using the droop Eqns. (4.28) and (4.29). The results are given in row 19 of Table 4.11.

Case II: Model B employing nonlinear V-P droop with dead band

This case study is similar to the previous study of Table 4.11 except VSC 2 which employs a nonlinear V-P droop with dead band. The droop control gains

TABLE 4.10

Study of European 1354-Bus System with 7-Terminal VSC-HVDC Network Incorporating Nonlinear DC Voltage Droop with Dead-Band Characteristics (Model B)

Base case power flow converged in seven iterations (NI=7)

$V_{1280} = 0.9554 \angle -35.2574$; $V_{1281} = 1.0354 \angle -37.3082$;

$V_{1286} = 1.0156 \angle -12.3760$; $V_{1292} = 1.0289 \angle -35.23066$;

$V_{1313} = 1.0351 \angle -37.4755$; $V_{1324} = 0.9972 \angle -29.0208$;

$V_{1353} = 0.9934 \angle -29.6628$;

AC-MTDC power flow with linear voltage droop control (VSCs connected to AC buses 1280, 1281, 1286, 1292, 1313, 1324 and 1353)

Given		Solution	
Quantities		**AC Buses**	**VSC**
V_{DCav}	3.02	$\theta_{1280} = -42.2332$;	$\theta_{sh1} = -51.5357$; $\theta_{sh2} = -34.9973$; $\theta_{sh3} = 10.4817$;
V_{1280}	0.98	$\theta_{1281} = -37.0767$;	$\theta_{sh4} = -34.3566$; $\theta_{sh5} = -35.6830$; $\theta_{sh6} = -27.4798$;
P_{sh2}	0.4	$\theta_{1286} = -12.1181$;	$\theta_{sh7} = -28.5106$; $m_1 = 1.0177$; $m_2 = 0.9905$;
Q_{sh2}	0.2	$\theta_{1292} = -35.4158$;	$m_3 = 0.9633$; $m_4 = 0.9765$; $m_5 = 0.9852$;
P_{sh3}	0.3	$\theta_{1313} = -37.2516$;	$m_6 = 0.9483$; $m_7 = 0.9507$;
Q_{sh3}	0.1	$\theta_{1324} = -28.6030$;	$V_{DC1} = 3.0208$; $V_{DC2} = 3.0198$; $V_{DC3} = 3.0199$;
P_{sh4}	0.2	$\theta_{1353} = 29.6316$;	$V_{DC4} = 3.0199$; $V_{DC5} = 3.0199$; $V_{DC6} = 3.0199$;
Q_{sh4}	0.1	$V_{1281} = 1.0372$;	$V_{DC7} = 3.0199$; $P_{DC1} = 1.6747$; $P_{DC2} = -0.4133$;
P_{sh5}	0.3	$V_{1286} = 1.0180$;	$P_{DC3} = -0.3124$; $P_{DC4} = -0.2119$; $P_{DC5} = -0.3125$;
Q_{sh5}	0.15	$V_{1292} = 1.0325$;	$P_{DC6} = -0.2119$; $P_{DC7} = -0.2121$;
P_{sh6}	0.2	$V_{1313} = 1.0367$;	Converter loss (%)
Q_{sh6}	0.09	$V_{1324} = 1.0031$;	$P_{loss1} = 3.39$; $P_{loss2} = 1.31$; $P_{loss3} = 1.23$;
P_{sh7}	0.2	$V_{1353} = 0.9996$;	$P_{loss4} = 1.19$; $P_{loss5} = 1.24$; $P_{loss6} = 1.19$;
Q_{sh7}	0.15	$NI = 7$;	$P_{loss7} = 1.20$;

Computation of references V_{DC}^* from droop equations

$V_{DC1} = 3.0208$; $V_{DC2} = 3.0198$;
$V_{DC3} = 3.0199$; $V_{DC4} = 3.0199$;
$V_{DC5} = 3.0199$; $V_{DC6} = 3.0199$;
$V_{DC7} = 3.0199$;
$V_{DCmax} = 3.02$; $V_{DChigh}^* = 3.015$;
$V_{DClow}^* = 3.01$; $V_{DCmin} = 3.009$;
$R_1 = 20$; $R_2 = 15$; $R_3 = 10$;
$R_4 = 10$; $R_5 = 10$; $R_6 = 15$;
$R_7 = 20$;

$V_{DC1}^* = 3.0211$; $V_{DC2}^* = 3.0202$; $V_{DC3}^* = 3.0202$;
$V_{DC4}^* = 3.0202$; $V_{DC5}^* = 3.0202$; $V_{DC6}^* = 3.0302$;
$V_{DC7}^* = 3.0202$; $I_{DC3}^* = -0.1068$; $I_{DC4}^* = -0.0735$;
$I_{DC5}^* = -0.1068$; $I_{DC6}^* = -0.0751$; $I_{DC7}^* = -0.0768$;
DC power
$P_{DC1}^* = 1.6679$; $P_{DC2}^* = -0.3412$;
$P_{DC3}^* = -0.3226$; $P_{DC4}^* = -0.2220$;
$P_{DC5}^* = -0.3227$; $P_{DC6}^* = -0.227$; $P_{DC7}^* = -0.2320$;
$NI = 4$

are identical to that of the study of Table 4.11. First, the AC-MTDC power flow is carried out. The results are shown in rows 3–17 of Table 4.12. Subsequently, a separate DC power flow is carried out to calculate the reference values for the droop control lines. The results are shown in row 19 of Table 4.12.

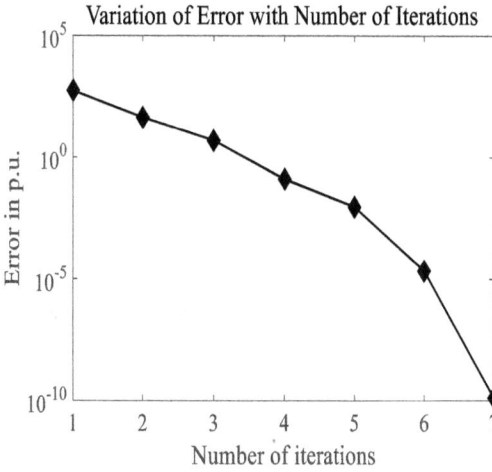

FIGURE 4.28   Convergence characteristic for the study of Table 4.9.

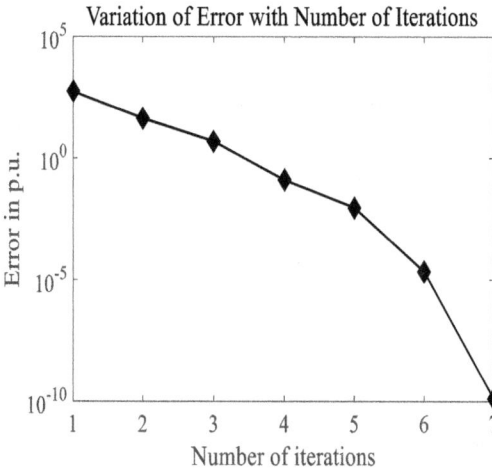

FIGURE 4.29   Convergence characteristic for the study of Table 4.10.

From the power-flow solution, it is observed that the converter connected at AC bus 272 operates at the point 'A' as shown in Figure 4.4.

The convergence characteristics of Tables 4.11 and 4.12 are shown in Figures 4.32 and 4.33, respectively. From Figures 4.32 and 4.33, it is observed that the proposed algorithm demonstrates excellent convergence characteristics, converging in six iterations. The bus voltage profiles for the studies of Tables 4.11 and 4.12 are shown in Figures 4.34 and 4.35, respectively. From Figures 4.34 and 4.35, it can be observed that the bus voltage profiles do not change much except at the AC terminal buses to which the converters are connected.

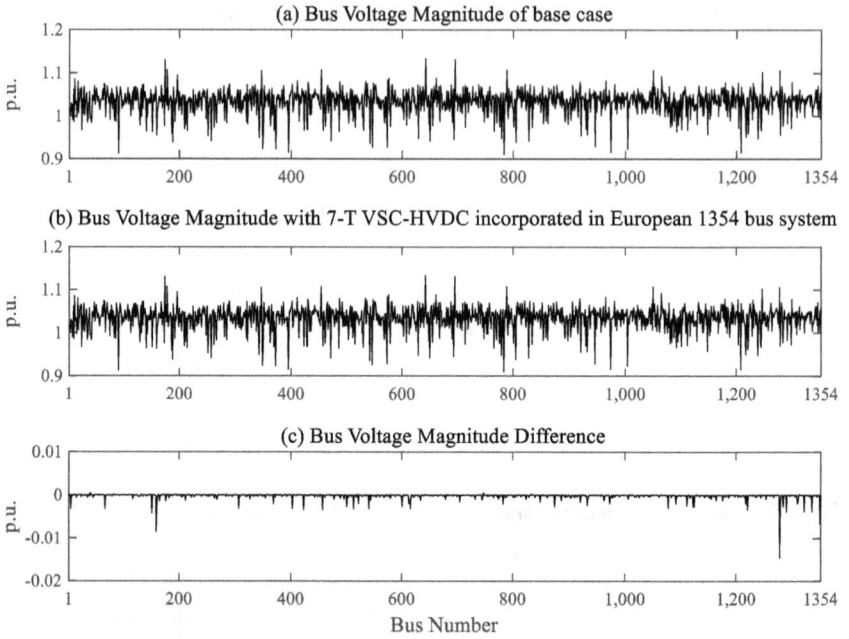

FIGURE 4.30   Bus voltage profile for the study of Table 4.9.

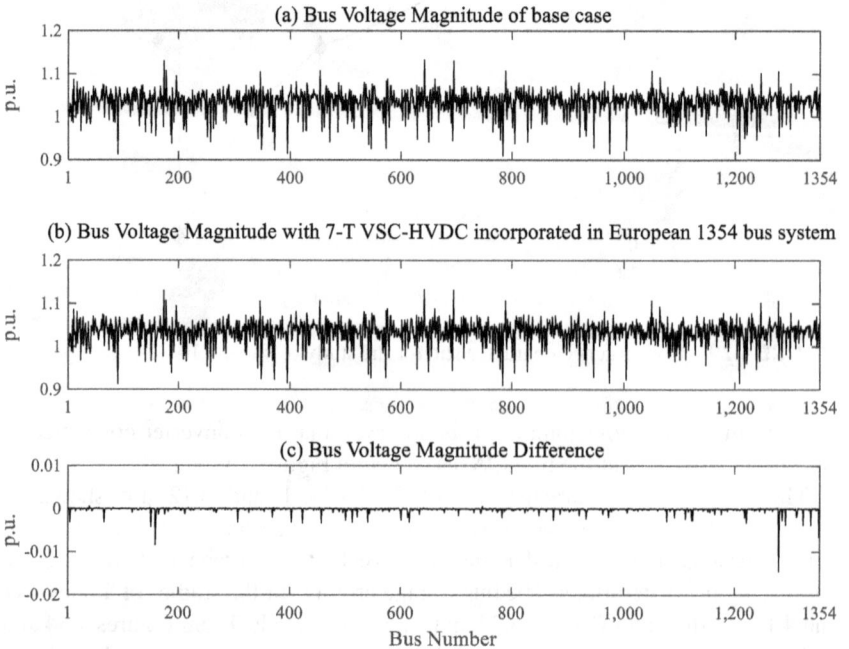

FIGURE 4.31   Bus voltage profile for the study of Table 4.10.

TABLE 4.11

Study of IEEE 300-Bus System with 5-Terminal VSC-HVDC Network Incorporating Linear DC Voltage Droop Characteristics (Model B)

Base case power flow (NI = 6)

$V_{266} = 1.011 \angle -11.24$; $V_{270} = 1.011 \angle -11.32$; $V_{271} = 0.998 \angle -17.67$;

$V_{272} = 0.981 \angle -19.46$; $V_{273} = 1.006 \angle -17.47$;

AC-MTDC power flow with linear voltage droop control (VSCs connected to AC buses 266, 270, 271, 272 and 273)

Given Quantities		AC Buses	VSC
		Solution	
V_{DCav}	3.0	$\theta_{266} = -11.5503$;	$\theta_{sh1} = -18.4481$; $\theta_{sh2} = -9.4048$; $\theta_{sh3} = -2.1709$;
V_{266}	1.03	$\theta_{270} = -11.5408$;	$\theta_{sh4} = -1.0782$; $\theta_{sh5} = -5.2494$; $m_1 = 1.0579$;
P_{sh2}	0.4	$\theta_{271} = -3.6012$;	$m_2 = 0.9803$; $m_3 = 1.0361$; $m_4 = 1.0384$;
Q_{sh2}	0.09	$\theta_{272} = -2.4997$;	$m_5 = 0.9876$; $V_{DC1} = 3.0009$; $V_{DC2} = 2.9997$;
P_{sh3}	0.3	$\theta_{273} = -6.8247$;	$V_{DC3} = 2.9998$; $V_{DC4} = 2.9998$; $V_{DC5} = 2.9998$;
Q_{sh3}	0.08	$V_{270} = 1.0299$;	$P_{DC1} = 1.3504$;
P_{sh4}	0.3	$V_{271} = 1.0909$;	Converter loss (%)
Q_{sh4}	0.05	$V_{272} = 1.0962$;	$P_{loss1} = 2.66$; $P_{loss2} = 1.29$; $P_{loss3} = 1.22$;
P_{sh5}	0.3	$V_{273} = 1.0391$;	$P_{loss4} = 1.23$; $P_{loss5} = 1.23$;
Q_{sh5}	0.08		
P_{DC2}	−0.4130		
P_{DC3}	−0.3123		
P_{DC4}	−0.3122		
P_{DC5}	−0.3124	NI = 6;	

Computation of references V_{DC}^* from droop equations

$V_{DC1} = 3.0009$; $V_{DC2} = 2.9997$; $V_{DC1}^* = 3.0031$; $V_{DC2}^* = 3.002$; $V_{DC3}^* = 3.0022$; $V_{DC4}^* = 3.0020$;

$V_{DC3} = 2.9998$; $V_{DC4} = 2.9998$; $V_{DC5}^* = 3.002$; $I_{DC3}^* = -0.0223$; $I_{DC4}^* = -0.1264$;

$V_{DC5} = 2.9998$; $R_1 = 20$; $R_2 = 15$; $I_{DC5}^* = -0.1372$; $P_{DC1}^* = 1.3056$; $P_{DC2}^* = -0.4468$;

$R_3 = 10$; $R_4 = 10$; $R_5 = 15$; $P_{DC3}^* = -0.0669$; $P_{DC4}^* = -0.3794$; $P_{DC5}^* = -0.4120$;

NI = 3

4.6.6 STUDIES WITH SEQUENTIAL POWER-FLOW MODEL OF EUROPEAN 1354-BUS TEST SYSTEM INTEGRATED WITH 7-TERMINAL MVDC GRID

Case I: Model B employing linear V-P and V-I droop characteristics

In this case study, the five-terminal MTDC grid is again integrated with the AC system at AC buses '1280', '1281', '1286', '1292' '1313', '1324' and '1353' and the sequential method of Model B is used to solve the network. In this study both the line active and reactive powers are specified. First, the AC-MTDC power flow is carried out. The specified values are given in rows 3–23 and column 1 of Table 4.13.

TABLE 4.12

Study of IEEE 300-Bus System with 5-Terminal VSC-HVDC Network Incorporating Nonlinear Voltage Droop Characteristics with Dead Band (Model B)

Base case power flow (NI = 6)

$V_{266} = 1.011 \angle -11.24$; $V_{270} = 1.011 \angle -11.32$; $V_{271} = 0.998 \angle -17.67$;

$V_{272} = 0.981 \angle -19.46$; $V_{273} = 1.006 \angle -17.47$;

AC-MTDC power flow with linear voltage droop control (VSCs connected to AC buses 266, 270, 271, 272 and 273)

Given Quantities		AC Buses	VSC
			Solution
V_{DCav}	3.02	$\theta_{266} = -11.4982$;	$\theta_{sh1} = -18.0960$; $\theta_{sh2} = -9.6028$; $\theta_{sh3} = -4.5960$;
V_{266}	1.02	$\theta_{270} = -11.5041$;	$\theta_{sh4} = -4.9446$; $\theta_{sh5} = -5.1988$; $m_1 = 0.9779$;
P_{sh2}	0.35	$\theta_{271} = -5.9761$;	$m_2 = 0.9657$; $m_3 = 1.0481$; $m_4 = 1.0549$; $m_5 = 0.9839$;
Q_{sh2}	0.1	$\theta_{272} = -5.8507$;	$V_{DC1} = 3.0208$; $V_{DC2} = 3.0198$; $V_{DC3} = 3.0198$;
P_{sh3}	0.3	$\theta_{273} = -6.760$;	$V_{DC4} = 3.0199$; $V_{DC5} = 3.0198$; $I_{DC3} = -1.034$;
Q_{sh3}	0.1	$V_{270} = 1.0203$;	$I_{DC4} = -0.0701$; $I_{DC5} = -0.1035$;
P_{sh4}	0.2	$V_{271} = 1.1094$;	DC power
Q_{sh4}	0.09	$V_{272} = 1.1179$;	$P_{DC1} = 1.1996$;
P_{sh5}	0.3	$V_{273} = 1.0401$;	Converter loss (%)
Q_{sh5}	0.1		$P_{loss1} = 2.0959$; $P_{loss2} = 1.2618$; $P_{loss3} = 1.22$;
P_{DC2}	−0.3627		$P_{loss4} = 1.1754$; $P_{loss5} = 1.2310$;
P_{DC3}	−0.3123		
P_{DC4}	−0.2118		
P_{DC5}	−0.3124		

NI = 6;

Computation of references V_{DC}^* from droop equations

$V_{DC1} = 3.0208$; $V_{DC2} = 3.0198$; $V_{DC1}^* = 3.0214$; $V_{DC2}^* = 3.0204$; $V_{DC3}^* = 3.0204$; $V_{DC4}^* = 3.0204$;

$V_{DC3} = 3.0198$; $V_{DC4} = 3.0199$; $V_{DC5}^* = 3.0204$; $P_{DC1} = 1.1996$; $P_{DC2} = -0.3627$; $P_{DC3} = -0.3123$;

$V_{DC5} = 3.0198$; $I_{DC3} = -1.034$; $P_{DC4} = -0.2118$; $P_{DC5} = -0.3124$; $P_{DC1}^* = 1.1882$; $P_{DC2}^* = -0.2913$;

$I_{DC4} = -0.0701$; $I_{DC5} = -0.1035$; $P_{DC3}^* = -0.3295$; $P_{DC4}^* = -0.2290$; $P_{DC5}^* - 0.3380$; $I_{DC1}^* = 0.3933$;

$V_{DCmax} = 3.02$; $V_{DChigh}^* = 3.015$; $I_{DC2}^* = -0.0965$; $I_{DC3}^* = -0.1091$; $I_{DC4}^* - 0.0758$; $I_{DC5}^* = -0.1119$;

$V_{DClow}^* = 3.01$; $V_{DCmin} = 3.009$;   NI = 4

$R_1 = 20$; $R_2 = 15$; $R_3 = 10$;

$R_4 = 10$; $R_5 = 15$;

The results of the AC-MTDC power flow are given in rows 3–23 and columns 2–3 of Table 4.13. The AC-MTDC power flow converges in seven number of iterations.

Subsequent to the AC-MTDC power flow, using the values of 'V_{DCa}' obtained (and hence, 'P_{DCa}' and 'I_{DCa}'), the DC voltage ('V_{DCa}^*') and thereafter, the power ('P_{DCa}^*') and current ('I_{DCa}^*') references for the droop lines of the VSCs are

Variation of Error with Number of iterations

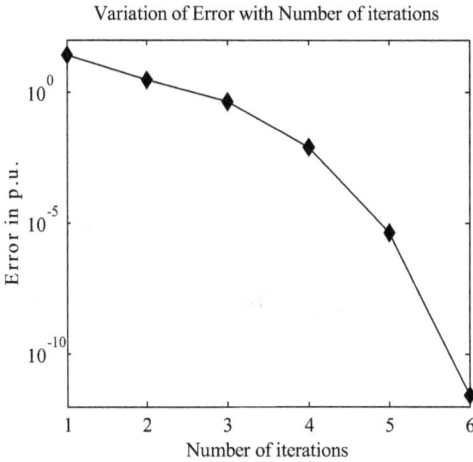

FIGURE 4.32 Convergence characteristic for the study of Table 4.11.

Variation of Error with Number of iterations

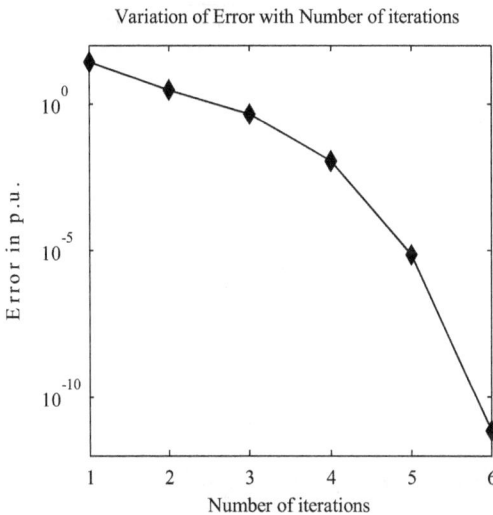

FIGURE 4.33 Convergence characteristic for the study of Table 4.12.

computed using the droop Eqns. (4.28) and (4.29). The results are given in row 25 of Table 4.13.

Case II: Model B employing nonlinear droop characteristics
    This case study is similar to the previous study of Table 4.13 except VSC 2 which employs a nonlinear V-P droop with dead band. The droop control gains are identical to that of the study of Table 4.13. First, the AC-MTDC power flow is carried out. The results are shown in rows 3–23 of Table 4.14. Subsequently,

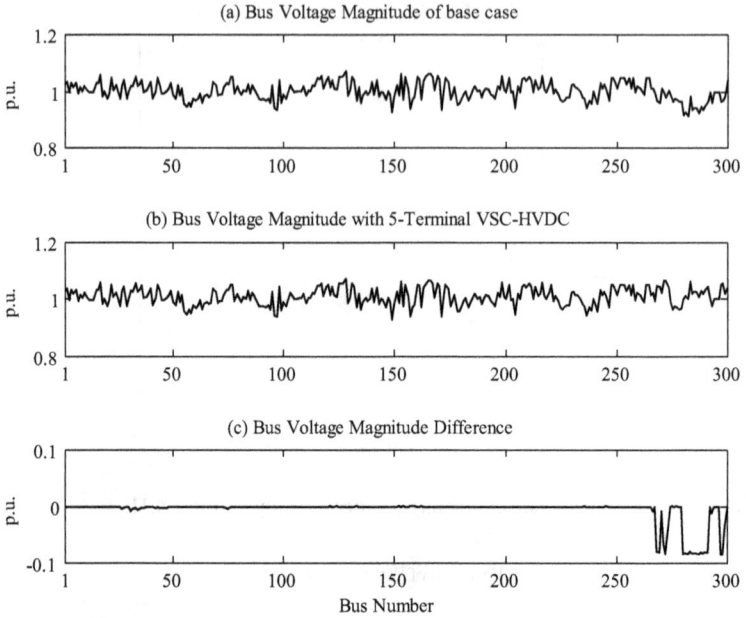

FIGURE 4.34 Bus voltage profile for the study of Table 4.11.

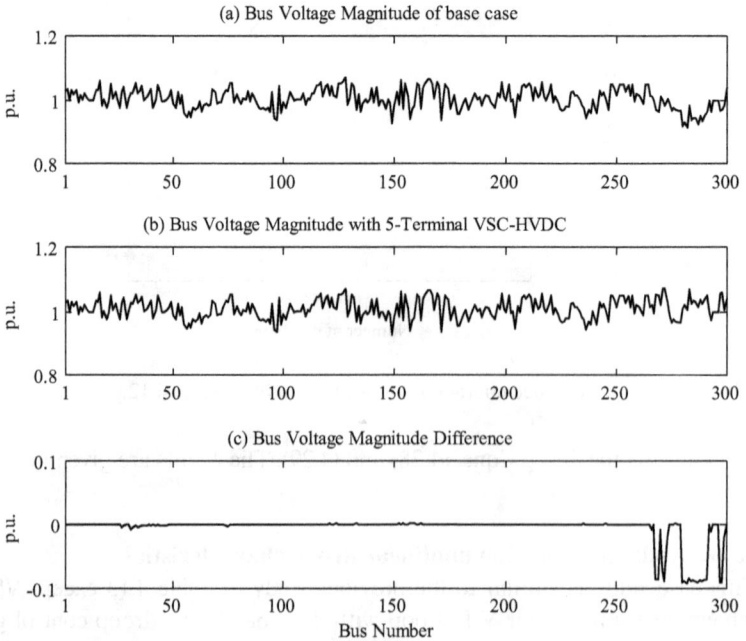

FIGURE 4.35 Bus voltage profile for the study of Table 4.12.

TABLE 4.13

Study of European 1354-Bus System with 7-Terminal VSC-HVDC Network Incorporating Linear DC Voltage Droop Characteristics (Model B)

Base case power flow converged in seven iterations (NI = 7)

$V_{1280} = 0.9554 \angle - 35.2574$; $V_{1281} = 1.0354 \angle - 37.3082$;

$V_{1286} = 1.0156 \angle - 12.3760$; $V_{1292} = 1.0289 \angle - 35.23066$;

$V_{1313} = 1.0351 \angle - 37.4755$; $V_{1324} = 0.9972 \angle - 29.0208$;

$V_{1353} = 0.9934 \angle - 29.6628$;

AC-MTDC power flow with linear voltage droop control (VSCs connected to AC buses 1280, 1281, 1286, 1292, 1313, 1324 and 1353)

Given Quantities		Solution	
		AC Buses	VSC
V_{DCav}	3.0	$\theta_{1280} = -42.2331$;	$\theta_{sh1} = -51.5353$; $\theta_{sh2} = -34.9973$; $\theta_{sh3} = 10.4817$;
V_{1280}	0.98	$\theta_{1281} = -37.0767$;	$\theta_{sh4} = -34.3565$; $\theta_{sh5} = -35.6829$; $\theta_{sh6} = -27.4798$;
P_{sh2}	0.4	$\theta_{1286} = -12.1181$;	$\theta_{sh7} = -28.5106$; $m_1 = 1.0244$; $m_2 = 0.9971$;
Q_{sh2}	0.2	$\theta_{1292} = -35.4157$;	$m_3 = 0.9698$; $m_4 = 0.9830$; $m_5 = 0.9917$;
P_{sh3}	0.3	$\theta_{1313} = -37.2515$;	$m_6 = 0.9546$; $m_7 = 0.9570$;
Q_{sh3}	0.1	$\theta_{1324} = -28.6030$;	$V_{DC1} = 3.008$; $V_{DC2} = 2.9998$; $V_{DC3} = 2.9999$;
P_{sh4}	0.2	$\theta_{1353} = 29.6316$;	$V_{DC4} = 2.9999$; $V_{DC5} = 2.9999$; $V_{DC6} = 2.9999$;
Q_{sh4}	0.1	$V_{1281} = 1.0372$;	$V_{DC7} = 2.9999$; $P_{DC1} = 1.6747$;
P_{sh5}	0.3	$V_{1286} = 1.0180$;	Converter loss (%)
Q_{sh5}	0.15	$V_{1292} = 1.0325$;	$P_{loss1} = 3.39$; $P_{loss2} = 1.31$; $P_{loss3} = 1.23$;
P_{sh6}	0.2	$V_{1313} = 1.0367$;	$P_{loss4} = 1.19$; $P_{loss5} = 1.24$; $P_{loss6} = 1.19$;
Q_{sh6}	0.09	$V_{1324} = 1.0031$;	$P_{loss7} = 1.20$;
P_{sh7}	0.2	$V_{1353} = 0.9996$;	
Q_{sh7}	0.15		
P_{DC2}	-0.4133		
P_{DC3}	-0.3124		
P_{DC4}	-0.2119		
P_{DC5}	-0.3125		
P_{DC6}	-0.2119		
P_{DC7}	-0.2121.	NI = 7	

Computation of references V_{DC}^* from droop equations

$V_{DC1} = 3.008$; $V_{DC2} = 2.9998$; $V_{DC1}^* = 3.0018$; $V_{DC2}^* = 3.0008$; $V_{DC3}^* = 3.0009$;

$V_{DC5} = 2.9999$; $V_{DC6} = 2.9999$; $V_{DC4}^* = 3.0010$; $V_{DC5}^* = 3.0009$; $V_{DC6}^* = 3.0009$;

$V_{DC7} = 2.9999$; $V_{DC7}^* = 3.0009$; $I_{DC4}^* = -0.0101$; $I_{DC5}^* = -0.1143$;

$R_1 = 20$; $R_2 = 15$; $R_3 = 10$; $I_{DC6}^* = -0.0857$; $I_{DC7}^* = -0.0907$;

$R_4 = 10$; $R_5 = 10$; $R_6 = 15$;   DC power

$R_7 = 20$; $P_{DC1}^* = 1.6543$; $P_{DC2}^* = -0.4286$;

$P_{DC3}^* = -0.3226$; $P_{DC4}^* = -0.0304$;

$P_{DC5}^* = -0.3430$; $P_{DC6}^* = -0.2572$; $P_{DC7}^* = -0.2721$;

NI = 3

TABLE 4.14
Study of European 1354-Bus System with 7-Terminal VSC-HVDC Network Incorporating Nonlinear DC Voltage Droop with Dead-Band Characteristics (Model B)

Base case power flow converged in seven iterations (NI = 7)

$V_{1280} = 0.9554 \angle -35.2574; V_{1281} = 1.0354 \angle -37.3082;$

$V_{1286} = 1.0156 \angle -12.3760; V_{1292} = 1.0289 \angle -35.23066;$

$V_{1313} = 1.0351 \angle -37.4755; V_{1324} = 0.9972 \angle -29.0208;$

$V_{1353} = 0.9934 \angle -29.6628;$

AC-MTDC power flow with linear voltage droop control (VSCs connected to AC buses 1280, 1281, 1286, 1292, 1313, 1324 and 1353)

Given Quantities		Solution	
		AC Buses	VSC
V_{DCav}	3.02	$\theta_{1280} = -42.2332;$	$\theta_{sh1} = -51.5357; \theta_{sh2} = -34.9973; \theta_{sh3} = 10.4817;$
V_{1280}	0.98	$\theta_{1281} = -37.0767;$	$\theta_{sh4} = -34.3566; \theta_{sh5} = -35.6830; \theta_{sh6} = -27.4798;$
P_{sh2}	0.4	$\theta_{1286} = -12.1181;$	$\theta_{sh7} = -28.5106; m_1 = 1.0177; m_2 = 0.9905;$
Q_{sh2}	0.2	$\theta_{1292} = -35.4158;$	$m_3 = 0.9633; m_4 = 0.9765; m_5 = 0.9852;$
P_{sh3}	0.3	$\theta_{1313} = -37.2516;$	$m_6 = 0.9483; m_7 = 0.9507;$
Q_{sh3}	0.1	$\theta_{1324} = -28.6030;$	$V_{DC1} = 3.0208; V_{DC2} = 3.0198; V_{DC3} = 3.0199;$
P_{sh4}	0.2	$\theta_{1353} = 29.6316;$	$V_{DC4} = 3.0199; V_{DC5} = 3.0199; V_{DC6} = 3.0199;$
Q_{sh4}	0.1	$V_{1281} = 1.0372;$	$P_{DC1} = 1.6747;$
P_{sh5}	0.3	$V_{1286} = 1.0180;$	Converter loss (%)
Q_{sh5}	0.15	$V_{1292} = 1.0325;$	$P_{loss1} = 3.39; P_{loss2} = 1.31; P_{loss3} = 1.23;$
P_{sh6}	0.2	$V_{1313} = 1.0367;$	$P_{loss4} = 1.19; P_{loss5} = 1.24; P_{loss6} = 1.19;$
Q_{sh6}	0.09	$V_{1324} = 1.0031;$	$P_{loss7} = 1.20;$
P_{sh7}	0.2	$V_{1353} = 0.9996;$	
Q_{sh7}	0.15		
P_{DC2}	−0.4133		
P_{DC3}	−0.3124		
P_{DC4}	−0.2119		
P_{DC5}	−0.3125		
P_{DC6}	−0.2119		
P_{DC7}	−0.2121	NI = 7;	

Computation of references V_{DC}^* from droop equations

$V_{DC1} = 3.0208; V_{DC2} = 3.0198;$

$V_{DC3} = 3.0199; V_{DC4} = 3.0199;$

$V_{DC5} = 3.0199; V_{DC6} = 3.0199;$

$V_{DC7} = 3.0199;$

$V_{DCmax} = 3.02; V_{DChigh} = 3.015;$

$V_{DClow}^* = 3.01; V_{DCmin} = 3.009;$

$R_1 = 20; R_2 = 15; R_3 = 10;$

$R_4 = 10; R_5 = 10; R_6 = 15;$

$R_7 = 20;$

$V_{DC1}^* = 3.0211; V_{DC2}^* = 3.0202; V_{DC3}^* = 3.0202;$

$V_{DC4}^* = 3.0202; V_{DC5}^* = 3.0202; V_{DC6}^* = 3.0302;$

$V_{DC7}^* = 3.0202; I_{DC3}^* = -0.1068; I_{DC4}^* = -0.0735;$

$I_{DC5}^* = -0.1068; I_{DC6}^* = -0.0751;$

$I_{DC7}^* = -0.0768;$

DC Power

$P_{DC1}^* = 1.6679; P_{DC2}^* = -0.3412;$

$P_{DC3}^* = -0.3226; P_{DC4}^* = -0.2220;$

$P_{DC5}^* = -0.3227; P_{DC6}^* = -0.227; P_{DC7}^* = -0.2320;$

NI = 4

a separate DC power flow is carried out to calculate the reference values for the droop control lines. The results are shown in row 25 of Table 4.14.

From the power-flow solution, it is observed that the converter connected at AC bus-1281 operates at the point 'A' as shown in Figure 4.4.

The convergence characteristics of Tables 4.13 and 4.14 are shown in Figures 4.36 and 4.37, respectively. From Figures 4.36 and 4.37, it is observed that the proposed algorithm demonstrates excellent convergence characteristics, converging in seven iterations. The bus voltage profiles for the studies of Tables 4.13 and 4.14 are shown in Figures 4.38 and 4.39, respectively. From Figures 4.38 and 4.39, it can be observed that the bus voltage profiles do not change much except at the AC terminal buses to which the converters are connected.

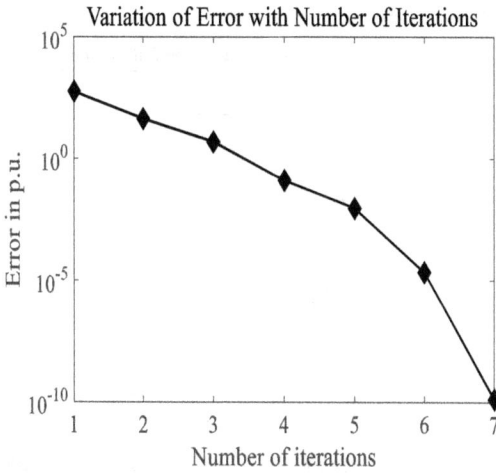

FIGURE 4.36 Convergence characteristic for the study of Table 4.13.

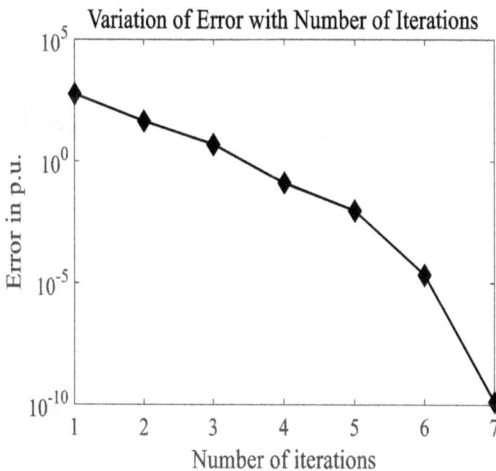

FIGURE 4.37 Convergence characteristic for the study of Table 4.14.

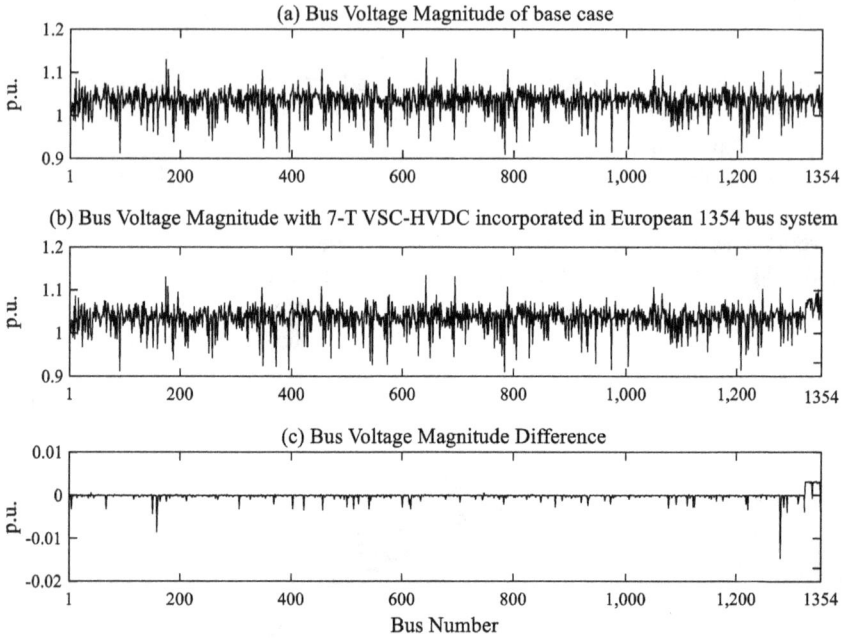

FIGURE 4.38   Bus voltage profile for the study of Table 4.13.

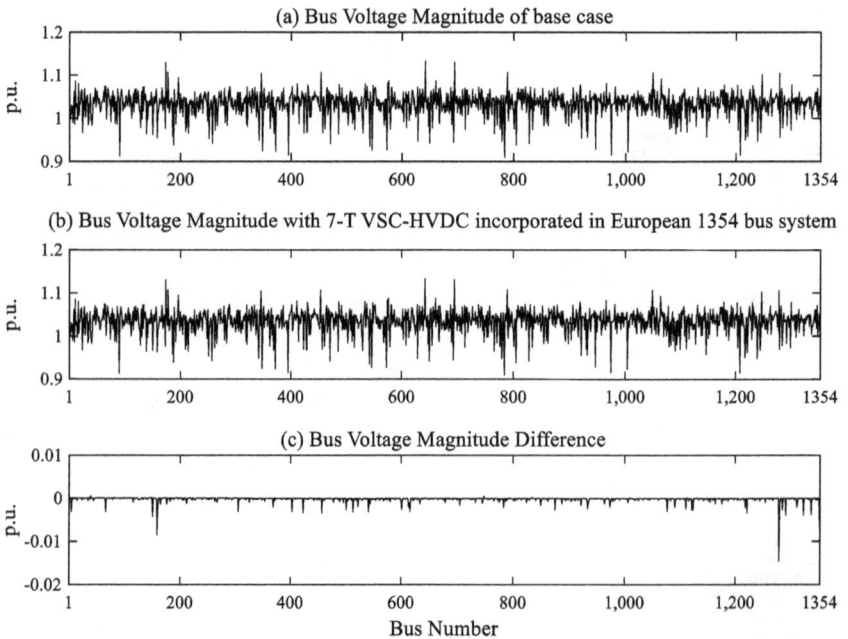

FIGURE 4.39   Bus voltage profile for the study of Table 4.14.

4.7   SUMMARY

In this chapter, a generalized approach for the development of both unified and sequential Newton power-flow models of integrated AC-MTDC systems employing DC voltage droop control is presented. In the proposed model, the modulation indices of the VSCs are obtained directly from the power-flow solution. Diverse MTDC grid control techniques including linear and nonlinear DC voltage droop control have been employed. It is observed that unlike droop model 'A', model 'B' facilitates the specification of both line end active and reactive power flows. The model displays excellent convergence characteristics, independent of the DC grid topology and the MTDC grid control technique employed. This validates the model.

5 Power-Flow Modelling of AC Power Systems Integrated with VSC-Based Multi-Terminal DC (AC-MVDC) Grids Incorporating Interline DC Power-Flow Controller (IDCPFC)

5.1 INTRODUCTION

One of the main challenges in VSC-based integrated AC–DC systems is the management of DC power flow within the DC grids. Although the VSCs control the power injections into a DC grid, the power flows within the DC grid depend upon the resistances of the DC links or cables. In this respect, DC power-flow control devices [40–45] have been conceptualised and developed, similar to Flexible AC Transmission Systems (FACTS) Controllers developed for AC grids. The interline DC power-flow controller (IDCPFC) reported in Refs. [45,53] is a DC power-flow controller which has been implemented for the power-flow management of MTDC grids. It is similar to the Interline Power-Flow Controller (IPFC) [53–55], which is a FACTS Controller [13].

Now, for planning, operation and control of VSC-based integrated AC–DC systems, their power-flow models are required. References [46,47,86–105] present some comprehensive power-flow models of VSC-based integrated AC–DC systems. However, none of these research works addresses the power-flow modelling of integrated AC–DC systems incorporating the IDCPFC.

This chapter presents the power-flow modelling of VSC-based integrated AC–DC systems employing IDCPFCs for power-flow management. Similar to Chapters 3 and 4, the VSC modulation index 'm' is considered as an unknown and can be obtained directly, from the power-flow solution. VSC losses are included in this model.

DOI: 10.1201/9781003252078-5

5.2 MODELLING OF AC-MVDC SYSTEMS INCORPORATING IDCPFCs

The basic assumptions adopted for this chapter are similar to those available in Section 3.2 (Chapter 3) of this book. Figure 5.1 shows an 'n' bus AC power system network integrated with a 'q' terminal VSC-MTDC grid. 'q' VSCs are used for integrating the VSC-MTDC grid with the AC network at AC buses 'i', '(i+1)', and so on, up to bus '(i+q−1)', through their respective converter transformers. Figure 5.1 also shows an IDCPFC incorporated within the MTDC grid for its power-flow management. The IDCPFC comprises 'z' (z≤q−1) variable DC voltage sources interconnected between DC bus '1' and 'z' other DC buses numbered as '2', '3', and so on, up to '(z+1)'. The equivalent circuit of Figure 5.1 is shown in Figure 5.2.

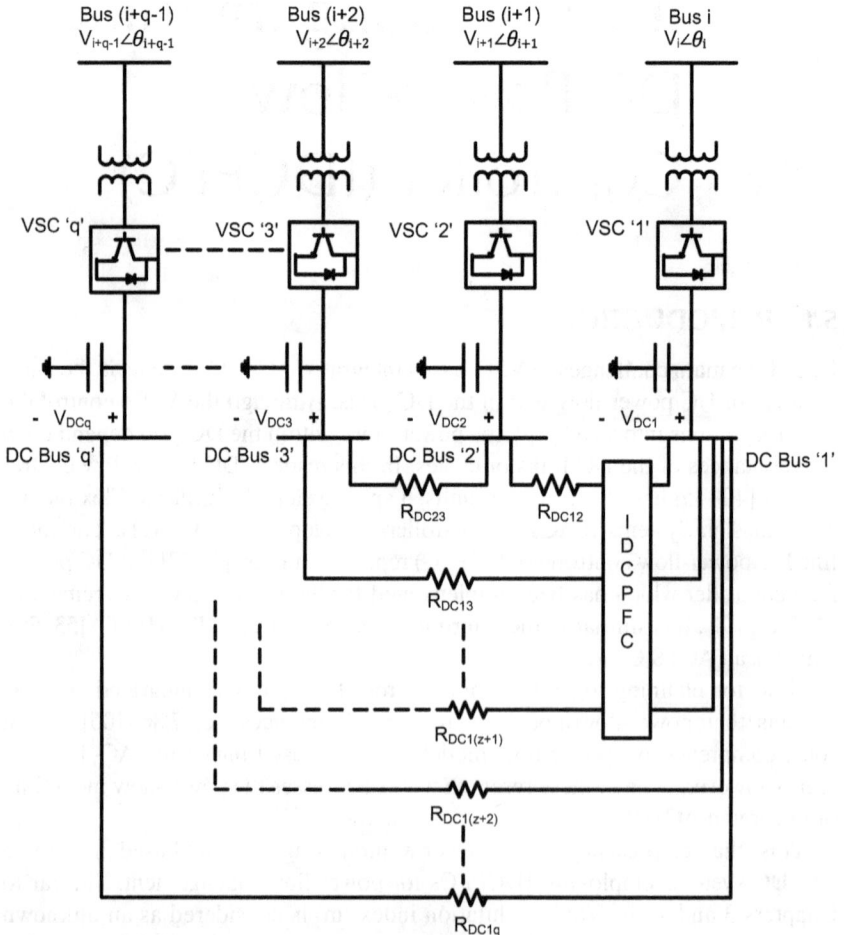

FIGURE 5.1   Schematic diagram of an AC-MTDC system incorporating an IDCPFC.

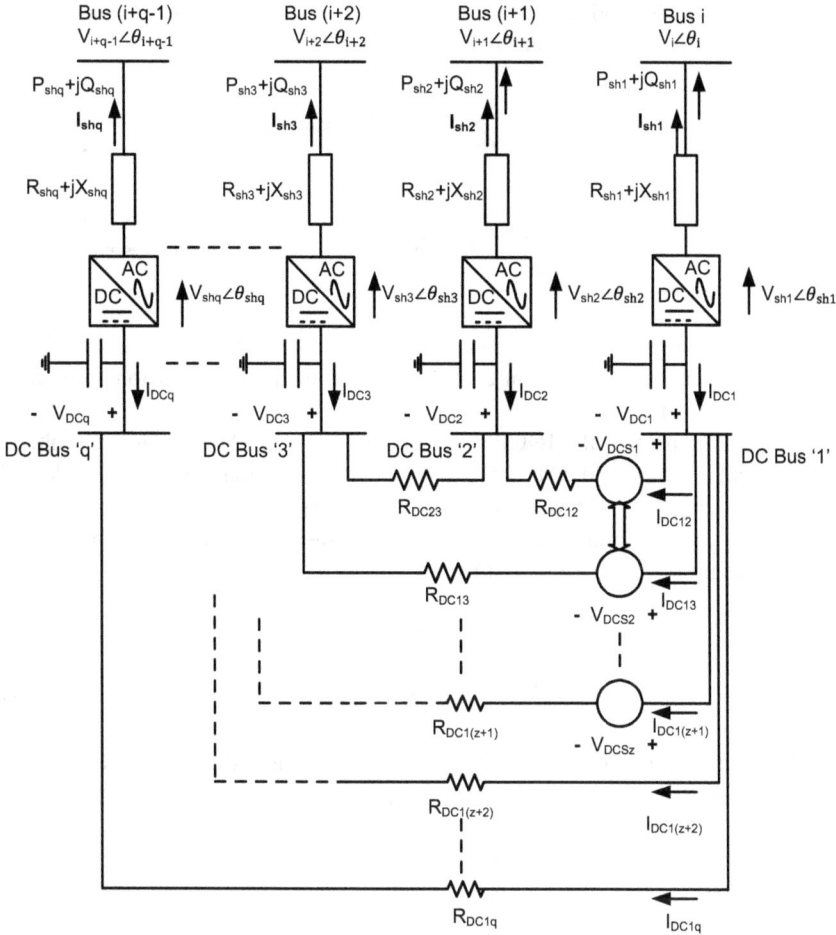

FIGURE 5.2 Equivalent circuit of AC-MTDC system incorporating an IDCPFC.

In Figure 5.2, the 'q' VSCs are represented by 'q' fundamental frequency, positive sequence voltage sources. The ath $(1 \leq a \leq q)$ VSC is connected to AC terminal bus '$(i+a-1)$' whose voltage is represented by the phasor $V_{i+a-1} = V_{i+a-1} \angle \theta_{i+a-1}$.

From Figure 5.2, the current in the link (not shown) connecting the ath VSC and its AC terminal bus is

$$\mathbf{I_{sha}} = \mathbf{y_{sha}} \left(\mathbf{V_{sha}} - \mathbf{V_{i+a-1}} \right) \tag{5.1}$$

where $V_{sha} = V_{sha} \angle \theta_{sha} = m_a$ c $V_{DCa} \angle \theta_{sha}$, $y_{sha} = 1/Z_{sha}$, $\mathbf{Z_{sha}} = R_{sha} + jX_{sha}$, R_{sha} and X_{sha} are the resistance and the leakage reactance of the ath converter transformer, respectively, 'm_a' is the VSC modulation index and the constant 'c' is representative of the VSC architecture [11].

From Figures 5.1 and 5.2, the net current injection at the AC bus '$(i+a-1)$' connected to the ath $(1 \leq a \leq p)$ converter can be written as

$$I_{i+a-1} = \sum_{k=1}^{n} Y_{(i+a-1)k} V_k - y_{sha} V_{sha} \qquad (5.2)$$

where $Y_{(i+a-1)(i+a-1)} = Y_{(i+a-1)(i+a-1)}^{old} + y_{sha}$ and $Y_{(i+a-1)(i+a-1)}^{old} = y_{(i+a-1)0} + \sum_{k=1, k \neq i+a-1}^{n} y_{(i+a-1)k}$

are the values of self-admittances of bus '$(i+a-1)$' with the ath VSC connected

and the original 'n' bus AC system without any VSC, respectively. Similarly, '$y_{(i+a-1)0}$' accounts for the shunt capacitances of all the transmission lines connected to bus '$(i+a-1)$'.

Now, in Figure 5.2, the IDCPFC comprises 'z' $(z \leq q-1)$ variable DC voltage sources and it is assumed, without any loss of generality, that the wth variable DC voltage source 'V_{DCsw}' $(1 \leq w \leq z)$ is connected in series with the link interconnecting the DC buses '1' and '$(1+w)$' $(1 \leq w \leq z, \forall z \leq q-1)$. Then, from Figure 5.2, the current and power in this link (connected between DC buses '1' and '$w+1$') are

$$I_{DC1(w+1)} = Y_{DC1(w+1)} \left[V_{DCw+1} - V_{DC1} + V_{DCsw} \right] \qquad (5.3)$$

$$P_{DC1(w+1)} = V_{DC1} I_{DC1(w+1)} \qquad (5.4)$$

Further, with the IDCPFC, the net DC current injection at the first DC bus can be written as

$$I_{DC1} = \sum_{u=1, u \neq 1}^{z+1} I_{DC1u} + \sum_{u=z+2, u \neq 1}^{q} I_{DC1u} \qquad (5.5)$$

Writing I_{DC1u} in the form of Eq. (5.3) and substituting in Eq. (5.5), we get

$$I_{DC1} = \sum_{u=1}^{q} Y_{DC1u} V_{DCu} + \sum_{u=1, u \neq 1}^{z+1} Y_{DC1u} V_{DCs(u-1)} \qquad (5.6)$$

In a similar manner, the net DC current injections at the other DC buses can also be written very easily. It can be shown that the net DC current injection at any arbitrary DC node 'u' $(1 \leq u \leq q)$ can be generalized as

$$I_{DCu} = \sum_{v=1}^{q} Y_{DCuv} V_{DCv} + \sum_{v=1, v \neq 1}^{z+1} Y_{DCuv} V_{DCs(v-1)} \qquad \text{if } u = 1$$

$$= \sum_{v=1}^{q} Y_{DCuv} V_{DCv} - Y_{DCu1} V_{DCs(u-1)} \qquad \text{if } 2 \leq u \leq z+1$$

$$= \sum_{v=1}^{q} Y_{DCuv} V_{DCv} \qquad \text{if } z+2 \le u \le q \qquad (5.7)$$

5.3 POWER-FLOW EQUATIONS OF INTEGRATED AC-MVDC SYSTEMS INCORPORATING IDCPFC

From Figure 5.2, at the AC bus '(i+a−1)' pertaining to the ath VSC, it can be shown that the net active and reactive power injections are

$$P_{i+a-1} = \sum_{k=1}^{n} V_{i+a-1} V_k Y_{(i+a-1)k} \cos\left[\theta_{i+a-1} - \theta_k - \phi_{(i+a-1)k}\right]$$
$$- m_a c\, V_{DCa} V_{i+a-1} y_{sha} \cos\left(\theta_{i+a-1} - \theta_{sha} - \phi_{sha}\right) \qquad (5.8)$$

$$Q_{i+a-1} = \sum_{k=1}^{n} V_{i+a-1} V_k Y_{(i+a-1)k} \sin\left[\theta_{i+a-1} - \theta_k - \phi_{(i+a-1)k}\right]$$
$$- m_a c\, V_{DCa} V_{i+a-1} y_{sha} \sin\left(\theta_{i+a-1} - \theta_{sha} - \phi_{sha}\right) \qquad (5.9)$$

where 'ϕ_{sha}' is the phase angle of y_{sha}.

In addition, from Figure 5.2, it can be shown that the active and reactive power flows at the terminal end of the link interconnecting the ath VSC to the AC bus '(i+a−1)' are

$$P_{sha} = m_a c\, V_{DCa} V_{i+a-1} y_{sha} \cos\left(\theta_{i+a-1} - \theta_{sha} - \phi_{sha}\right) - V_{i+a-1}^2 y_{sha} \cos\phi_{sha} \qquad (5.10)$$

$$Q_{sha} = m_a c\, V_{DCa} V_{i+a-1} y_{sha} \sin\left(\theta_{i+a-1} - \theta_{sha} - \phi_{sha}\right) + V_{i+a-1}^2 y_{sha} \sin\phi_{sha} \qquad (5.11)$$

Now, if all the VSCs as well as the IDCPFC are lossless, the AC–DC power balance equation for the ath ($1 \le a \le q$) VSC can be written using Eq. (5.7) as

$$\left(m_a c\, V_{DCa}\right)^2 y_{sha} \cos\phi_{sha} - m_a c\, V_{DCa} V_{i+a-1} y_{sha} \cos\left(\theta_{sha} - \theta_{i+a-1} - \phi_{sha}\right)$$

$$= -\sum_{v=1}^{q} V_{DCa} V_{DCv} Y_{DCav} \qquad \text{if } z+2 \le a \le q$$

$$= -\sum_{v=1}^{q} V_{DCa} V_{DCv} Y_{DCav} + V_{DCa} V_{DCs(a-1)} Y_{DCa1} \qquad \text{if } 2 \le a \le z+1$$

$$= -\sum_{v=1}^{q} V_{DCa} V_{DCv} Y_{DCav} - \sum_{v=1,\, v \ne 1}^{z+1} V_{DCa} V_{DCs(v-1)} Y_{DCav} \qquad \text{if } a = 1$$

If VSC losses are considered, the above equation for the ath $(1 \leq a \leq q)$ VSC becomes

$$\left(m_a c\, V_{DCa}\right)^2 y_{sha} \cos\phi_{sha} - m_a c\, V_{DCa} V_{i+a-1} y_{sha} \cos\left(\theta_{sha} - \theta_{i+a-1} - \phi_{sha}\right)$$

$$= -\sum_{v=1}^{q} V_{DCa} V_{DCv} Y_{DCav} - P_{lossa} \qquad\qquad \text{if } (z+2) \leq a \leq q$$

$$= -\sum_{v=1}^{q} V_{DCa} V_{DCv} Y_{DCav} + V_{DCa} V_{DCs(a-1)} Y_{DCa1} - P_{lossa} \qquad \text{if } 2 \leq a \leq z+1$$

$$= -\sum_{v=1}^{q} V_{DCa} V_{DCv} Y_{DCav} - \sum_{v=1,\, v\neq 1}^{z+1} V_{DCa} V_{DCs(v-1)} Y_{DCav} - P_{lossa} \qquad \text{if } a = 1$$

$$\text{or, } f_{1a} = 0 \qquad \forall\, a,\ 1 \leq a \leq q \tag{5.12}$$

where P_{lossa} represents the losses [18, 92] of the ath VSC as already detailed in Chapter 3 {Eq. (3.10)}.

Equation (5.12) represents 'q' independent equations. The detailed derivation of Eq. (5.12) is shown in Appendix A.

Now, in the AC-MTDC system (Figure 5.1) with 'q' VSCs, if it is assumed that the rth $(1 \leq r \leq q)$ VSC is used for voltage control of its corresponding AC bus, we have

$$V_{i+a-1}^{sp} - V_{i+a-1}^{cal} = 0 \qquad \forall a, 1 \leq a \leq q,\ a = r \tag{5.13}$$

Also, not more than '(q–1)' line active and reactive power flows {Eqs. (5.10) and (5.11)} can be specified, which give us '(2q–2)' independent equations given as

$$P_{sha}^{sp} - P_{sha}^{cal} = 0 \tag{5.14}$$

$$Q_{sha}^{sp} - Q_{sha}^{cal} = 0 \tag{5.15}$$

$$\forall a,\ 1 \leq a \leq q,\ a \neq r.$$

Instead of PQ control mode, if a VSC operates in the PV one, Eq. (5.15) changes to

$$V_{i+a-1}^{sp} - V_{i+a-1}^{cal} = 0 \qquad \forall a,\ 1 \leq a \leq q,\ a \neq r \tag{5.16}$$

Further, the net reactive power injection at AC bus '(i+r–1)' can be specified as its voltage is controlled by the rth VSC. Thus, we get

$$Q_{i+a-1}^{sp} - Q_{i+a-1}^{cal} = 0 \qquad \forall a,\ 1 \leq a \leq q,\ a = r \tag{5.17}$$

In Eqs. (5.13)–(5.17), V_{i+a-1}^{sp}, Q_{i+a-1}^{sp}, P_{sha}^{sp} and Q_{sha}^{sp} are specified values while V_{i+a-1}^{cal}, Q_{i+a-1}^{cal}, P_{sha}^{cal} and Q_{sha}^{cal} are calculated values {using Eqs. (5.9)–(5.11)}.

Now, in the DC network as shown in Figure 5.2, the IDCPFC is represented by 'z' variable DC voltage sources (incorporated in series with 'z' DC links). Hence, the inclusion of the IDCPFC in the integrated AC-MTDC system introduces 'z' additional unknowns. To solve them would require 'z' specified or known quantities. It is important to note that if the IDCPFC is considered lossless, from Figure 5.2, the power delivered by the IDCPFC is

$$P_{IDCPFC} = \left[V_{DCs1} \left\{ -I_{DC12} \right\} + V_{DCs2} \left\{ -I_{DC13} \right\} + \cdots V_{DCsz} \left\{ -I_{DC1(z+1)} \right\} \right] = 0 \quad (5.18)$$

Equation (5.18) represents a single, independent equation. Thus, additional '(z–1)' equations are required for a complete solution of the IDCPFC variables, which is similar to the degree of freedom of an IPFC [45–47]. These '(z–1)' equations are obtained from the control objectives of the IDCPFC. Reference [43] has implemented DC link power-flow control with an IDCPFC. In this chapter, both current and power-flow controls of the DC link(s) have been considered.

With DC link current control, the line currents in all the 'z' DC links (containing the 'z' variable DC voltage sources of the IDCPFC) except one can be controlled. If it is presumed that the line current in the DC link containing the yth $(1 \le y \le z)$ variable DC voltage source 'V_{DCsy}' is not being controlled by the IDCPFC, the control equations for the rest of the '(z–1)' DC links are

$$I_{DC1(w+1)}^{sp} - I_{DC1(w+1)}^{cal} = 0 \quad (5.19)$$

In a similar manner, the control equations for the power flow in the '(z–1)' DC links are

$$P_{DC1(w+1)}^{sp} - P_{DC1(w+1)}^{cal} = 0 \quad (5.20)$$

Equations (5.19) and (5.20) can be generalised as

$$f_{2w} = 0 \quad \forall w, 1 \le w \le z-1, w \ne y, z \le q-1 \quad (5.21)$$

It may be noted that if the IDCPFC employs current control for some of the DC links and power-flow control for the rest, Eq. (5.21) would comprise both Eqs. (5.19) and (5.20).

In the above equations, $I_{DC1(w+1)}^{sp}$ and $P_{DC1(w+1)}^{sp}$ are the specified values of the current and power flow in the DC link between the DC buses '1' and '(1+w)' (and containing the variable DC voltage source 'V_{DCsw}'), respectively, while $I_{DC1(w+1)}^{cal}$ and $P_{DC1(w+1)}^{cal}$ are their calculated value obtained using Eqs. (5.3) and (5.4), respectively.

5.4   IMPLEMENTATION OF POWER-FLOW IN INTEGRATED AC-MVDC SYSTEMS INCORPORATING IDCPFC

If it is assumed that there are 'g' generators connected at the first 'g' buses of the 'n' bus AC system with bus 1 being the slack bus, then, for the AC-MTDC system incorporating an IDCPFC with 'z' variable DC voltage sources and following DC slack bus control, the unified AC-MTDC power-flow problem is of the form

Compute: θ, **V**, **X**

Given: **P**, **Q**, **R**

with

$$\theta = \left[\theta_2 \cdots \theta_n\right]^T , \mathbf{V} = \left[V_{g+1} \cdots V_n\right]^T, \theta_{sh} = \left[\theta_{sh1} \cdots \theta_{shq}\right]^T, \mathbf{m} = \left[m_1 \cdots m_q\right]^T,$$

$$\mathbf{V_{DC}} = \left[V_{DC2} \cdots V_{DCq}\right]^T$$

$$\mathbf{V_{DCs}} = \left[V_{DCs1} \cdots V_{DCsz}\right]^T \text{ and } \mathbf{X} = \left[\theta_{sh}^T \quad \mathbf{m}^T \quad \mathbf{V_{DC}^T} \quad \mathbf{V_{DCs}^T}\right]^T$$

$$\mathbf{P} = [P_2 \cdots P_n]^T, \mathbf{Q} = \left[Q_{g+1} \cdots Q_n\right]^T, \mathbf{P_{sh}} = [P_{sh2} \cdots P_{shq}], \mathbf{Q_{sh}} = \left[Q_{sh2} \cdots Q_{shq}\right]$$

$$\mathbf{f_1} = \left[f_{11} \cdots f_{1q}\right], \mathbf{f_2} = \left[f_{21} \cdots f_{2(z-1)}\right]$$

and $\mathbf{R} = \left[\ \mathbf{P_{sh}} \quad \mathbf{Q_{sh}} \quad V_{i+r-1} \quad P_{IDCPFC} \quad \mathbf{f_1} \quad \mathbf{f_2}\right]^T$

It is presumed that in this model, DC slack bus control (master slave control) is adopted ('V_{DC1}' is specified). Also, the master VSC 'r' controls the voltage magnitude of the AC bus '(i+r−1)' unlike the other '(q−1)' slave VSCs, which control the line active as well as reactive power flows.

The Newton power-flow equation can be written as

$$\mathbf{J}\left[\Delta\theta^T \quad \Delta\mathbf{V}^T \quad \Delta\theta_{sh}^T \quad \Delta\mathbf{m}^T \quad \Delta\mathbf{V_{DC}^T}\Delta\mathbf{V_{DCs}^T}\right]^T = \left[\Delta\mathbf{P}^T \quad \Delta\mathbf{Q}^T \quad \Delta\mathbf{R}^T\right]^T \qquad (5.22)$$

where **J** is the power-flow Jacobian.

It may be noted that instead of the DC slack bus voltage 'V_{DC1}', if the average voltage of all the DC terminals 'V_{DCav}' is specified, the following modifications are required.

$$\mathbf{V_{DC}} = \left[V_{DC1} \cdots V_{DCq}\right]^T, \mathbf{R} = \left[\ \mathbf{P_{sh}} \quad \mathbf{Q_{sh}} \quad V_{i+r-1} \quad V_{DCav} \quad P_{IDCPFC} \quad \mathbf{f_1} \quad \mathbf{f_2}\right]^T.$$

The individual elements of **J** have to be appropriately modified for the above case.

If 'x' IDCPFCs are present, '$\mathbf{V_{DCs}}$' gets enlarged (with 'x*z' elements) and 'P_{IDCPFC}' is replaced by a vector '$\mathbf{P_{IDCPFC}}$' having 'x' elements, each governed by an equation similar to Eq. (5.18). Also, the individual elements of **J** are appropriately modified.

Figure 5.3 depicts the flow chart of the proposed approach with IDCPFC.

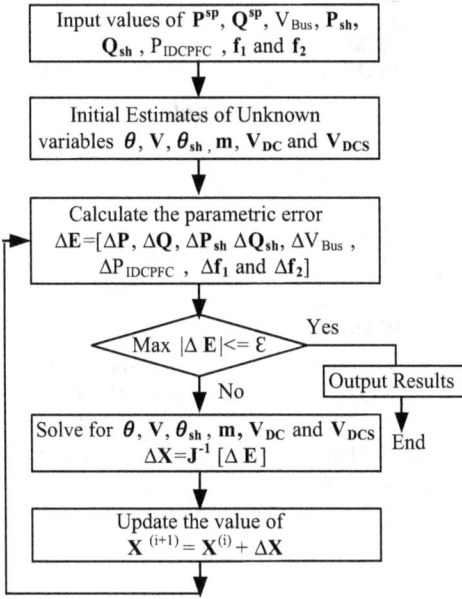

FIGURE 5.3   Flow chart of the proposed approach with IDCPFC.

5.5   CASE STUDIES AND RESULTS

For validation of the above model, a large number of studies were carried out by employing diverse DC voltage control strategies on MTDC grids integrated with the IEEE 300-bus and European 1354-bus test systems [115,116]. For all the VSC coupling transformers, $R_{sha} = 0.001$ p.u. and $X_{sha} = 0.1$ p.u. ($\forall$ a, $1 \le a \le q$). For interconnections between DC terminals, $R_{DCuv} = 0.01$ p.u. ($\forall u, v, 1 \le u \le q, 1 \le v \le q$, $u \ne v$), throughout the chapter [93]. The converter loss constants 'a_1', 'b_1' and 'c_1' are chosen to be 0.011, 0.003 and 0.0043, respectively [18,102,105]. The initial values of all the variable DC voltage sources (V_{DCsy} $\forall y$, $1 \le y \le z$) were chosen as 0.001 p.u. In all occurrences, a termination error tolerance of 10^{-8} p.u. was selected. 'NI' denotes the number of iterations. In all the results given in Tables 5.1–5.8, the values of bus voltage magnitudes, current magnitudes, active and reactive powers and droop control gains are denoted in p.u., while phase angles of voltage phasors are denoted in degrees.

5.5.1   STUDY OF 3-TERMINAL VSC-HVDC NETWORK INCORPORATING IDCPFC IN IEEE 300 BUS SYSTEM

Case I: DC link current control using IDCPFC

In this study, at first, a three-terminal VSC-MTDC grid is integrated with the IEEE-300 bus test system at buses 266, 270 and 271 and the AC-MTDC power flow is carried out. For this analysis, DC slack bus control is assumed ('V_{DC1}' is specified).

TABLE 5.1

Study of IEEE 300-Bus System with 3-Terminal VSC-HVDC Network Incorporating IDCPFC in DC Current Control Mode

AC-MTDC power flow without any IDCPFC (VSCs connected to AC buses 266, 270 and 271)

$P_{sh2} = 0.3$; $Q_{sh2} = 0.1$; $P_{sh3} = 0.2$; $Q_{sh3} = 0.09$; $V_{DC1} = 3.0$; $V_{DC2} = 2.9991$;

$V_{DC3} = 2.9992$; $I_{DC12} = 0.093$; NI=6;

AC-MTDC Power Flow with IDCPFC

Given Quantities		AC Buses	VSCs	
			Solution	
V_{266}	1.02	$V_{270} = 1.0202$;	$\theta_{sh1} = -14.2881$; $\theta_{sh2} = -9.7713$;	
V_{DC1}	3.0	$V_{271} = 1.0619$;	$\theta_{sh3} = -11.6866$; $V_{DC2} = 2.9991$;	
P_{sh2}	0.3	$\theta_{266} = -11.3851$;	$V_{DC3} = 2.9989$; $m_1 = 0.9894$;	
Q_{sh2}	0.1	$\theta_{270} = -11.4005$;	$m_2 = 0.9721$; $m_3 = 1.0099$;	
P_{sh3}	0.2	$\theta_{271} = -12.6900$;		
Q_{sh3}	0.09		**IDCPFC**	
			$V_{DCs1} = -0.000254$; $V_{DCs2} = 0.00056$;	
			$I_{DC13} = 0.0548$; $P_{DC12} = 0.36$;	
IDCPFC			$P_{DC13} = 0.1645$;	
I_{DC12}	0.12		DC power	Converter loss (%)
			$P_{DC1} = 0.5245$;	$P_{loss1} = 1.44$;
			$P_{DC2} = -0.3124$;	$P_{loss2} = 1.23$;
			$P_{DC3} = -0.2118$;	$P_{loss3} = 1.18$;
			NI=6;	

The power-flow solution is shown in the first row of Table 5.1, with the line current in the DC link between DC buses 1 and 2 computed to be 0.093 p.u. Subsequently, an IDCPFC (having two variable DC voltage sources) is incorporated in the three-terminal VSC-MTDC grid integrated with the IEEE 300-bus test network and the AC-MTDC power flow is again carried out. The IDCPFC is used to control the line current in the DC link between DC buses 1 and 2 to a specified value of 0.12 p.u. (the DC link current without any IDCPFC is 0.093 p.u.). The specified quantities for this study are shown in rows 3–11 and columns 1–2 of Table 5.1. The power-flow solution with the IDCPFC is shown in rows 3–11 and columns 3–4 of Table 5.1.

Subsequently, a study is conducted on the same AC-MTDC network but with the VSCs now connected to AC buses 268, 272 and 273. The AC-MTDC power-flow solution without any IDCPFC is shown in the first row of Table 5.2, with the current in the DC link between the DC buses 1 and 3 computed to be 0.1098 p.u. Subsequently, an IDCPFC (having two variable DC voltage sources) is again incorporated in the three-terminal VSC-MTDC grid integrated with the IEEE 300-bus test network and the AC-MTDC power flow is again carried out. The IDCPFC maintains the DC current between DC buses 1 and 3 to a value of 0.12 p.u. (the DC link current without any IDCPFC is 0.1098 p.u.). The specified quantities for this study are shown in rows 3–11 and columns 1–2 of Table 5.2. The AC-MTDC power-flow solution with the IDCPFC is shown in rows 3–11 and columns 3–4 of Table 5.2.

TABLE 5.2

Study of IEEE 300-Bus System with 3-Terminal VSC-HVDC Network Incorporating IDCPFC in DC Current Control Mode

AC-MTDC power flow without any IDCPFC (VSCs connected to AC buses 268, 272 and 273)

$P_{sh2} = 0.35$; $Q_{sh2} = 0.1$; $P_{sh3} = 0.3$; $Q_{sh3} = 0.1$; $V_{DC1} = 3.0$; $V_{DC2} = 2.9988$; $V_{268} = 0.98$; $V_{DC3} = 2.9989$; $I_{DC13} = 0.1098$; NI=6;

AC-MTDC Power Flow with IDCPFC

Given Quantities		Solution	
		AC Buses	VSCs
V_{268}	0.98	$V_{272} = 1.0086$;	$\theta_{sh1} = -43.2336$; $\theta_{sh2} = -29.2839$;
V_{DC1}	3.0	$V_{273} = 1.0366$;	$\theta_{sh3} = -5.1793$; $V_{DC2} = 2.9988$;
P_{sh2}	0.35	$\theta_{268} = -39.1739$;	$V_{DC3} = 2.9989$; $m_1 = 0.9418$;
Q_{sh2}	0.1	$\theta_{272} = -31.2289$;	$m_2 = 0.9615$; $m_3 = 0.9874$;
P_{sh3}	0.3	$\theta_{273} = -6.7581$;	
Q_{sh3}	0.1		IDCPFC
			$V_{DCs1} = 0.00016$; $V_{DCs2} = -0.00014$;
IDCPFC			$I_{DC12} = 0.0815$; $I_{DC13} = 0.0933$; $P_{DC12} = 0.2444$;
I_{DC13}	0.12		DC power $\qquad$ Converter loss (%)
			$P_{DC1} = 0.6754$; $\qquad$ $P_{loss1} = 1.54$;
			$P_{DC2} = -0.3628$; $\qquad$ $P_{loss2} = 1.26$;
			$P_{DC3} = -0.3124$; $\qquad$ $P_{loss3} = 1.23$;
			NI=6;

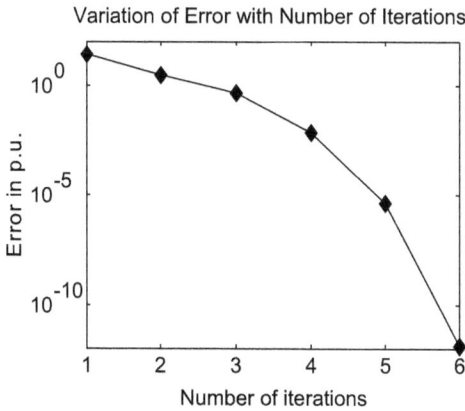

FIGURE 5.4 Convergence characteristic for the base case power flow in IEEE 300-bus system.

The convergence characteristics corresponding to the base case, study of row 1 of Table 5.1 (without IDCPFC) and rows 3–5 of Table 5.1 (with IDCPFC) are shown in Figures 5.4–5.6, respectively.

Variation of Error with Number of Iterations

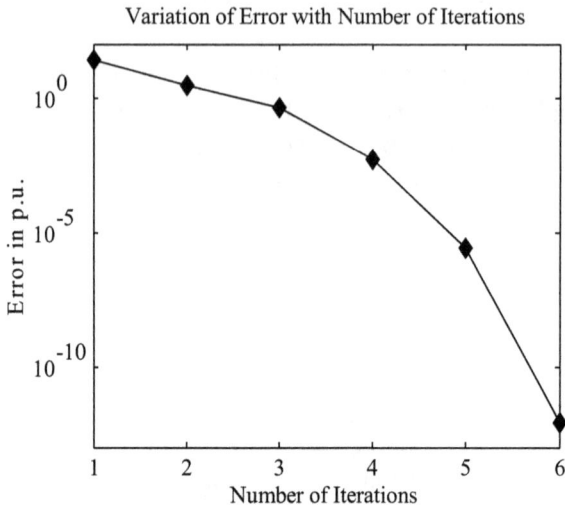

FIGURE 5.5   Convergence characteristic for the case study of row 1 in Table 5.1.

Variation of Error with Number of Iterations

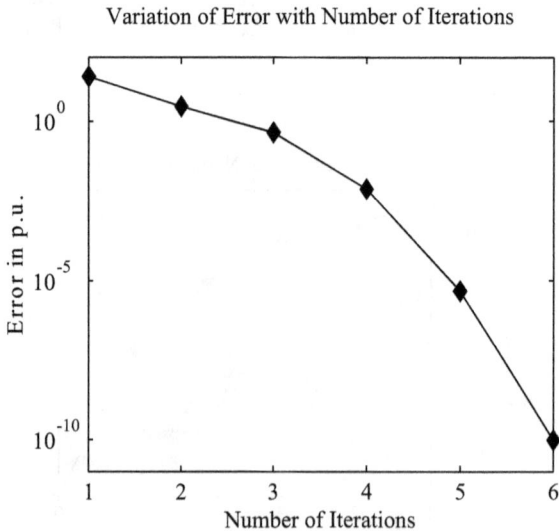

FIGURE 5.6   Convergence characteristic for the case study of rows 3–11 in Table 5.1.

Similarly, the convergence characteristic plots for the study of row 1 of Table 5.2 (without IDCPFC) and rows 3–5 of Table 5.2 (with IDCPFC) are shown in Figures 5.7 and 5.8, respectively. From Figures 5.4–5.8, it is observed that the AC-MTDC power-flow solutions with the IDCPFC demonstrate quadratic convergence characteristics, similar to the base case power-flow. Also, the convergence

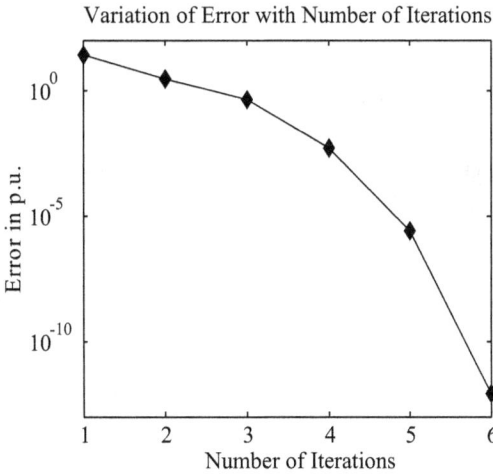

FIGURE 5.7   Convergence characteristic for the case study of row 1 in Table 5.2.

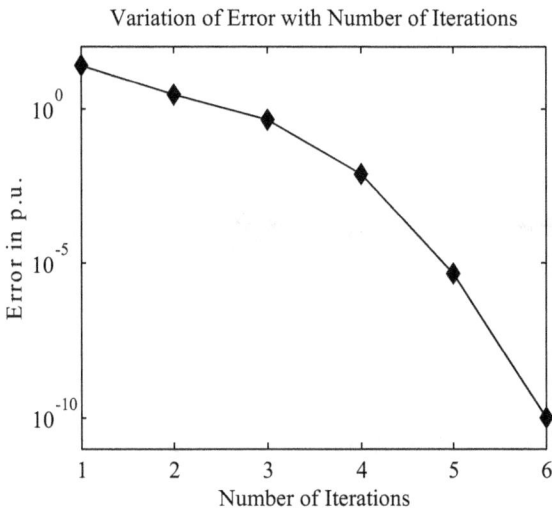

FIGURE 5.8   Convergence characteristic for the case study of rows 3–11 in Table 5.2.

pattern is independent of the MTDC grid location and the IDCPFC operating point specifications.

The bus voltage profiles for the studies of Tables 5.1 and 5.2 are depicted in Figures 5.9 and 5.10, respectively. From Figures 5.9 and 5.10, it is observed that the bus voltage profiles of the AC-MTDC system including IDCPFC do not alter much from that of the base case except the AC buses at which the VSCs are connected.

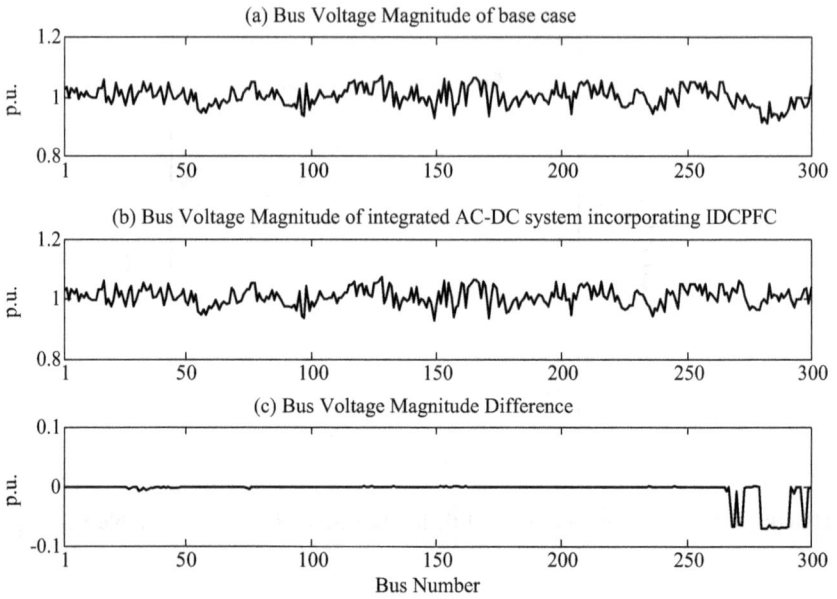

FIGURE 5.9   Bus voltage profile for the study of Table 5.1.

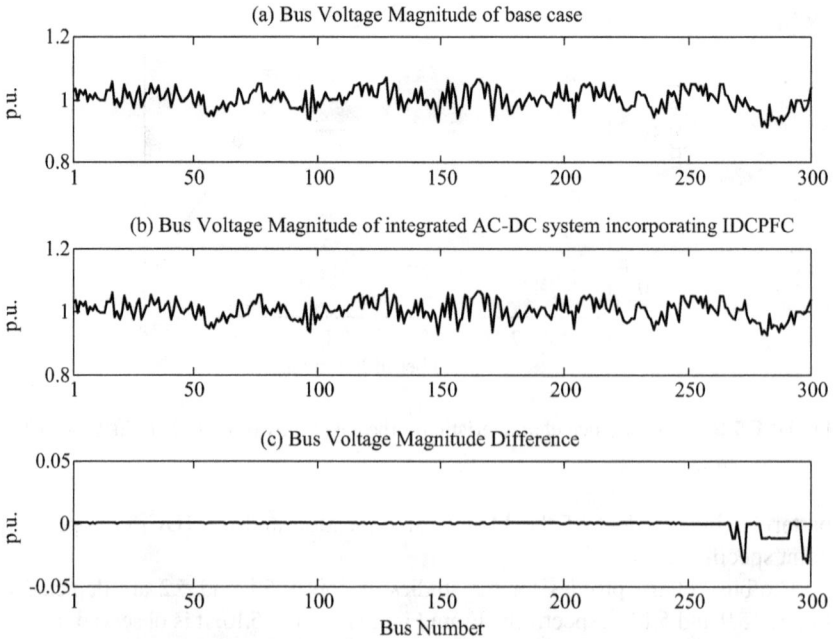

FIGURE 5.10   Bus voltage profile for the study of Table 5.2.

TABLE 5.3

Study of IEEE 300-Bus System with 3-Terminal VSC-HVDC Network Incorporating IDCPFC in DC Power Control Mode

AC-MTDC power flow without any IDCPFC (VSCs connected to AC buses 266, 270 and 271)

$P_{sh2} = 0.3$; $Q_{sh2} = 0.1$; $P_{sh3} = 0.2$; $Q_{sh3} = 0.09$; $V_{DC1} = 3.0$; $V_{DC2} = 2.9991$;

$V_{DC3} = 2.9992$; $P_{DC13} = 0.2454$; NI=6;

AC-MTDC Power Flow with IDCPFC

Given Quantities		AC Buses	VSCs	
			Solution	
V_{266}	1.02	$V_{270} = 1.0202$;	$\theta_{sh1} = -14.2879$; $\theta_{sh2} = -9.7712$;	
V_{DC1}	3.0	$V_{271} = 1.0619$;	$\theta_{sh3} = -11.6866$; $V_{DC2} = 2.9990$;	
P_{sh2}	0.3	$\theta_{266} = -11.3850$;	$V_{DC3} = 2.9992$; $m_1 = 0.9894$;	
Q_{sh2}	0.1	$\theta_{270} = -11.4005$;	$m_2 = 0.9721$; $m_3 = 1.0097$;	
P_{sh3}	0.2	$\theta_{271} = -12.6899$;	IDCPFC	
Q_{sh3}	0.09		$V_{DCs1} = 0.00018$; $V_{DCs2} = -0.00016$;	
			$I_{DC12} = 0.116$; $I_{DC13} = 0.0933$; $P_{DC12} = 0.3481$;	
IDCPFC			DC power	Converter loss (%)
P_{DC13}	0.28		$P_{DC1} = 0.5244$;	$P_{loss1} = 1.44$;
			$P_{DC2} = -0.3124$;	$P_{loss2} = 1.23$;
				$P_{loss3} = 1.18$;
			NI=6;	

Case II: DC link power control using IDCPFC

In this study, at first, a three-terminal VSC-MTDC grid is integrated with the IEEE 300-bus test system at buses 266, 270 and 271 and the AC-MTDC power flow is carried out. For this analysis, DC slack bus control is assumed ('V_{DC1}' is specified). The power-flow solution is shown in the first row of Table 5.3, with the sending end line power in the DC link between DC buses 1 and 3 computed to be 0.2454 p.u. Subsequently, an IDCPFC (having two variable DC voltage sources) is incorporated in the three-terminal VSC-MTDC grid integrated with the IEEE 300-bus test network and the AC-MTDC power flow is again carried out. The IDCPFC is used to control the sending end power in the DC link between DC buses 1 and 3 to a specified value of 0.28 p.u. (the sending end DC link power without any IDCPFC is 0.2454 p.u.). The specified quantities are shown in rows 3–11 and columns 1–2 of Table 5.3. The AC-MTDC power-flow solution with the IDCPFC is shown in rows 3–11 and columns 3–4 of Table 5.3.

Subsequently, a study is executed on the same three-terminal AC-MTDC network but with the three VSCs connected to AC buses 268, 272 and 273. The AC-MTDC power-flow solution without any IDCPFC is shown in the first row of Table 5.4, with the sending end power flow in the DC link between the DC buses 1 and 2 computed to be 0.3461 p.u. Subsequently, an IDCPFC (having two variable DC voltage sources) is incorporated in the three-terminal VSC-MTDC grid

TABLE 5.4

Study of IEEE 300-Bus System with 3-Terminal VSC-HVDC Network Incorporating IDCPFC in DC Power Control Mode

AC-MTDC power flow without any IDCPFC (VSCs connected to AC buses 268, 272 and 273)

$P_{sh2} = 0.35$; $Q_{sh2} = 0.1$; $P_{sh3} = 0.3$; $Q_{sh3} = 0.1$; $V_{DC1} = 3.0$; $V_{DC2} = 2.9988$; $V_{268} = 0.98$;

$V_{DC3} = 2.9989$; $P_{DC12} = 0.3461$; NI=6;

AC-MTDC Power Flow with IDCPFC

Given Quantities		AC Buses	VSCs
			Solution
V_{268}	0.98	$V_{272} = 1.0086$;	$\theta_{sh1} = -43.2335$; $\theta_{sh2} = -29.2837$;
V_{DC1}	3.0	$V_{273} = 1.0366$;	$\theta_{sh3} = -5.1792$; $V_{DC2} = 2.9989$;
P_{sh2}	0.35	$\theta_{268} = -39.1738$;	$V_{DC3} = 2.9989$; $m_1 = 0.9418$;
Q_{sh2}	0.1	$\theta_{272} = -31.2288$;	$m_2 = 0.9615$; $m_3 = 0.9874$;
P_{sh3}	0.3	$\theta_{273} = -6.7581$;	
Q_{sh3}	0.1		**IDCPFC**
			$V_{DCs1} = -0.000064$; $V_{DCs2} = 0.000074$;
IDCPFC			$I_{DC12} = 0.12$; $P_{DC13} = 0.3571$; $I_{DC13} = 0.119$;
P_{DC12}	0.36		DC power Converter loss (%)
			$P_{DC1} = 0.6754$; $P_{loss1} = 1.54$;
			$P_{DC2} = -0.3628$; $P_{loss2} = 1.26$;
			$P_{DC3} = -0.3124$; $P_{loss3} = 1.23$;
			NI=6;

integrated with the IEEE 300-bus test network and the AC-MTDC power flow is again carried out. The IDCPFC is used to control the sending end power in the DC link between DC buses 1 and 2 to a specified value of 0.36 p.u. (the sending end DC link power without any IDCPFC is 0.3461 p.u.). The specified quantities are shown in rows 3–11 and columns 1–2 of Table 5.4. The AC-MTDC power-flow solution with the IDCPFC is shown in rows 3–11 and columns 3–4 of Table 5.4.

The convergence characteristics corresponding to the studies of rows 3–11 of Table 5.3 (with IDCPFC) and rows 3–11 of Table 5.4 (with IDCPFC) are shown in Figures 5.11 and 5.12, respectively. From Figures 5.11 and 5.12, it is observed that the AC-MTDC power-flow solutions with the IDCPFC demonstrate quadratic convergence characteristics, similar to the base case power flow.

The bus voltage profiles for the studies of Tables 5.3 and 5.4 are shown in Figures 5.13 and 5.14, respectively. Again, from Figures 5.13 and 5.14, it is observed that the bus voltage profiles do not change much from that of the base case except at the AC terminal buses to which the VSCs are connected.

5.5.2 STUDY OF 7-TERMINAL VSC-HVDC NETWORK INCORPORATING IDCPFC IN EUROPEAN 1354 BUS SYSTEM

Case I: DC current control by using IDCPFC

In this study, at first, the power flow of seven-terminal VSC-HVDC system incorporating in European 1354 bus system is executed. The VSC-HVDC network

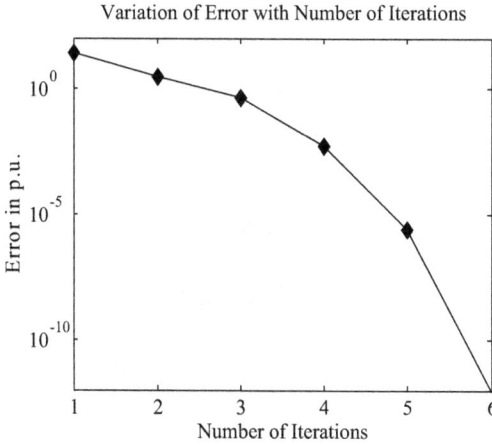

FIGURE 5.11   Convergence characteristic for the case study of rows 3–11 in Table 5.3.

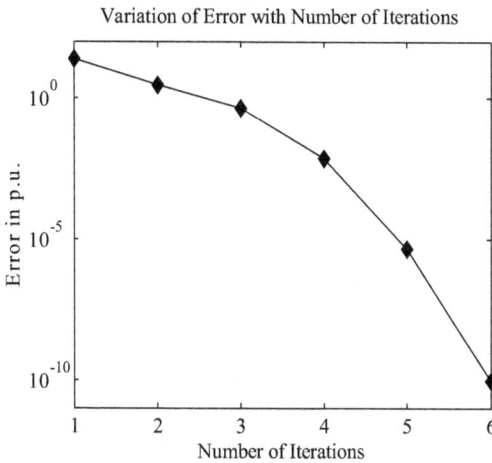

FIGURE 5.12   Convergence characteristic for the case study of rows 3–11 in Table 5.4.

is connected between AC buses 1280, 1281, 1286, 1292, 1313, 1324 and 1353. The results are given in the first row of Table 5.5. The line current in the DC link between DC buses 1 and 2 is calculated to a value of 0.0996 p.u. Then a power flow of seven-terminal DC network includes an IDCPFC with two variable DC voltage sources in European 1354-bus system are executed. The IDCPFC is used to control the line current in the DC link between DC buses 1 and 2 to a specified value of 0.11 p.u. (the DC link current without any IDCPFC is 0.0996 p.u.). For this analysis, DC slack bus control is assumed ('V_{DC1}' is specified). The specified quantities for this study are shown in rows 3–19 and columns 1–2 of Table 5.5. The power-flow results are shown in rows 3–19 and columns 3–4 of Table 5.5.

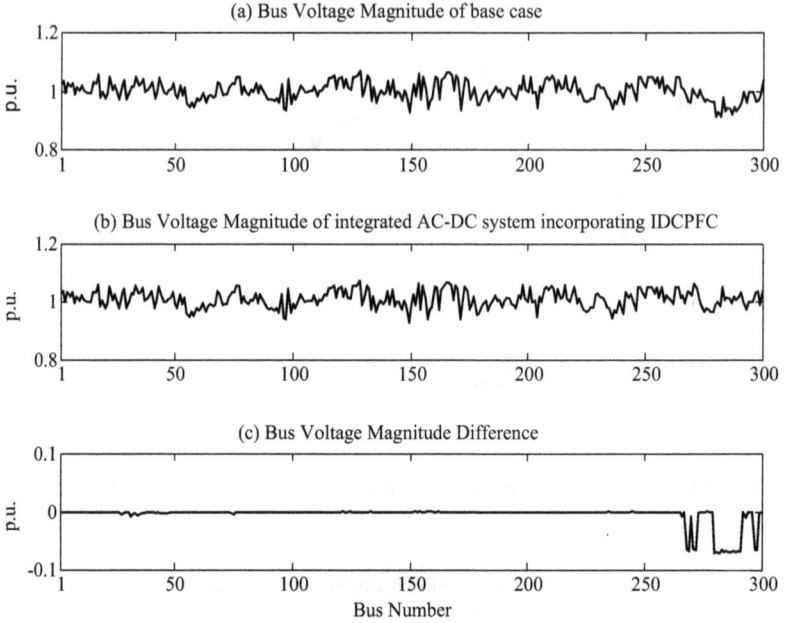

FIGURE 5.13 Bus voltage profile for the study of Table 5.3.

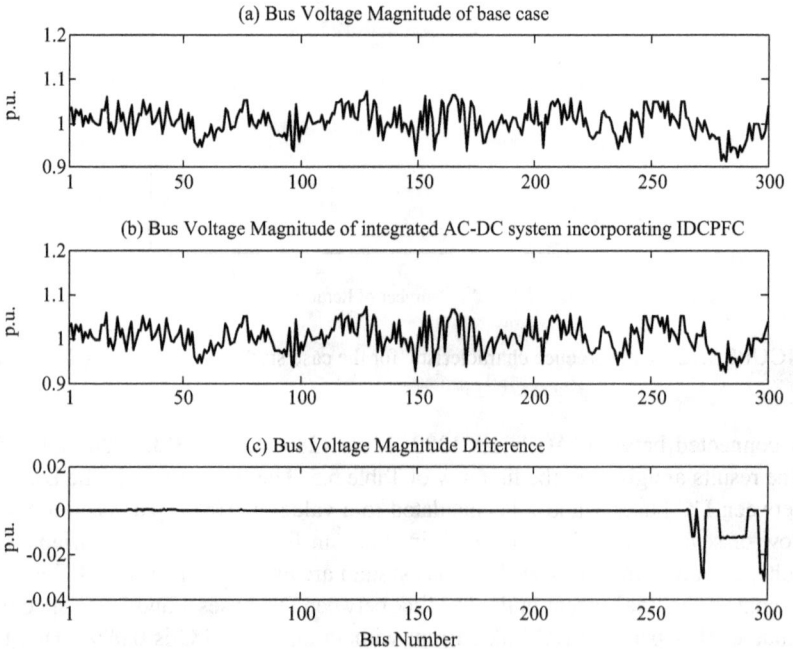

FIGURE 5.14 Bus voltage profile for the study of Table 5.4.

TABLE 5.5
Study of European 1354-Bus System with 7-Terminal VSC-HVDC Network Incorporating IDCPFC in DC Current Control Mode

Without any IDCPFC

$P_{sh2} = 0.3$; $Q_{sh2} = 0.08$; $P_{sh3} = 0.2$; $Q_{sh3} = 0.09$; $P_{sh4} = 0.3$; $Q_{sh4} = 0.08$; $P_{sh5} = 0.4$;

$Q_{sh5} = 0.07$; $P_{sh6} = 0.2$; $P_{sh7} = 0.3$;

$V_{1280} = 0.98$; $V_{1324} = 1.01$; $V_{1353} = 1.02$; $V_{DC1} = 3.0$; $V_{DC2} = 2.990$; $V_{DC3} = 2.9991$;

$V_{DC4} = 2.9990$; $V_{DC5} = 2.9990$; $V_{DC6} = 2.9991$; $V_{DC7} = 2.9990$;

$I_{DC12} = 0.0996$; NI=7;

AC-MTDC power flow with IDCPFC (VSCs connected to AC buses 1280, 1281, 1286, 1292, 1313, 1324 and 1353)

Given Quantities		Solution		
		AC Buses	VSCs	
V_{1280}	0.98	$\theta_{1280} = -42.5836$;	$\theta_{sh1} = -52.4238$; $\theta_{sh2} = -35.4367$; $\theta_{sh3} = -11.1185$;	
V_{1324}	1.01	$\theta_{1281} = -37.0197$;	$\theta_{sh4} = -33.5743$; $\theta_{sh5} = -35.0660$; $\theta_{sh6} = -27.5064$;	
V_{1353}	1.02	$\theta_{1286} = -12.2104$;	$\theta_{sh7} = -27.8653$;	
		$\theta_{1292} = -35.1696$;	$V_{DC2} = 2.9987$; $V_{DC3} = 2.9986$; $V_{DC4} = 2.9988$;	
		$\theta_{1313} = -37.1813$;	$V_{DC5} = 2.9987$; $V_{DC6} = 2.9988$; $V_{DC7} = 2.9988$;	
V_{DC1}	3.0	$\theta_{1324} = -28.5904$;	$m_1 = 1.0302$; $m_2 = 0.9856$;	
		$\theta_{1353} = -29.3645$;	$m_3 = 0.9685$; $m_4 = 0.9818$; $m_5 = 0.9847$;	
P_{sh2}	0.3	$V_{1281} = 1.0365$;	$m_6 = 0.9754$; $m_7 = 1.0333$;	
Q_{sh2}	0.08	$V_{1286} = 1.0176$;		
P_{sh3}	0.2	$V_{1292} = 1.0325$;	**IDCPFC**	
Q_{sh3}	0.09	$V_{1313} = 1.0362$;	$V_{DCs1} = 0.000158$; $V_{DCs2} = 0.0015$;	
P_{sh4}	0.3		$P_{DC13} = 0.4217$; $I_{DC13} = 0.1406$; $P_{DC12} = 0.3777$;	
Q_{sh4}	0.08		DC power	Converter loss (%)
P_{sh5}	0.4		$P_{DC1} = 1.7797$;	$P_{loss1} = 3.6313$;
Q_{sh5}	0.07		$P_{DC2} = -0.3124$;	$P_{loss2} = 1.2285$;
P_{sh6}	0.2		$P_{DC3} = -0.2119$;	$P_{loss3} = 1.1846$;
P_{sh7}	0.3		$P_{DC4} = -0.3124$;	$P_{loss4} = 1.2291$;
IDCPFC			$P_{DC5} = -0.4130$;	$P_{loss5} = 1.2836$;
			$P_{DC6} = -0.2124$;	$P_{loss6} = 1.2338$;
I_{DC12}	0.11		$P_{DC7} = -0.3168$;	$P_{loss7} = 1.6198$;

IN=7;

Case II: DC power control by using IDCPFC

This study is similar to the study of Table 5.5, but the line power in DC link between buses 1 and 3 is specified instead of DC line current. First, the power flow of seven-terminal VSC-HVDC system incorporating in European 1354-bus system is executed without any IDCPFC and the line power in DC link between buses 1 and 3 is calculated to a value of 0.2454 p.u. The results are given in first row of Table 5.6. The IDCPFC is used to control the line power in the DC link between DC buses 1 and 3 to a specified value of 0.28 p.u. (the DC link power without any IDCPFC is 0.2454 p.u.). For ease of analysis, the first converter acts

TABLE 5.6

Study of European 1354-Bus System with 7-Terminal VSC-HVDC Network Incorporating IDCPFC in DC Power Control Mode

Without any IDCPFC

$P_{sh2} = 0.3$; $Q_{sh2} = 0.08$; $P_{sh3} = 0.2$; $Q_{sh3} = 0.09$; $P_{sh4} = 0.3$; $Q_{sh4} = 0.08$; $P_{sh5} = 0.4$;

$\quad Q_{sh5} = 0.07$; $P_{sh6} = 0.2$; $P_{sh7} = 0.3$;

$V_{1280} = 0.98$; $V_{1324} = 1.01$; $V_{1353} = 1.02$; $V_{DC1} = 3.0$; $V_{DC2} = 2.990$; $V_{DC3} = 2.9991$;

$V_{DC4} = 2.9990$; $V_{DC5} = 2.9990$; $V_{DC6} = 2.9991$; $V_{DC7} = 2.9990$;

$P_{DC13} = 0.2845$; NI=7;

AC-MTDC power flow with IDCPFC (VSCs connected to AC buses 1280, 1281, 1286, 1292, 1313, 1324 and 1353)

Given Quantities		AC Buses	VSCs
			Solution
V_{1280}	0.98	$\theta_{1280} = -42.5836$;	$\theta_{sh1} = -52.4238$; $\theta_{sh2} = -35.4367$; $\theta_{sh3} = -11.1185$;
V_{1324}	1.01	$\theta_{1281} = -37.0197$;	$\theta_{sh4} = -33.5743$; $\theta_{sh5} = -35.0660$; $\theta_{sh6} = -27.5064$;
V_{135}	1.02	$\theta_{1286} = -12.2104$;	$\theta_{sh7} = -27.8653$;
		$\theta_{1292} = -35.1696$;	$V_{DC2} = 2.9986$; $V_{DC3} = 2.9988$; $V_{DC4} = 2.9988$;
		$\theta_{1313} = -37.1813$;	$V_{DC5} = 2.9987$; $V_{DC6} = 2.9988$; $V_{DC7} = 2.9988$;
		$\theta_{1324} = -28.5904$;	$m_1 = 1.0302$; $m_2 = 0.9856$;
V_{DC1}	3.0	$\theta_{1353} = -29.3645$;	
		$V_{1281} = 1.0365$;	$m_3 = 0.9685$; $m_4 = 0.9818$; $m_5 = 0.9847$;
		$V_{1286} = 1.0176$;	$m_6 = 0.9754$; $m_7 = 1.0333$;
P_{sh2}	0.3	$V_{1292} = 1.0325$;	
Q_{sh2}	0.08	$V_{1313} = 1.0362$;	**IDCPFC**
P_{sh3}	0.2		$V_{DCs1} = 0.0015$; $V_{DCs2} = 0.00008$;
Q_{sh3}	0.09		$I_{DC12} = -0.0065$; $I_{DC13} = 0.11$; $P_{DC12} = -0.0194$;
P_{sh4}	0.3		
Q_{sh4}	0.08		**DC power**	**Converter loss (%)**
P_{sh5}	0.4		$P_{DC1} = 1.7797$;	$P_{loss1} = 3.6313$;
Q_{sh5}	0.07		$P_{DC2} = -0.3124$;	$P_{loss2} = 1.2285$;
P_{sh6}	0.2		$P_{DC3} = -0.2119$;	$P_{loss3} = 1.1846$;
P_{sh7}	0.3		$P_{DC4} = -0.3124$;	$P_{loss4} = 1.2291$;
IDCPFC			$P_{DC5} = -0.4130$;	$P_{loss5} = 1.2836$;
			$P_{DC6} = -0.2124$;	$P_{loss6} = 1.2338$;
P_{DC13}	0.33		$P_{DC7} = -0.3168$;	$P_{loss7} = 1.6198$;
			IN=7;

as a master converter. The specified quantities for this study are shown in rows 3–19 and columns 1–2 of Table 5.6. The power-flow results are shown in rows 3–19 and columns 3–4 of Table 5.6.

The convergence characteristic plots for the study of base case, row 1 of Table 5.5 (without IDCPFC), rows 3–19 of Table 5.5 (DC link current control) and rows 3–19 of Table 5.6 (DC link power control) are shown in Figures 5.15–5.18, respectively. From Figures 5.15–5.18, it is observed that the convergence characteristics including IDCPFC are similar to the base case quadratic convergence characteristics. The bus voltage profiles for the studies of Tables 5.5 and 5.6 are

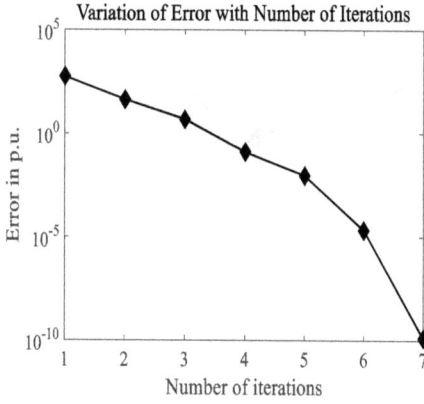

FIGURE 5.15   Convergence characteristic of the base case power flow in the European 1354-bus system.

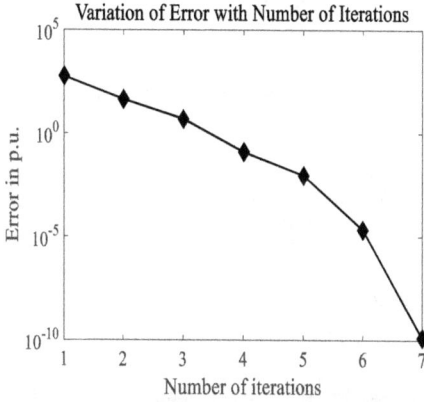

FIGURE 5.16   Convergence characteristic for the study of first row in Table 5.5.

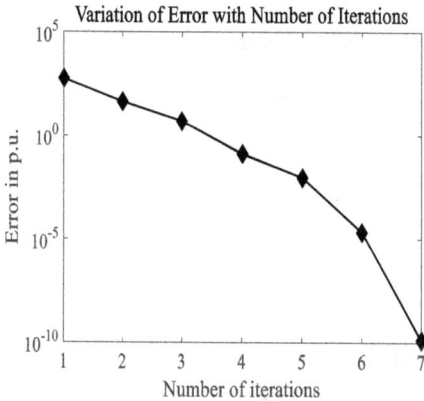

FIGURE 5.17   Convergence characteristic for the study of 3–19 rows in Table 5.5.

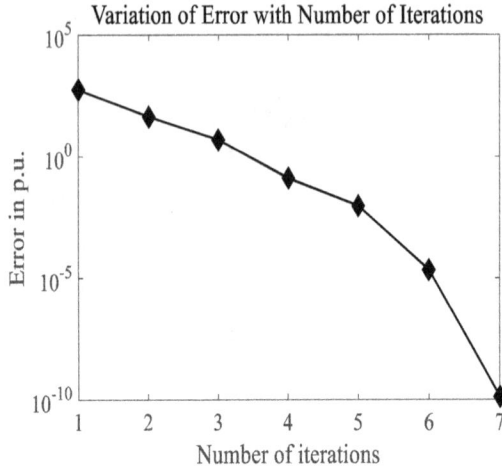

FIGURE 5.18   Convergence characteristic for the study of rows 3–19 in Table 5.6.

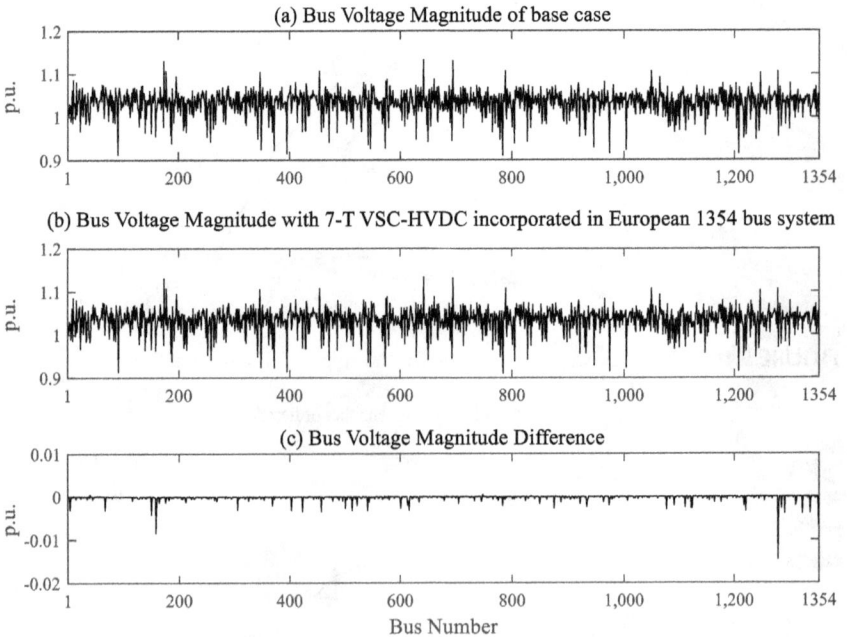

FIGURE 5.19   Bus voltage profile for the study of Table 5.5.

depicted in Figures 5.19 and 5.20, respectively. From Figures 5.19 and 5.20, it is observed that the voltage profile of the network including IDCPFC does not alter much from base case except the terminals of converter-connected AC buses.

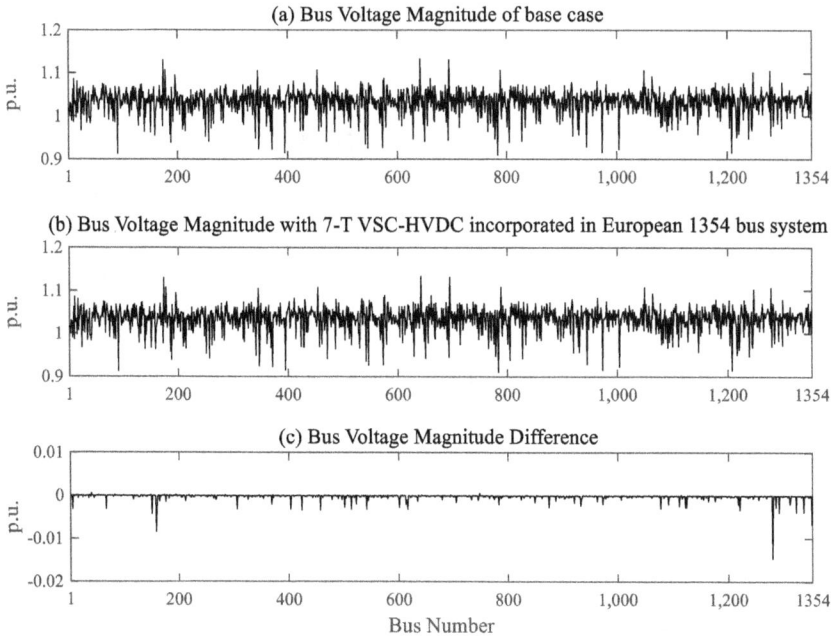

FIGURE 5.20   Bus voltage profile for the study of Table 5.6.

5.6   SUMMARY

This chapter presents a unified Newton power-flow model of VSC-based integrated AC–DC systems incorporating IDCPFCs. The IDCPFC is a DC power-flow controller used for the power-flow management of DC grids. The proposed model has been implemented with IDCPFC in three-terminal MTDC grids integrated with the IEEE 300-bus test network. Also, a seven-terminal MTDC grid integrated with the European 1354-bus test network is considered for the implementation of proposed model. The convergence characteristics validate the model.

SUMMARY

This chapter ...

6 Power-Flow Modelling of AC Power Systems Integrated with VSC-Based Multi-Terminal DC (AC-MVDC) Grids Incorporating Renewable Energy Sources

6.1 INTRODUCTION

Due to the ever-increasing demand of electrical energy amid environmental concerns, both the industry and the academia have been focusing on the harnessing of renewable energy sources, particularly the integration of offshore renewable energy sources [19–21]. In order to integrate such remotely located energy sources with the AC grid, multi-terminal VSC-based HVDC (MVDC) systems have been proposed over conventional HVDC which can connect different offshore stations at the same voltage or at different voltages.

Now, for planning, operation and control of AC-MVDC systems integrated with renewable energy sources, their power-flow models are required. As already discussed in the previous chapters, MVDC systems can employ DC slack-bus control or DC voltage droop control for operation. A power-flow model should include both these control strategies for completeness.

6.2 MODELLING OF AC-MVDC SYSTEMS INCORPORATING RENEWABLE ENERGY SOURCES

The assumptions adopted for modelling are similar to those in Chapter 3 of this book. Figure 6.1 shows a 'p' bus HVDC grid integrated with an 'n' bus AC power system network. The interfacing of the HVDC grid with the AC network takes place at 'q' (q≤p) AC buses through 'q' VSCs and their respective converter transformers. The DC sides of these 'q' VSCs are connected in the PTP configuration and constitute part of the larger 'p' bus DC grid. The rest of the '(p–q)'

DOI: 10.1201/9781003252078-6

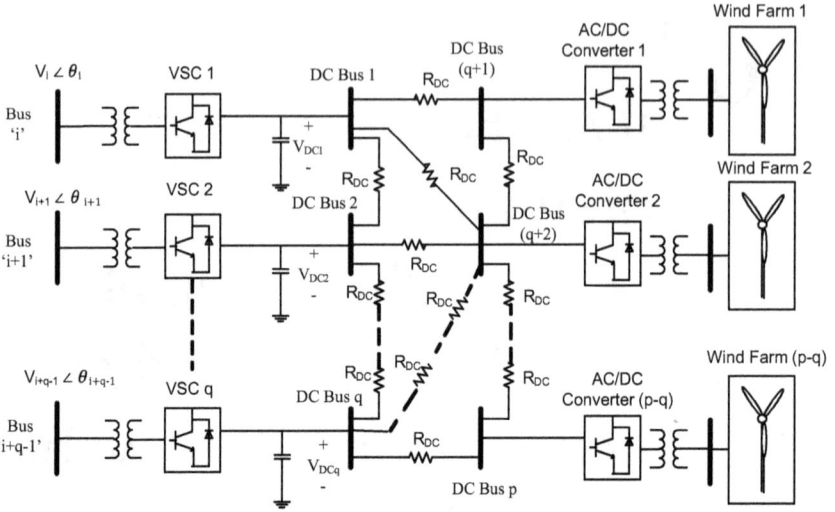

FIGURE 6.1 Schematic diagram of hybrid AC-MVDC system with offshore wind farms.

DC buses are appropriately interfaced (through AC/DC converters) with offshore wind farms (OWFs). Without loss of generality, it is assumed that the 'q' VSCs are connected to AC buses 'i', '(i+1)', and so on, up to bus '(i+q−1)', through 'q' converter transformers and the '(p−q)' OWFs are interfaced with DC buses '(q+1)', '(q+2)', and so on, up to DC bus 'p'.

Figure 6.2 shows the equivalent circuit for the network shown in Figure 6.1. In Figure 6.2, each of the 'q' VSCs is represented as a fundamental frequency, positive sequence voltage source. Thus, V_{sha} represents the voltage phasor pertaining to the ath ($1 \le a \le q$) VSC. Each converter transformer is represented by its leakage impedance. The ath ($1 \le a \le q$) VSC is connected to AC bus '(i+a−1)' whose voltage is represented by the phasor $V_{i+a-1} = V_{i+a-1} \angle \theta_{i+a-1}$, through the ath converter transformer. All the '(p−q)' DC buses with renewable energy sources are represented as DC power injections.

In Figure 6.2, let R_{sha} and X_{sha} be the resistance and the leakage reactance of the converter transformer of the ath ($1 \le a \le q$) VSC, respectively. Also, let $y_{sha} = 1/Z_{sha}$, where $Z_{sha} = R_{sha} + jX_{sha}$. Then, from Figure 6.2, the current in the link (not shown) connecting the ath VSC to its AC terminal bus '(i+a−1)' is

$$I_{sha} = y_{sha} \left(V_{sha} - V_{i+a-1} \right) \tag{6.1}$$

where $V_{sha} = V_{sha} \angle \theta_{sha} = m_a c V_{DCa} \angle \theta_{sha}$. In the above equation, 'm_a' and 'V_{DCa}' are the modulation index and the DC side voltage of the ath VSC, respectively, while 'θ_{sha}' is the phase angle of V_{sha}. 'c' is a constant which depends on the VSC architecture [11].

Again, from Figure 6.2, the net current injection at the AC bus '(i+a−1)' connected to the ath ($1 \le a \le q$) VSC can be written as

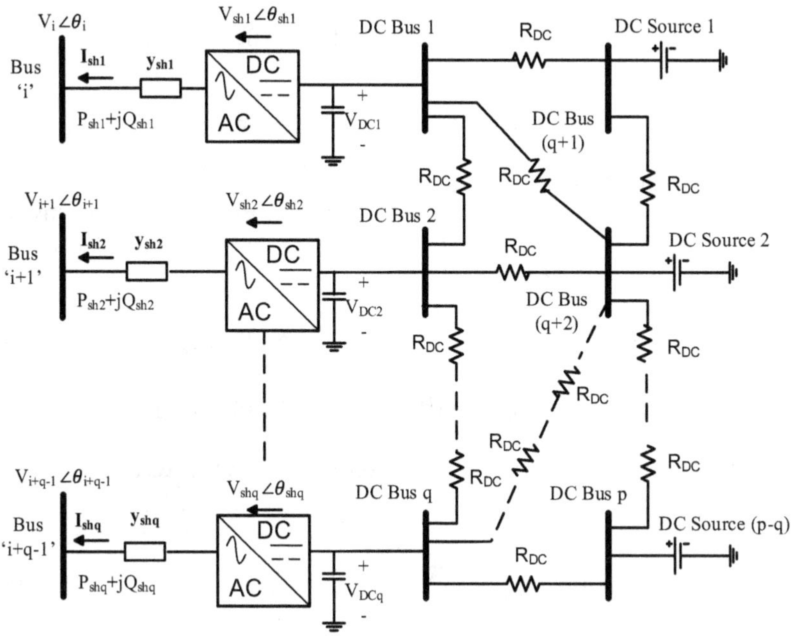

FIGURE 6.2   Equivalent circuit of hybrid AC-MVDC system with offshore wind farms.

$$I_{i+a-1} = \sum_{k=1}^{n} Y_{(i+a-1)k} V_k - y_{sha} V_{sha} \qquad (6.2)$$

where $Y_{(i+a-1)(i+a-1)} = Y^{old}_{(i+a-1)(i+a-1)} + y_{sha}$ and $Y^{old}_{(i+a-1)(i+a-1)} = y_{(i+a-1)0} + \sum_{k=1, k \neq i+a-1}^{n} y_{(i+a-1)k}$

are the values of self-admittances for the bus '$(i+a-1)$' with the ath VSC connected and in the original 'n' bus AC system without any VSC, respectively.

6.3 POWER-FLOW EQUATIONS OF INTEGRATED AC-MVDC SYSTEMS WITH RENEWABLE ENERGY SOURCES

Now, from Figure 6.2, at the AC bus '$(i+a-1)$' pertaining to the ath VSC, it can be shown that the net active and reactive power injections are

$$P_{i+a-1} = \sum_{k=1}^{n} V_{i+a-1} V_k Y_{(i+a-1)k} \cos\left[\theta_{i+a-1} - \theta_k - \phi_{(i+a-1)k}\right]$$
$$- m_a c V_{DCa} V_{i+a-1} y_{sha} \cos\left(\theta_{i+a-1} - \theta_{sha} - \phi_{sha}\right) \qquad (6.3)$$

$$Q_{i+a-1} = \sum_{k=1}^{n} V_{i+a-1} V_k Y_{(i+a-1)k} \sin\left[\theta_{i+a-1} - \theta_k - \phi_{(i+a-1)k}\right]$$

$$- m_a c V_{DCa} V_{i+a-1} y_{sha} \sin\left(\theta_{i+a-1} - \theta_{sha} - \phi_{sha}\right)$$

(6.4)

where 'ϕ_{sha}' is the phase angle of y_{sha}.

In addition, from Figure 6.2, it can be shown using Eq. (6.2) that the active and reactive power flows at the terminal end of the link interconnecting the ath VSC to the AC bus '$(i+a-1)$' are

$$P_{sha} = m_a c\, V_{DCa} V_{i+a-1} y_{sha} \cos\left(\theta_{i+a-1} - \theta_{sha} - \phi_{sha}\right) - V_{i+a-1}^2 y_{sha} \cos\phi_{sha} \quad (6.5)$$

$$Q_{sha} = m_a c\, V_{DCa} V_{i+a-1} y_{sha} \sin\left(\theta_{i+a-1} - \theta_{sha} - \phi_{sha}\right) + V_{i+a-1}^2 y_{sha} \sin\phi_{sha} \quad (6.6)$$

Also, from Figure 6.2, by virtue of the power balance on the AC and DC sides of the ath VSC,

$$\mathrm{Re}\left(V_{sha}\, I_{sha}^*\right) + \sum_{v=1}^{p} V_{DCa} V_{DCv} Y_{DCav} = -P_{lossa} \quad (6.7)$$

Substitution of Eq. (6.1) in Eq. (4.7) gives

$$\left(m_a c\, V_{DCa}\right)^2 y_{sha} \cos\phi_{sha} - m_a c\, V_{DCa} V_{i+a-1} y_{sha} \cos\left(\theta_{sha} - \theta_{i+a-1} - \phi_{sha}\right)$$

$$+ \sum_{v=1}^{p} V_{DCa} V_{DCv} Y_{DCav} + P_{lossa} = 0$$

$$\text{or,}\quad f_{1a} = 0 \quad \forall\ 1 \le a \le q \quad (6.8)$$

Thus, for 'q' VSCs, 'q' independent equations are obtained.

In Eqs. (6.7) and (6.8), $Y_{DCav} = -\dfrac{1}{R_{DCav}}$, where '$R_{DCav}$' is the resistance of the DC link between DC buses 'a' and 'v'. Also, 'P_{lossa}' represents the losses of the ath VSC as already detailed in Chapter 3 and is again given below:

$$P_{lossa} = a_1 + b_1 I_{sha} + c_1 I_{sha}^2 \quad (6.9)$$

where 'a_1', 'b_1' and 'c_1' are loss factors [18,102] and

$$I_{sha} = y_{sha}\left[V_{i+a-1}^2 + \left(m_a c\, V_{DCa}\right)^2 - 2 V_{i+a-1} m_a c V_{DCa} \cos\left(\theta_{i+a-1} - \theta_{sha}\right)\right]^{1/2} \quad (6.10)$$

The derivation of Eq. (6.10) is given in Appendix A.

Now, in the AC-MTDC system (Figure 6.2) with 'q' VSCs, if it is assumed that the rth ($1 \le r \le q$) VSC is used for voltage control of its corresponding AC bus, we have

$$V^{sp}_{i+a-1} - V^{cal}_{i+a-1} = 0 \qquad \forall a, 1 \le a \le q, \quad a = r \qquad (6.11)$$

Also, not more than '(q–1)' line active and reactive power flows {Eqs. (6.5) and (6.6)} can be specified, which give us '(2q–2)' independent equations given as

$$P^{sp}_{sha} - P^{cal}_{sha} = 0 \qquad (6.12)$$

$$Q^{sp}_{sha} - Q^{cal}_{sha} = 0 \qquad (6.13)$$

$$\forall a, 1 \le a \le q, \quad a \ne r.$$

Instead of PQ control mode, if a VSC operates in the PV one, Eq. (6.13) changes to

$$V^{sp}_{i+a-1} - V^{cal}_{i+a-1} = 0 \qquad \forall a, 1 \le a \le q, \quad a \ne r \qquad (6.14)$$

Further, the net reactive power injection at AC bus '(i+r–1)' can be specified as its voltage is controlled by the rth VSC. Thus, we get

$$Q^{sp}_{i+a-1} - Q^{cal}_{i+a-1} = 0 \qquad \forall a, 1 \le a \le q, \quad a = r \qquad (6.15)$$

In Eqs. (6.11)–(6.15), V^{sp}_{i+a-1}, Q^{sp}_{i+a-1}, P^{sp}_{sha} and Q^{sp}_{sha} are specified values, while V^{cal}_{i+a-1}, Q^{cal}_{i+a-1}, P^{cal}_{sha} and Q^{cal}_{sha} are calculated values {using Eqs. (6.4)–(6.6)}.

6.4 MODELLING OF INTEGRATED AC-MVDC SYSTEMS WITH RENEWABLE ENERGY SOURCES EMPLOYING DC SLACK-BUS CONTROL

In Figure 6.1, if it is assumed that 'g' generators are connected at the first 'g' buses of the 'n' bus AC system with bus 1 being the slack bus, then the Newton-Raphson power-flow model of the 'n' bus AC system integrated with a 'p' terminal DC network with '(p–q)' offshore wind farms is written as

Unified method employing DC slack-bus control

Compute: θ, **V**, **X**

Given: **P, Q, R**

with

$$\theta = [\theta_2 \cdots \theta_n]^T, \ \mathbf{V} = \left[V_{g+1} \cdots V_n \right]^T, \ \mathbf{\theta_{sh}} = \left[\theta_{sh1} \cdots \theta_{shq} \right]^T, \ \mathbf{m} = \left[m_1 \cdots m_q \right]^T,$$

$$\mathbf{V_{DC}} = \left[V_{DC2} \cdots V_{DCp} \right]^T$$

$$\mathbf{X} = \begin{bmatrix} \boldsymbol{\theta}_{sh}^T & \mathbf{m}^T & \mathbf{V}_{DC}^T \end{bmatrix}^T$$

$$\mathbf{P} = \begin{bmatrix} P_2 \cdots P_n \end{bmatrix}^T, \mathbf{Q} = \begin{bmatrix} Q_{g+1} \cdots Q_n \end{bmatrix}^T, \mathbf{P}_{sh} = \begin{bmatrix} P_{sh2} \cdots P_{shq} \end{bmatrix}, \mathbf{Q}_{sh} = \begin{bmatrix} Q_{sh2} \cdots Q_{shq} \end{bmatrix}$$

$$\mathbf{f}_1 = \begin{bmatrix} f_{11} \cdots f_{1q} \end{bmatrix}, \mathbf{P}_{DCWF} = \begin{bmatrix} P_{DCWF1} \cdots P_{DCWF(p-q)} \end{bmatrix}, \mathbf{R} = \begin{bmatrix} \mathbf{P}_{sh} & \mathbf{Q}_{sh} & V_{i+r-1} & \mathbf{f}_1 & \mathbf{P}_{DCWF} \end{bmatrix}^T$$

In this model, DC slack-bus control is assumed with the master VSC 'r' controlling the voltage magnitude of its AC terminal bus '(i+r–1)' unlike the other '(q–1)' slave VSCs which control the line active and reactive power flows.

The Newton power-flow equation is

$$\mathbf{J} \begin{bmatrix} \Delta\boldsymbol{\theta}^T & \Delta\mathbf{V}^T & \Delta\boldsymbol{\theta}_{sh}^T & \Delta\mathbf{m}^T & \Delta\mathbf{V}_{DC}^T \end{bmatrix}^T = \begin{bmatrix} \Delta\mathbf{P}^T & \Delta\mathbf{Q}^T & \Delta\mathbf{R}^T \end{bmatrix}^T \qquad (6.16)$$

where $\mathbf{J}$ is the power-flow Jacobian.

In Eq. (6.16), '$\Delta\mathbf{P}$', '$\Delta\mathbf{Q}$' and '$\Delta\mathbf{R}$' represent the mismatch vectors while $\Delta\boldsymbol{\theta}$, $\Delta\mathbf{V}$, $\Delta\boldsymbol{\theta}_{sh}$, $\Delta\mathbf{m}$ and $\Delta\mathbf{V}_{DC}$ represent the correction vectors.

Sequential method employing DC slack-bus control

In sequential method employing DC slack-bus control (already described in Chapter 3), the effect of the 'q' VSCs represented as 'q' equivalent loads on the secondary sides of the converter transformers. This is shown in Figure 6.3. The secondaries of the 'q' converter transformers are shown connected to 'q' fictitious

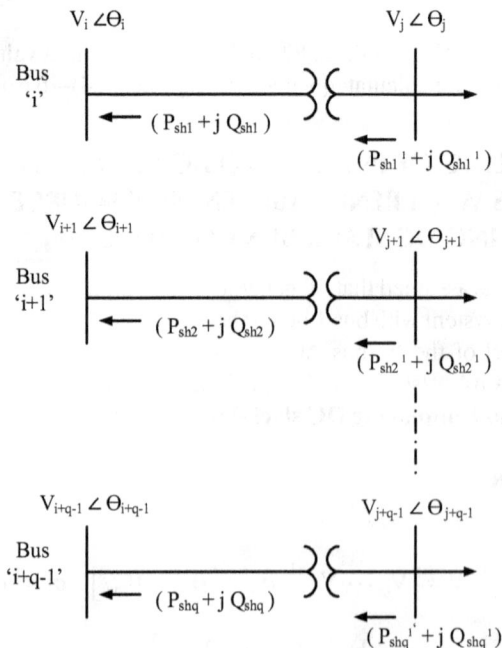

FIGURE 6.3 Representation of VSCs as equivalent complex load powers.

AC buses 'j', '(j+1)', and so on, up to bus '(j+q-1)', as shown in the figure. The effect of the 'q' VSCs are represented as 'q' equivalent complex loads at these fictitious AC buses 'j' to '(j+q-1)'. In this respect, it may be noted from Figure 6.3 that '(q-1)' complex powers ('S_{sh2}' to 'S_{shq}') are specified only in the terminal ends of the lines connected to their AC terminal buses {buses '(i+1)' to '(i+q-1)'}. Let $S_{sha} = P_{sha} + jQ_{sha}$ ($2 \leq a \leq q$) be the complex power specified in the terminal end of the line connected to the AC bus '(i+a-1)' through the ath coupling transformer. Then, from Figure 6.3, the active component of the equivalent complex load power at the fictitious AC bus '(j+a-1)' {which represents the effect of the ath VSC} can be expressed as

$$P_{sha}' = P_{sha} + I_{sha}^2 \, R_{sha} = P_{sha}^{sp} + \left(\frac{P_{sha}^{sp\,2} + Q_{sha}^{sp\,2}}{V_{i+a-1}^2} \right) R_{sha} \tag{6.17}$$

where P_{sha}^{sp} is the specified active power in the terminal end of the line connected to the AC bus '(i+a-1)'.

But from Figure 6.2, the active power delivered by the ath ($1 \leq a \leq q$) VSC at its AC terminal is

$$P_{sha}' = Re\left(V_{sha} \, I_{sha}^* \right) = m_a^2 \, c^2 \, V_{DCa}^2 \, y_{sha} \, \cos\phi_{sha}$$
$$- m_a c V_{DCa} V_{i+a-1} y_{sha} \cos\left(\theta_{sha} - \theta_{i+a-1} - \phi_{sha} \right) \tag{6.18}$$

Substituting Eq. (6.17) in (6.18), we get

$$m_a^2 \, c^2 \, V_{DCa}^2 \, y_{sha} \, \cos\phi_{sha} - m_a c \, V_{DCa} V_{i+a-1} y_{sha} \cos\left(\theta_{sha} - \theta_{i+a-1} - \phi_{sha} \right)$$

$$- P_{sha}^{sp} - \left(\frac{P_{sha}^{sp\,2} + Q_{sha}^{sp\,2}}{V_{i+a-1}^2} \right) R_{sha} = 0$$

$$\text{or,} \quad f_{2a} = 0 \tag{6.19}$$

In a similar manner, the reactive component of the equivalent complex load power at the fictitious AC bus '(j+a-1)' {which represents the effect of the ath VSC} can be expressed as

$$-m_a^2 c^2 V_{DCa}^2 y_{sha} \sin\phi_{sha} - m_a c \, V_{DCa} V_{i+a-1} y_{sha} \sin\left(\theta_{sha} - \theta_{i+a-1} - \phi_{sha} \right)$$

$$- Q_{sha}^{sp} - \left(\frac{P_{sha}^{sp\,2} + Q_{sha}^{sp\,2}}{V_{i+a-1}^2} \right) X_{sha} = 0$$

$$\text{or,} \quad f_{3a} = 0 \tag{6.20}$$

If the ath ($1 \leq a \leq q$) VSC operates in the PV control mode, 'Q_{sha}' is not specified, and hence, Eq. (6.20) is modified to {using Eq. (A.8) of Appendix A}

$$P'_{sha} - P^{sp}_{sha} - I^2_{sha}R_{sha} = 0$$

or, $\quad m_a^2 c^2 V_{DCa}^2 y_{sha} \cos\phi_{sha} - m_a c V_{DCa} V_{i+a-1} y_{sha} \cos(\theta_{sha} - \theta_{i+a-1} - \phi_{sha}) - P^{sp}_{sha}$

$$-\left[V^2_{i+a-1} y^2_{sha} + m_a^2 c^2 V^2_{DCa} y^2_{sha} - 2m_{aghg} c V_{DCa} V_{i+a-1} \cos(\theta_{i+a-1} - \theta_{sha}) \right] R_{sha} = 0$$

or, $\quad f_{4a} = 0$ $\qquad\qquad$ (6.21)

AC Network Solution

Corresponding to Figure 6.1, the Newton power-flow equation for the sequential solution of the AC power system variables can be written as

Solve: $\boldsymbol{\theta}$, **V**, **X**
Specified: **P**, **Q**, **R**

where

$$\boldsymbol{\theta} = \left[\theta_2 \cdots \theta_n\right]^T, \ \mathbf{V} = \left[V_{g+1} \cdots V_n\right]^T, \ \boldsymbol{\theta}_{sh} = \left[\theta_{sh1} \cdots \theta_q\right]^T, \ \mathbf{m} = \left[m_1 \cdots m_q\right]^T$$

and $\mathbf{X} = \left[\boldsymbol{\theta}_{sh}^T, \ \mathbf{m}^T\right]^T$

$$\mathbf{P} = \left[P_2 \cdots P_n\right]^T, \ \mathbf{Q} = \left[Q_{g+1} \cdots Q_n\right]^T,$$

$$\mathbf{f}_2 = \left[f_{22} \cdots f_{2q}\right], \ \mathbf{f}_3 = \left[f_{32} \cdots f_{3q}\right] \text{ and } \mathbf{R} = \left[V_{i+r-1}, \ \mathbf{f}_2, \ \mathbf{f}_3, \ f_{11}\right]^T$$

Thus, the basic Newton power-flow equation is given below:

$$
\begin{bmatrix}
& \mathbf{J}_{old} & & \dfrac{\partial \mathbf{P}}{\partial \boldsymbol{\theta}_{sh}} & \dfrac{\partial \mathbf{P}}{\partial \mathbf{m}} \\[2mm]
& & & \dfrac{\partial \mathbf{Q}}{\partial \boldsymbol{\theta}_{sh}} & \dfrac{\partial \mathbf{Q}}{\partial \mathbf{m}} \\[2mm]
\dfrac{\partial \mathbf{R}}{\partial \boldsymbol{\theta}} & \dfrac{\partial \mathbf{R}}{\partial \mathbf{V}} & & \dfrac{\partial \mathbf{R}}{\partial \boldsymbol{\theta}_{sh}} & \dfrac{\partial \mathbf{R}}{\partial \mathbf{m}}
\end{bmatrix}
\begin{bmatrix}
\Delta\boldsymbol{\theta} \\ \Delta\mathbf{V} \\ \Delta\boldsymbol{\theta}_{sh} \\ \Delta\mathbf{m}
\end{bmatrix}
=
\begin{bmatrix}
\Delta\mathbf{P} \\ \Delta\mathbf{Q} \\ \Delta\mathbf{R}
\end{bmatrix}
\qquad (6.22)
$$

In Eq. (6.22), $\mathbf{J}_{old}$ is the conventional load flow (without incorporating HVDC link) Jacobian sub-block given as follows:

$$
\mathbf{J}_{old} =
\begin{bmatrix}
\dfrac{\partial \mathbf{P}}{\partial \boldsymbol{\theta}} & \dfrac{\partial \mathbf{P}}{\partial \mathbf{V}} \\[2mm]
\dfrac{\partial \mathbf{Q}}{\partial \boldsymbol{\theta}} & \dfrac{\partial \mathbf{Q}}{\partial \mathbf{V}}
\end{bmatrix}
$$

Also, in Eq. (6.22), '$\Delta\mathbf{P}$', '$\Delta\mathbf{Q}$' and '$\Delta\mathbf{R}$' represent the mismatch vectors. In addition, $\Delta\boldsymbol{\theta}$, $\Delta\mathbf{V}$, $\Delta\boldsymbol{\theta}_{sh}$ and $\Delta\mathbf{m}$ represent correction vectors. In the above formulation, it is assumed that all the slave converters operate in the PQ control mode. However, it may be noted that if they are made to operate in the PV control mode, the corresponding elements of the correction and mismatch vectors have to be modified accordingly. Some typical elements of Eq. (6.22) are given in Appendix A.

DC network solution

Again, corresponding to Figure 6.1, the Newton power-flow equation for the sequential solution of the DC variables of the 'q' bus MVDC system can be written as

Solve: $\mathbf{V_{DC}}$
Specified: $\mathbf{R}$
$\mathbf{R} = [\mathbf{f}_s, \mathbf{P_{DCWF}}]$

where

$$\mathbf{V_{DC}} = \begin{bmatrix} V_{DC2} \cdots V_{DCq} \end{bmatrix}, \mathbf{f}_5 = \begin{bmatrix} f_{12} \cdots f_{1q} \end{bmatrix} \text{ and } \mathbf{P_{DCWF}} = \begin{bmatrix} P_{DCWF1} \cdots P_{DCWF(P-q)} \end{bmatrix}$$

Thus the basic Newton power-flow equation is given below

$$\left[\frac{\partial \mathbf{R}}{\partial \mathbf{V_{DC}}} \right] [\Delta \mathbf{V_{DC}}] = [\Delta \mathbf{R}] \tag{6.23}$$

Also, in Eq. (6.23), '$\Delta\mathbf{R}$' is the mismatch vector whereas $\Delta\mathbf{V_{DC}}$ represents the correction vector comprising the DC voltages of all DC buses except the DC slack bus.

The flow charts of the unified and sequential NR power-flow methods of AC-MVDC system employing DC slack-bus control are given in Figures 6.4 and 6.5, respectively.

6.5 MODELLING OF AC-MVDC SYSTEMS WITH RENEWABLE ENERGY SOURCES EMPLOYING DC VOLTAGE DROOP CONTROL

6.5.1 Types of DC Voltage Droop Control

As already mentioned in Chapter 4, in DC voltage droop control [26–31], multiple converters participate in the DC voltage control scheme. Droop control comprises either linear voltage droop characteristics like voltage-power (V-P) or voltage-current (V-I) droops or nonlinear voltage droop characteristics with dead-bands and limits.

1. Voltage-Power (V-P) Droop
   If the ath VSC follows a linear V-P droop characteristic, its rectifying power can be expressed as

$$P_{DCa} = R_a \left(V_{DCa}^* - V_{DCa} \right) + P_{DCa}^* \tag{6.24}$$

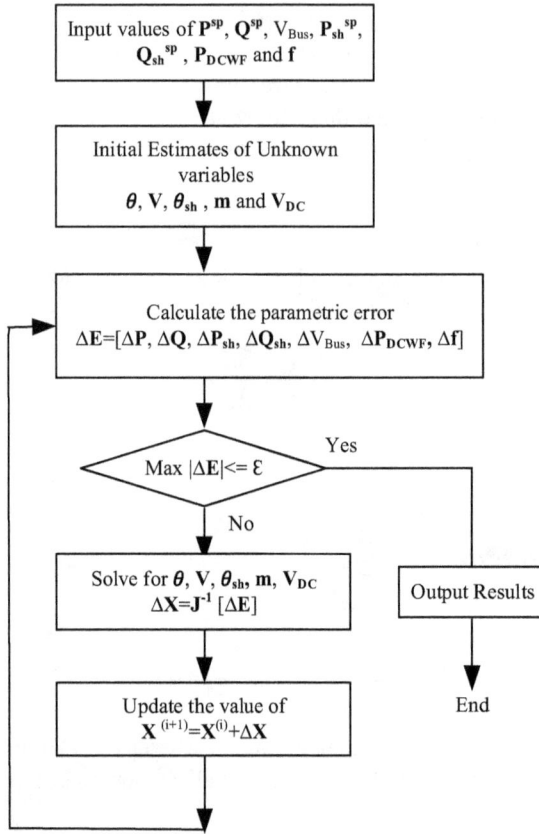

FIGURE 6.4 Flow chart of the unified power-flow algorithm.

2. Voltage-Current (V-I) Droop

If the ath VSC follows a linear V-I droop characteristic; the net DC current injection at its terminal can be expressed as

$$I_{DCa} = R_a \left(V_{DCa}^* - V_{DCa} \right) + I_{DCa}^* \qquad (6.25)$$

Thus, the rectifying power of the VSC can be expressed as

$$P_{DCa} = V_{DCa} \left[R_a \left(V_{DCa}^* - V_{DCa} \right) + I_{DCa}^* \right] \qquad (6.26)$$

Figure 4.3 depicts typical linear V-P and V-I droop characteristics for any arbitrary VSC 'a'.

In Eqs. (6.24)–(6.26), 'V_{DCa}^*', 'P_{DCa}^*' and 'I_{DCa}^*' represent the DC voltage, power and current references of the droop characteristics, respectively, while 'R_a' is the droop control gain.

Input values of $\mathbf{P^{sp}}$, $\mathbf{Q^{sp}}$, V_{Bus}, $\mathbf{P_{sh}}$, $\mathbf{Q_{sh}}$, V_{DC1}, f_1, f_2, P_{DCWF} and f_3

Initial estimates of unknown variables θ, V, θ_{sh}, m and V_{DC}

Calculate the parametric Error for AC network
$\Delta E_{AC} = [\Delta \mathbf{P}, \Delta \mathbf{Q}, \Delta V_{Bus}, \Delta f_{11}, \Delta f_2$ and Δf_3]

Solve for θ, V, θ_{sh}, and m
$\Delta \mathbf{X_{AC}} = \mathbf{J}^{-1} [\Delta \mathbf{E_{AC}}]$

Update the value of
$\mathbf{X_{AC}}^{(I+1)} = \mathbf{X_{AC}}^{(I)} + \Delta \mathbf{X_{AC}}$

Calculate the parametric Error for DC network
$\Delta \mathbf{E_{DC}} = [\Delta f_{12}....\Delta f_{1q}, \Delta P_{DCWF}]$

Max [|$\Delta \mathbf{E_{AC}}$|, | $\Delta \mathbf{E_{DC}}$|] <= ε

Output results

Solve for V_{DC}
$\Delta \mathbf{X_{DC}} = \mathbf{J}^{-1} [\Delta \mathbf{E_{DC}}]$

End

Update the value of
$\mathbf{X_{DC}}^{(I+1)} = \mathbf{X_{DC}}^{(I)} + \Delta \mathbf{X_{DC}}$

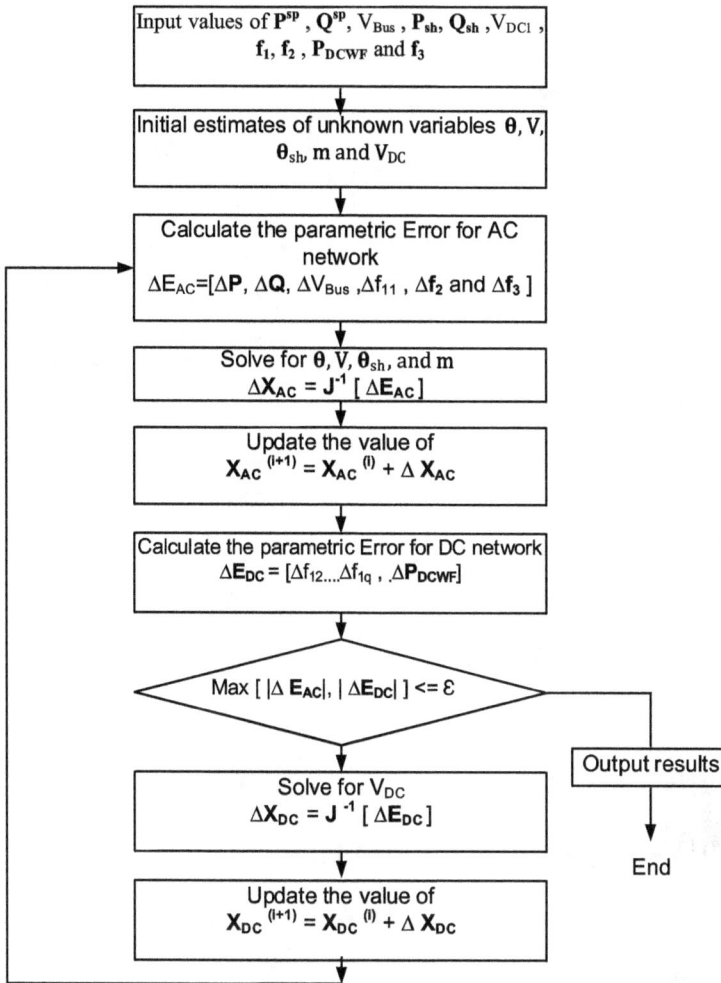

FIGURE 6.5 Flow chart of the sequential power-flow algorithm.

Computation of 'V_{DCa}^{*}' and 'P_{DCa}^{*}'

As already mentioned in Section 4.4 of Chapter 4, the values of the DC power and current references 'P_{DCa}^{*}' and 'I_{DCa}^{*}' in Eqs. (6.24) and (6.25) for all the 'q' converters are either pre-specified or obtained from a DC power flow.

3. Voltage-Power (V-P) Droop with dead band

As already described in Chapter 4, if the ath VSC ($1 \le a \le q$) follows a nonlinear voltage droop characteristics with dead band and voltage limits as shown in Figure 6.6, the droop characteristic can be expressed as

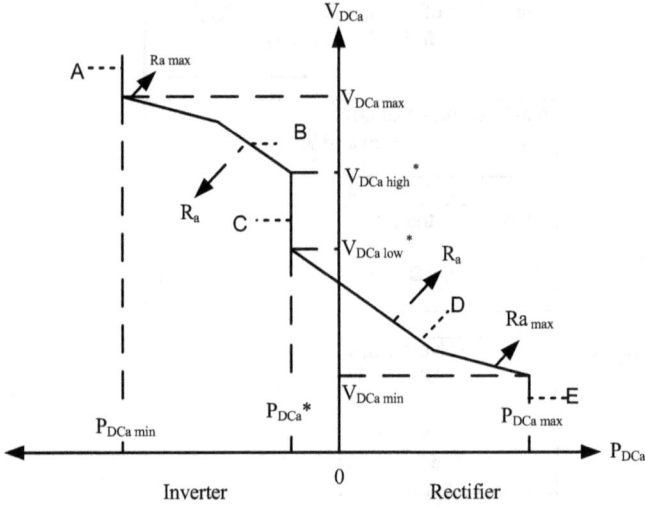

FIGURE 6.6 Nonlinear voltage droop characteristic of the ath VSC with dead band and voltage limits.

$$P_{DCa} = R_{a\,max}\left(V_{DCa\,max} - V_{DCa}\right) + \left[R_a\left(V^*_{DCa\,high} - V_{DCa\,max}\right) + P^*_{DCa}\right]$$

$$\text{for } V_{DCa} \geq V_{DCa\,max} \tag{6.27}$$

$$= R_a\left(V^*_{DCa\,high} - V_{DCa}\right) + P^*_{DCa} \qquad \text{for } V^*_{DCa\,high} < V_{DCa} < V_{DCa\,max} \tag{6.28}$$

$$= 0.\left(V^*_{DCa\,high} - V_{DCa}\right) + P^*_{DCa} \qquad \text{for } V^*_{DCa\,low} \leq V_{DCa} \leq V^*_{DCa\,high} \tag{6.29}$$

$$= R_a\left(V^*_{DCa\,low} - V_{DCa}\right) + P^*_{DCa} \qquad \text{for } V^*_{DCa\,min} < V_{DCa} < V^*_{DCa\,low} \tag{6.30}$$

$$= R_{a\,max}\left(V_{DCa\,min} - V_{DCa}\right) + \left[R_a\left(V^*_{DCa\,low} - V_{DCa\,min}\right) + P^*_{DCa}\right] \quad \text{for } V_{DCa} \leq V_{DCa\,min}$$

$$\tag{6.31}$$

4. DC voltage margin control

As already discussed in Chapter 4, the V-P characteristic pertaining to the DC voltage margin control is shown in Figure 6.7.

The inclusion of linear and nonlinear DC voltage droop control is accommodated in the AC-MTDC model with renewable energy sources as shown below.

6.5.2 IMPLEMENTATION OF DC VOLTAGE DROOP CONTROL IN INTEGRATED AC-MVDC SYSTEMS INTERFACED WITH OFFSHORE WIND FARMS

Let us assume now that all the 'q' VSCs in the AC-MTDC system shown in Figures 6.1 and 6.2 operate on droop control. To simplify matters, let all the 'q'

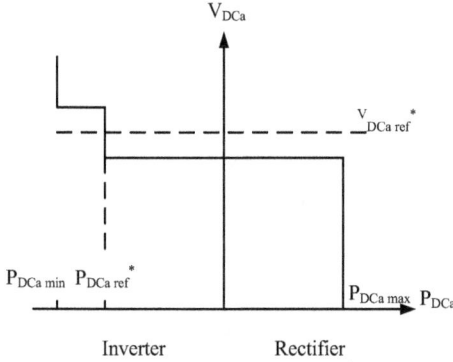

FIGURE 6.7   Voltage margin characteristic of the ath VSC.

VSCs follow linear V-P droops. Then, for the ath VSC ($1 \leq a \leq q$), from Eq. (6.24), we have

$$P_{DCa} = V_{DCa}I_{DCa} = \sum_{v=1}^{p} V_{DCa}V_{DCv}Y_{DCav} = R_a\left(V_{DCa}^* - V_{DCa}\right) + P_{DCa}^*$$

$$\text{or,} \quad \sum_{v=1}^{p} V_{DCa}V_{DCv}Y_{DCav} + R_a V_{DCa} - R_a V_{DCa}^* - P_{DCa}^* = 0 \qquad (6.32)$$

$$\text{or,} \quad f_{2a} = 0 \quad \forall a, \ 1 \leq a \leq q \qquad (6.33)$$

Equation (6.33) represents 'q' independent equations.

Now, two distinctly different models can be realized depending on whether the values of 'V_{DCa}^*' and 'P_{DCa}^*' are specified or not. These are elaborated below.

Model 'A': Values of 'V_{DCa}^*' and 'P_{DCa}^*' are known a priori

In some cases, the values of 'V_{DCa}^*' and 'P_{DCa}^*' for all the 'q' converters are pre-specified, having been obtained from a DC power-flow, or otherwise. In such cases, the 'q' independent droop equations represented by Eq. (6.33) along with '(p−q)' specified values of rectified wind farm power injections $\left(P_{DCWFs}\right)$ are sufficient to compute the values of the DC bus voltages 'V_{DCa}' directly, and subsequently, the DC bus power injections 'P_{DCa}'. Now, as already described in Section 4.5 of Chapter 4, once 'P_{DCa}' are known, the active powers 'P_{sha}' in the lines joining the '(q−1)' VSCs to their corresponding AC buses cannot be specified for the AC-MTDC power-flow. This is not in line with practical considerations which are targeted to maintain a specified 'P_{sha}'. This is a major drawback of the model.

Now, under the assumption that there are 'g' generators connected at the first 'g' buses of the 'n' bus AC system with bus 1 being the slack bus, the unified AC-MTDC power-flow problem corresponding to model 'A' is of the form

Compute: θ, **V**, **X**

Given: **P**, **Q**, **R**

where

$$\theta = \begin{bmatrix} \theta_2 \cdots \theta_n \end{bmatrix}^T, \ \mathbf{V} = \begin{bmatrix} V_{g+1} \cdots V_n \end{bmatrix}^T, \ \theta_{sh} = \begin{bmatrix} \theta_{sh1} \cdots \theta_{shq} \end{bmatrix}^T, \ \mathbf{m} = \begin{bmatrix} m_1 \cdots m_q \end{bmatrix}^T,$$

$$\mathbf{X} = \begin{bmatrix} \theta_{sh}^T & \mathbf{m}^T \end{bmatrix}^T$$

$$\mathbf{P} = \begin{bmatrix} P_2 \cdots P_n \end{bmatrix}^T, \ \mathbf{Q} = \begin{bmatrix} Q_{g+1} \cdots Q_n \end{bmatrix}^T, \ \mathbf{Q}_{sh} = \begin{bmatrix} Q_{sh2} \cdots Q_{shq} \end{bmatrix}, \ \mathbf{f}_1 = \begin{bmatrix} f_{11} \cdots f_{1q} \end{bmatrix},$$

$$\mathbf{R} = \begin{bmatrix} \mathbf{Q}_{sh} & V_{i+r-1} & \mathbf{f}_1 \end{bmatrix}^T$$

For this model, it is presumed that VSC 'r' controls the voltage magnitude of the AC bus '(i+r–1)' unlike the other '(q–1)' VSCs, which control the line reactive power flows.

The Newton power-flow equation is

$$\mathbf{J} \begin{bmatrix} \Delta\theta^T & \Delta\mathbf{V}^T & \Delta\theta_{sh}^T & \Delta\mathbf{m}^T \end{bmatrix}^T = \begin{bmatrix} \Delta\mathbf{P}^T & \Delta\mathbf{Q}^T & \Delta\mathbf{R}^T \end{bmatrix}^T \quad (6.34)$$

where **J** is the power-flow Jacobian.

In Eq. (6.34), '$\Delta\mathbf{P}$', '$\Delta\mathbf{Q}$' and '$\Delta\mathbf{R}$' represent the mismatch vectors, while $\Delta\theta$, $\Delta\mathbf{V}$, $\Delta\theta_{sh}$ and $\Delta\mathbf{m}$ represent the correction vectors. The elements of '**J**' can be obtained very easily from Eq. (6.34).

Thus, in Model 'A', if 'V_{DCa}^*' and 'P_{DCa}^*' are known, 'V_{DCa}' ($1 \leq a \leq p$) can be solved using Eq. (6.32) and '(p–q)' specified values of rectified wind farm power injections ('P_{DCWFs}'), independent of the AC-MTDC power flow {Eq. (6.34)}. Subsequent to the AC-MTDC power flow, the rest of the unknowns are also solved and the line active power flows 'P_{sha}' can be computed {using Eq. (6.5)}.

To summarize, Model 'A' addresses the problem 'given the DC voltage and power (or current) references of the VSC droop lines and the target line reactive power flows, what should be the line active power flow values?'

Figure 6.8 depicts the flow chart of the Newton-Raphson (NR) power-flow algorithm for droop control Model A.

Model 'B': Values of 'V_{DC}^*' and 'P_{DC}^*' are not known a priori

If the DC voltage ('V_{DC}^*') and power ('P_{DC}^*') reference values of the 'q' VSCs are not known, the DC bus voltages 'V_{DCa}' ($1 \leq a \leq p$) and hence the DC bus power injections 'P_{DCa}' cannot be computed from the '(p–q)' specified values of rectified wind farm power injections ('P_{DCWFs}') only (it is assumed that q $\neq$ 0). This enables the line active power-flow values 'P_{sh}' to be specified control objectives, which is in line with practical MTDC control. This is an advantage over model 'A'.

For the above modelling strategy, the unified AC-MTDC power-flow problem is of the form

Unified method of solution for model B

Compute: θ, **V**, **X**

Given: **P**, **Q**, **R**

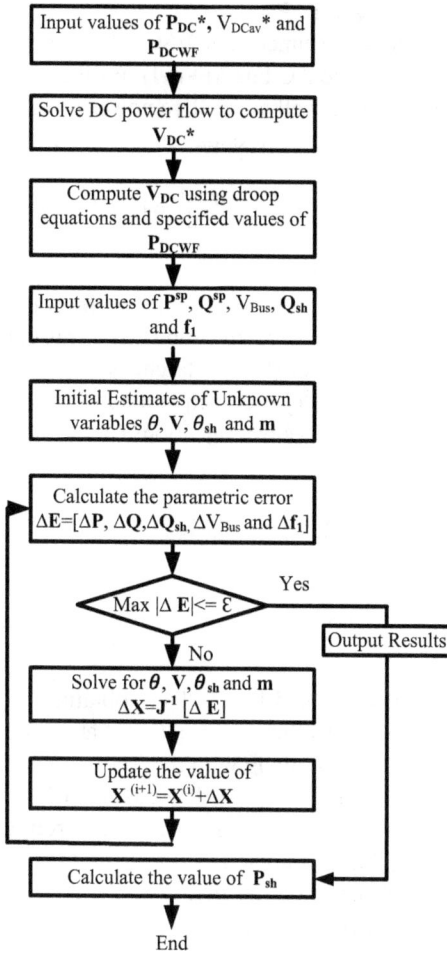

FIGURE 6.8   Flow chart of the NR power-flow algorithm (Model A).

with

$$\theta = \left[\theta_2 \cdots \theta_n\right]^T, \mathbf{V} = \left[V_{g+1} \cdots V_n\right]^T, \theta_{sh} = \left[\theta_{sh1} \cdots \theta_{shq}\right]^T, \mathbf{m} = \left[m_1 \cdots m_q\right]^T,$$

$$\mathbf{V}_{DC} = \left[V_{DC1} \cdots V_{DCp}\right]^T$$

$$\mathbf{X} = \left[\theta_{sh}^T \quad \mathbf{m}^T \quad \mathbf{V}_{DC}^T\right]^T$$

$$\mathbf{P} = \left[P_2 \cdots P_n\right]^T, \mathbf{Q} = \left[Q_{g+1} \cdots Q_n\right]^T, \mathbf{P}_{sh} = [P_{sh2} \cdots P_{shq}], \mathbf{Q}_{sh} = \left[Q_{sh2} \cdots Q_{shq}\right],$$

$$\mathbf{f}_1 = \left[f_{11} \cdots f_{1q}\right], \mathbf{P}_{DCWF} = \left[P_{DCWF1} \cdots P_{DCWF(p-q)}\right]$$

and $\mathbf{R} = \begin{bmatrix} \mathbf{P}_{sh} & \mathbf{Q}_{sh} & V_{i+r-1} & V_{DCav} & f_1 & P_{DCWF} \end{bmatrix}^T$

For this model too, it is presumed that VSC 'r' is employed for the control of the voltage magnitude of the AC bus '(i+r–1)' unlike the other '(q–1)' VSCs, which control the line active as well as reactive power flows.

The unified AC-MTDC power-flow equation is

$$\mathbf{J}\begin{bmatrix} \Delta\boldsymbol{\theta}^T & \Delta\mathbf{V}^T & \Delta\boldsymbol{\theta}_{sh}^T & \Delta\mathbf{m}^T & \Delta\mathbf{V}_{DC}^T \end{bmatrix}^T = \begin{bmatrix} \Delta\mathbf{P}^T & \Delta\mathbf{Q}^T & \Delta\mathbf{R}^T \end{bmatrix}^T \quad (6.35)$$

where $\mathbf{J}$ is the power-flow Jacobian.

The values of 'V_{DCa}' ($1 \le a \le p$) are now obtained from the AC-MTDC power flow {Eq. (6.35)}, and the DC bus power injections 'P_{DCa}' are computed. Thereafter, the DC voltage ('V_{DCa}^*') and power (P_{DCa}^*) references of the droop lines of the 'q' VSCs can be computed from the 'q' droop equations along with the '(p–q)' specified values of rectified wind farm power injections ('P_{DCWFs}'), as elaborated below.

From Eq. (6.24),

$$P_{DCa}^* + R_a V_{DCa}^* - R_a V_{DCa} - P_{DCa} = 0$$

$$\text{or,} \quad \sum_{v=1}^{p} V_{DCa}^* V_{DCv}^* Y_{DCav} + R_a V_{DCa}^* - R_a V_{DCa} - P_{DCa} = 0 \quad (6.36)$$

The values of 'V_{DCa}' ($1 \le a \le p$) and hence 'P_{DCa}' obtained from the AC-MTDC power flow {Eq. (6.35)} are substituted in Eq. (6.36) above and the 'p' voltages 'V_{DCa}^*' are solved using Eq. (6.36) along with the '(p–q)' specified values of rectified wind farm power injections ('P_{DCWFs}'). The voltages ('V_{DCa}^*') corresponding to the 'q' VSCs are taken as their DC voltage references. From the values of 'V_{DCa}^*', the power references 'P_{DCa}^*' are computed.

From Eq. (6.35), it can be observed that both the AC and DC variables are solved together and are known as the unified method. In a similar manner, the AC and DC variables can also be solved sequentially. This is known as the sequential method. The sequential method is detailed below.

Sequential method of solution for model B
AC network solution
Compute: $\boldsymbol{\theta}$, $\mathbf{V}$, $\mathbf{X}$
Given: $\mathbf{P}$, $\mathbf{Q}$, $\mathbf{R}$
with

$$\boldsymbol{\theta} = \begin{bmatrix} \theta_2 \cdots \theta_n \end{bmatrix}^T, \mathbf{V} = \begin{bmatrix} V_{g+1} \cdots V_n \end{bmatrix}^T, \boldsymbol{\theta}_{sh} = \begin{bmatrix} \theta_{sh1} \cdots \theta_{shq} \end{bmatrix}^T, \mathbf{m} = \begin{bmatrix} m_1 \cdots m_q \end{bmatrix}^T,$$

$$\mathbf{X} = \begin{bmatrix} \boldsymbol{\theta}_{sh}^T & \mathbf{m}^T \end{bmatrix}^T$$

$$\mathbf{P} = \begin{bmatrix} P_2 \cdots P_n \end{bmatrix}^T, \mathbf{Q} = \begin{bmatrix} Q_{g+1} \cdots Q_n \end{bmatrix}^T, \mathbf{P}_{sh} = [P_{sh2} \cdots P_{shq}], \mathbf{Q}_{sh} = \begin{bmatrix} Q_{sh2} \cdots Q_{shq} \end{bmatrix},$$

and $\mathbf{R} = \begin{bmatrix} \mathbf{P}_{sh} & \mathbf{Q}_{sh} & V_{i+r-1} & f_{11} \end{bmatrix}^T$

For this model too, it is presumed that VSC 'r' is employed for the control of voltage magnitude of the AC bus '(i+r−1)' unlike the other '(q−1)' VSCs, which control the line active as well as reactive power flows.

The AC-MTDC power-flow equation is

$$\mathbf{J}\left[\, \Delta\boldsymbol{\theta}^T \ \ \Delta\mathbf{V}^T \ \ \Delta\boldsymbol{\theta}_{sh}^T \ \ \Delta\mathbf{m}^T \ \right]^T = \left[\, \Delta\mathbf{P}^T \ \ \Delta\mathbf{Q}^T \ \ \Delta\mathbf{R}^T \,\right]^T \qquad (4.29)$$

where $\mathbf{J}$ is the power-flow Jacobian.

DC network solution

The power-flow solution pertaining to DC network solution can be formed as

Solve: $\mathbf{V}_{DC}$

Specified: $\mathbf{V}_{DCavg}$, $\mathbf{f}_6$, $\mathbf{P}_{DCWF}$

where

$$\mathbf{V}_{DC} = \left[\, \mathbf{V}_{DC1} \cdots \mathbf{V}_{DCq} \,\right], \ \mathbf{f}_6 = \mathbf{P}_{DC2} \cdots \mathbf{P}_{DCq}, \ \mathbf{P}_{DCWF} = \left[\, \mathbf{P}_{DCWF1} \cdots \mathbf{P}_{DCWF(p-q)} \,\right]$$

and $\mathbf{R}_{DC} = \left[\, \mathbf{V}_{DCavg} \ \mathbf{f}_6 \ \mathbf{P}_{DCWF} \,\right]^T$

The Newton power-flow equation for the above is

$$\mathbf{J}_{DC} \ \Delta\mathbf{V}_{DC}^T = \Delta\mathbf{R}_{DC} \qquad (4.30)$$

In Eq. (4.30), '$\Delta\mathbf{R}_{DC}$' and '$\Delta\mathbf{V}_{DC}$' are the mismatch and correction vectors, respectively.

Similar to the unified method, the DC reference values can now be calculated using droop equations.

Figures 6.9 and 6.10 depict the flow charts of the proposed unified and sequential Newton-Raphson (NR) power-flow algorithms for droop control model 'B'.

6.6 CASE STUDIES AND RESULTS

For validation of the above models, a large number of studies were carried out by employing diverse DC voltage droop control strategies on MTDC grids integrated with European 1354-bus networks [116]. Offshore wind farms (OWFs) are appropriately interfaced at '(p−q)' buses in the MTDC grid. In all the occurrences, the VSC constant was selected as $c = \dfrac{1}{2\sqrt{2}}$ [11]. Also, for all the VSC coupling transformers, $R_{sha} = 0.001$ p.u. and $X_{sha} = 0.1$ p.u. ($\forall a, 1 \le a \le q$). The converter loss constants 'a_1', 'b_1' and 'c_1' were chosen as 0.011, 0.003 and 0.0043, respectively [18,105]. For interconnections between DC terminals, $R_{DCuv} = 0.01$ p.u. ($\forall u, v, 1 \le u \le q, 1 \le v \le q, u \ne v$) has been selected throughout the chapter [93]. In all occurrences, a termination error tolerance of 10^{-8} p.u. was selected. 'NI' denotes the number of iterations. In all the results given in Tables 6.1–6.9, values of bus voltage magnitudes, current magnitudes, active and reactive powers and droop control gains are denoted in p.u. while phase angles of voltage phasors are denoted in degrees.

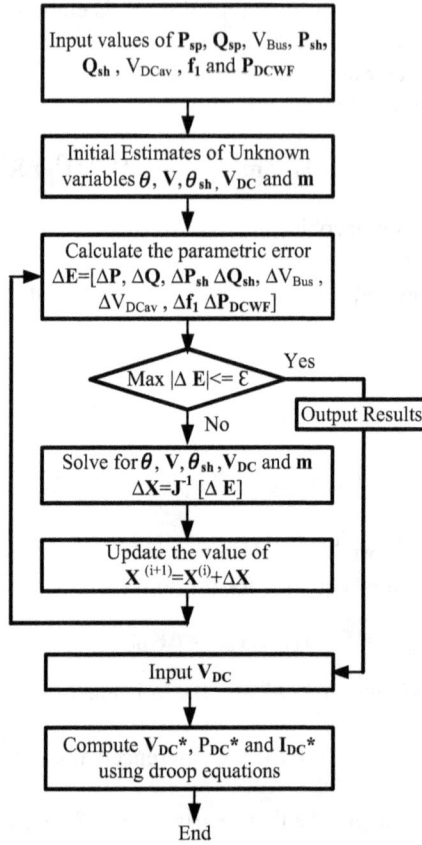

FIGURE 6.9   Flow chart of the unified NR power-flow algorithm (Model B).

6.6.1 Study with Unified Power-Flow Model of European 1354-Bus Test System Integrated with 7-Terminal MVDC Network Employing DC Slack-Bus Control and Interfaced with Offshore Wind Farms

Firstly, a base case power flow (without any MTDC network connected) was carried out. The results are shown in row 1 of Table 6.1. Subsequently, a seven-terminal VSC-HVDC network with DC slack-bus control is integrated with the European 1354-bus system at AC buses 1280, 1281, 1286, 1292 and 1313. Two offshore wind farms are connected with DC buses '6' and '7' of the seven-terminal HVDC network. The master converter is connected to bus no. 1280 and operates in bus voltage control mode. The slave converters are connected to AC buses 1281, 1286, 1292 and 1313 and operate in the PQ control mode. The specified quantities are shown in rows 4–6 and column 1 of Table 6.1. The power-flow solution is shown in rows 4–6 and column 2–3 of Table 6.1.

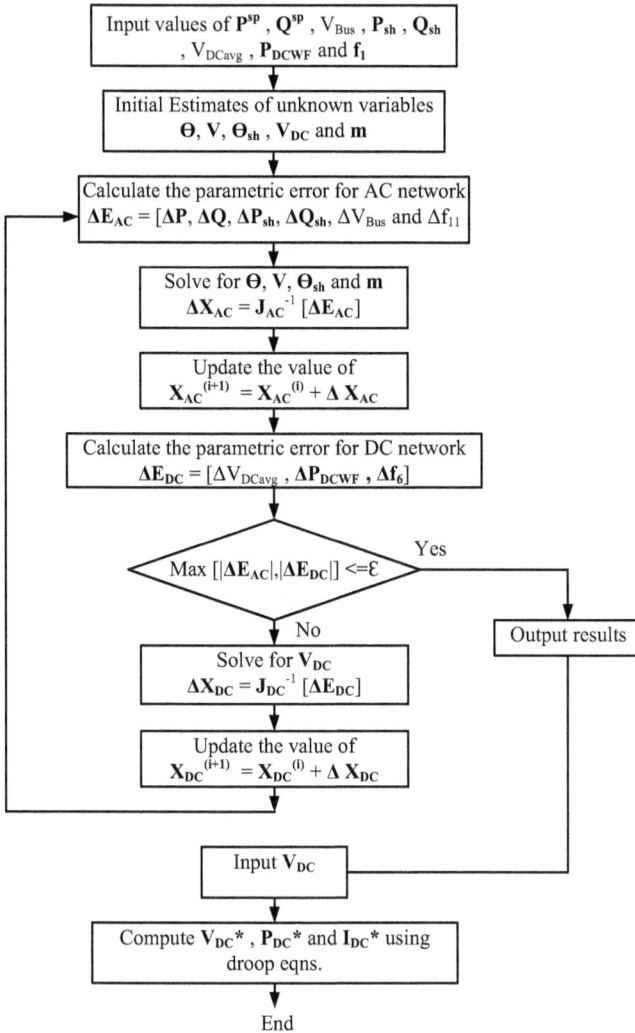

FIGURE 6.10   Flow chart of the sequential NR power-flow algorithm (Model B).

The convergence characteristic plots for the studies corresponding to row 1 of Table 6.1 (base case) and rows 4–6 of Table 6.1 are shown in Figures 6.11 and 6.12, respectively. From Figures 6.11 and 6.12, it can be observed that the proposed AC-MTDC model with offshore wind farms demonstrates quadratic convergence characteristics, similar to the base case power-flow. The bus voltage profile corresponding to this study is shown in Figure 6.13. From Figure 6.13, it is observed that the bus voltage profile for the AC-MTDC power-flow is similar to the base case except at the terminals at which the VSCs are connected.

TABLE 6.1

Study of European 1354-Bus System with 7-Terminal VSC-HVDC Network Employing DC Slack-Bus Control and Incorporating OWFs

Base case power flow converged in seven iterations (NI=7)

$V_{1280} = 0.9554 \angle -35.2574$; $V_{1281} = 1.0354 \angle -37.3082$; $V_{1286} = 1.0156 \angle -12.3760$;

$V_{1292} = 1.0289 \angle -35.23066$; $V_{1313} = 1.0351 \angle -37.4755$;

AC-MTDC Power-Flow with OWFs (Slack-Bus Control)

Given Quantities	Power-Flow Solution	
Master converter	Master converter	
$V_{1280} = 0.98$;	$\theta_{sh1} = -42.0550$; $m_1 = 0.9824$;	
$V_{DC1} = 3.0$;	Slave converters	
	$V_{DC2} = 2.9994$; $V_{DC3} = 2.9994$; $V_{DC4} = 2.9995$; $V_{DC5} = 2.9994$;	
	$m_2 = 0.9844$; $m_3 = 0.9652$; $m_4 = 0.9770$; $m_5 = 0.9808$;	
	$\theta_{sh2} = -33.3004$; $\theta_{sh3} = -9.7072$; $\theta_{sh4} = -32.6959$; $\theta_{sh5} = -33.9965$;	
Slave converters	AC terminal buses	Wind farm terminals
$P_{sh2} = 0.5$;	$\theta_{1280} = -37.6895$;	$V_{DC6} = 2.9999$;
$Q_{sh2} = 0.06$;	$V_{1281} = 1.0365 \angle -35.9458$;	$V_{DC7} = 2.9998$;
$P_{sh3} = 0.4$;	$V_{1286} = 1.0175 \angle -11.9055$;	
$Q_{sh3} = 0.05$;	$V_{1292} = 1.0315 \angle -34.3022$;	
$P_{sh4} = 0.3$;	$V_{1313} = 1.0361 \angle -36.1221$;	
$Q_{sh4} = 0.04$;		
$P_{sh5} = 0.4$;		
$Q_{sh5} = 0.03$;		
Rectifying power of	DC power	Converter loss (%)
offshore wind farms	$P_{DC1} = 0.7523$;	$P_{loss1} = 1.82$;
$P_{DCWF1} = 0.5$;	$P_{DC2} = -0.5137$;	$P_{loss2} = 1.35$;
$P_{DCWF2} = 0.4$;	$P_{DC3} = -0.4130$;	$P_{loss3} = 1.29$;
	$P_{DC4} = -0.3123$;	$P_{loss4} = 1.23$;
	$P_{DC5} = -0.4130$;	$P_{loss5} = 1.28$;
	NI = 7	

6.6.2 STUDY WITH UNIFIED POWER-FLOW MODEL OF EUROPEAN 1354 BUS TEST SYSTEM INTEGRATED WITH 7-TERMINAL MVDC NETWORK EMPLOYING DC VOLTAGE DROOP CONTROL AND INTERFACED WITH OFFSHORE WIND FARMS

Case I: Model A employing linear V-P and V-I droop characteristics with OWFs

This study is similar to the study of Table 6.1 except that the VSCs are operated in the linear DC voltage droop control mode. The VSCs connected to AC buses 1280, 1281 and 1286 are operated in V-P droop, while the VSC connected

TABLE 6.2
Study of European 1354-Bus System with 7-Terminal VSC-HVDC Network Incorporating OWFs (Model A)

Base case power flow converged in seven iterations (NI=7)

$V_{1280} = 0.9554 \angle -35.2574$; $V_{1281} = 1.0354 \angle -37.3082$; $V_{1286} = 1.0156 \angle -12.3760$;

$V_{1292} = 1.0289 \angle -35.23066$; $V_{1313} = 1.0351 \angle -37.4755$;

DC Power Flow

Given Quantities	Solution
$V^*_{DCav} = 3.02$; $P^*_{DC2} = 0.5$;	$V^*_{DC1} = 3.0192$; $V^*_{DC2} = 3.0202$; $V^*_{DC3} = 3.0202$; $V^*_{DC4} = 3.0201$;
$P^*_{DC3} = 0.4$; $P^*_{DC4} = 0.3$;	$V^*_{DC5} = 3.0202$; $P^*_{DC1} = 1.5995$; $I^*_{DC4} = 0.0993$; $I^*_{DC5} = 0.1324$;
$P^*_{DC5} = 0.4$;	NI=5;

Computation of V_{DC} from Droop Equation

Given Quantities	Solution
$V^*_{DC1} = 3.0192$; $V^*_{DC2} = 3.0202$;	$V_{DC1} = 3.0284$; $V_{DC2} = 3.0294$;
$V^*_{DC3} = .0202$;	$V_{DC3} = 3.0294$; $V_{DC4} = 3.0292$;
$V^*_{DC4} = 3.0201$; $V^*_{DC5} = 3.0202$;	$V_{DC5} = 3.0293$; $V_{DC6} = 3.0295$;
$P^*_{DC1} = 1.5995$; $I^*_{DC4} = 0.0993$;	$V_{DC7} = 3.0295$;
$I^*_{DC5} = 0.1324$;	NI=4;
$R_1 = 20$; $R_2 = 15$; $R_3 = 10$;	
$R_4 = 15$; $R_5 = 10$;	
$P_{DCWF1} = 0.6$; $P_{DCWF2} = 0.5$;	

AC-MTDC power flow with linear voltage droop control (VSCs connected to AC buses 1280, 1281, 1292 and 1313)

Given Quantities	Solution		
		AC Buses	VSC
$V_{DC1} = 3.0284$;	$\theta_{1280} = -27.6972$;		$\theta_{sh1} = -7.4114$;
$V_{DC2} = 3.0294$;	$V_{1281} = 1.0365 \angle -6.6495$;		$\theta_{sh2} = -38.6253$;
$V_{DC3} = 3.0294$;	$V_{1286} = 1.0175 \angle -12.5431$;		$\theta_{sh3} -14.3087$;
$V_{DC4} = 3.0292$;	$V_{1292} = 1.0315\angle -33.8577$;		$\theta_{sh4} -33.3333$;
$V_{DC5} = 3.0293$;	$V_{1313} = 1.0361\angle -36.7961$;		$\theta_{sh5} = -37.5226$;
$V_{DC6} = 3.0295$;	$P_{sh1} = 1.7487$;		$m_1 = 0.9338$;
$V_{DC7} = 3.0295$;	$P_{sh2} = -0.3752$;		$m_2 = 0.9858$; $m_3 = 0.9583$;
$V_{1280} = 0.98$;	$P_{sh3} = -0.3204$;		$m_4 = 0.9731$; $m_5 = 0.9807$;
$Q_{sh2} = 0.0192$;	$P_{sh4} = 0.0995$;		
$Q_{sh3} = 0.2$;	$P_{sh5} = -0.1365$;		
$Q_{sh4} = 0.1$;	DC power		Converter loss (%)
$Q_{sh5} = 0.15$;	$P_{DC1} = -1.7830$; $P_{DC2} = 0.3621$;		$P_{loss1} = 3.01$; $P_{loss2} = 1.29$;
	$P_{DC3} = 0.3078$; $P_{DC4} = -0.1110$;		$P_{loss3} = 1.25$; $P_{loss4} = 1.15$;
	$P_{DC5} = 0.1247$;		$P_{loss5} = 1.18$;
	NI=7		

TABLE 6.3

Study of European 1354-Bus System with 7-Terminal VSC-HVDC Network Employing Linear DC Voltage Droop Characteristics and Incorporating OWFs (Model B)

Base case power flow converged in seven iterations (NI = 7)

$V_{1280} = 0.9554 \angle -35.2574$; $V_{1281} = 1.0354 \angle -37.3082$; $V_{1286} = 1.0156 \angle -12.3760$;

$V_{1292} = 1.0289 \angle -35.23066$; $V_{1313} = 1.0351 \angle -37.4755$;

(VSCs connected to AC buses 1280, 1281, 1286, 1292 and 1313)

Power flow of AC-MTDC system with OWFs

Given	Solution	
Quantities	AC Buses	VSCs
V_{DCav} 3.0	$\theta_{1280} = -35.9687$;	$\theta_{sh1} = -38.0780$; $\theta_{sh2} = -33.3122$;
V_{1280} 0.98	$V_{1281} = 1.0367 \angle -35.9449$;	$\theta_{sh3} = -9.6939$; $\theta_{sh4} = -32.5091$;
P_{sh2} 0.5	$V_{1286} = 1.0182 \angle -11.8789$;	$\theta_{sh5} = -35.0800$; $m_1 = 0.9687$; $m_2 = 0.9881$;
Q_{sh2} 0.1	$V_{1292} = 1.0326 \angle -34.1039$;	$m_3 = 0.9694$; $m_4 = 0.9816$; $m_5 = 0.9846$;
P_{sh3} 0.4	$V_{1313} = 1.0362 \angle -36.1347$;	$V_{DC1} = 3.0002$; $V_{DC2} = 2.9998$;
Q_{sh3} 0.09		$V_{DC3} = 2.9998$; $V_{DC4} = 2.9999$; $V_{DC5} = 2.9999$;
P_{sh4} 0.3		Wind farm terminals
Q_{sh4} 0.08		$V_{DC6} = 3.0003$; $V_{DC7} = 3.0002$;
P_{sh5} 0.2		Converter loss (%)
Q_{sh5} 0.07		$P_{loss1} = 1.44$; $P_{loss2} = 1.35$; $P_{loss3} = 1.29$;
Rectifying power		$P_{loss4} = 1.23$; $P_{loss5} = 1.18$;
of offshore wind		DC power
farms		$P_{DC1} = 0.3513$; $P_{DC2} = -0.5138$; $P_{DC3} = -0.4131$;
$P_{DCWF1} = 0.6$; NI = 7;		$P_{DC4} = -0.3124$; $P_{DC5} = -0.2119$;
$P_{DCWF2} = 0.5$;		
Computation of references V^*_{DC} from droop eqns.		
$V_{DC1} = 3.0002$; $V_{DC2} = 2.9998$; $V_{DC3} = 2.9998$;		$V^*_{DC1} = 2.9910$; $V^*_{DC2} = 2.9905$; $V^*_{DC3} = 2.9905$;
$V_{DC4} = 2.9999$; $V_{DC5} = 2.9999$; $R_1 = 20$; $R_2 = 15$;		$V^*_{DC4} = 2.9907$; $V^*_{DC5} = 2.907$;
$R_3 = 10$; $R_4 = 15$; $R_5 = 10$;		NI = 3;

to AC buses 1292 and 1313 operate in the V-I droop. Two offshore wind farms with rectifying powers $P_{DCWF1} = 0.6$ p.u. and $P_{DCWF2} = 0.5$ p.u. are injected into the DC grid at terminals 6 and 7, respectively. The droop control gains of VSCs 1, 2, 3, 4 and 5 are set to 20, 15, 10, 15 and 10, respectively [24].

At first, a DC load flow is carried out to calculate the reference values of the droop lines of the five VSCs. The results are shown in rows 3–4 of Table 6.2. Subsequently, the DC grid bus voltages are computed by using the three droop equations and two values of offshore wind farm power injections (P_{DCWF1} and P_{DCWF2}). The results are given in rows 5–7 of Table 6.2. Finally, an

TABLE 6.4

Study of European 1354-Bus System with 7-Terminal VSC-HVDC Network Employing Nonlinear DC Voltage Droop Control and Incorporating OWFs (Model A)

Base case power flow converged in seven iterations (NI=7)

$V_{1280} = 0.9554 \angle -35.2574$; $V_{1281} = 1.0354 \angle -37.3082$; $V_{1286} = 1.0156 \angle -12.3760$;

$V_{1292} = 1.0289 \angle -35.23066$; $V_{1313} = 1.0351 \angle -37.4755$;

DC Power Flow to Calculate DC Reference Values

Specified Quantities	Power-Flow Solution
$V^*_{DCav} = 3.02$; $P^*_{DC2} = -0.5$;	$V^*_{DC1} = 3.0208$; $V^*_{DC2} = 3.0198$; $V^*_{DC3} = 3.0198$;
$P^*_{DC3} = -0.4$; $P^*_{DC4} = -0.4$;	$V^*_{DC4} = 3.0198$; $V^*_{DC5} = 3.0199$; $P^*_{DC1} = 1.6005$;
$P^*_{DC5} = -0.3$;	$I^*_{DC4} = -0.1325$; $I^*_{DC5} = -0.0993$;
	NI = 3

AC-MTDC power flow with linear voltage droop and nonlinear droop with dead band (VSCs connected to AC buses 1280, 1281, 1286, 1292 and 1313)

Specified Quantities		Power-Flow Solution	
Control Parameters		**DC Terminal Buses**	
$V_{DCmax} = 3.014$; $V^*_{DChigh} = 3.013$;		$V_{DC1} = 3.0252$; $V_{DC2} = 3.0242$; $V_{DC3} = 3.0241$;	
$V^*_{DClow} = 3.012$; $V_{DCmin} = 3.009$;		$V_{DC4} = 3.0242$; $V_{DC5} = 3.0243$; $V_{DC6} = 3.0247$;	
$R_1 = 20$; $R_2 = 15$; $R_3 = 10$;		$V_{DC7} = 3.0246$;	
$R_4 = 15$; $R_5 = 10$;		DC power	Converter loss (%)
$R_{max} = 30$;		$P_{DC1} = 1.5123$;	$P_{loss1} = 3.03$;
		$P_{DC2} = -0.5664$;	$P_{loss2} = 1.39$;
		$P_{DC3} = -0.7135$;	$P_{loss3} = 1.50$;
		$P_{DC4} = -0.5984$;	$P_{loss4} = 1.41$;
		$P_{DC5} = -0.4333$;	$P_{loss5} = 1.29$;
Converter	**Control Mode**	VSCs	
1,2	V-P droop	$\theta_{sh1} = -49.4866$; $\theta_{sh2} = -33.3392$; $\theta_{sh3} = -7.8296$;	
3	V-P droop with dead band	$\theta_{sh4} = -31.0834$; $\theta_{sh5} = -34.2051$; $m_1 = 1.0074$;	
4,5	V-I droop	$m_2 = 0.9786$; $m_3 = 0.9608$; $m_4 = 0.9747$; $m_5 = 0.9748$;	
$V_{266} = 0.98$; $Q_{sh2} = 0.08$;		AC terminal buses	
$Q_{sh3} = 0.06$; $Q_{sh4} = 0.07$;		$\theta_{1280} = -41.0210$; $V_{1281} = 1.0367 \angle -36.2532$;	
$Q_{sh5} = 0.05$; $P_{DCWF1} = 0.5$;		$V_{1286} = 1.0182 \angle -11.6516$; $V_{1292} = 1.0326 \angle -34.1885$;	
$P_{DCWF2} = 0.3$;		$V_{1313} = 1.0362 \angle -36.4320$;	
		NI = 7	

AC-MTDC load flow is carried out. The power-flow solution is shown in rows 8–10 of Table 6.2.

The convergence characteristic corresponding to the study of Table 6.2 is shown in Figure 6.14. From Figure 6.14, it is observed that the AC-MTDC power-flow model with offshore wind farms demonstrates a quadratic convergence

TABLE 6.5

Study of European 1354-Bus System with 7-Terminal VSC-HVDC Network Employing DC Voltage Margin Control and Incorporating OWFs (Model A)

Base case power flow converged in seven iterations (NI=7)

$V_{1280} = 0.9554 \angle -35.2574$; $V_{1281} = 1.0354 \angle -37.3082$; $V_{1286} = 1.0156 \angle -12.3760$;

$V_{1292} = 1.0289 \angle -35.23066$; $V_{1313} = 1.0351 \angle -37.4755$;

DC Power Flow to Calculate DC Reference Values

Specified Quantities	Power-Flow Solution
$V^*_{DCav} = 3.02$; $P^*_{DC2} = -0.5$;	$V^*_{DC1} = 3.0207$; $V^*_{DC2} = 3.0198$; $V^*_{DC3} = 3.0198$;
$P^*_{DC3} = -0.4$; $P^*_{DC4} = -0.2$;	$V^*_{DC4} = 3.0199$; $V^*_{DC5} = 3.0199$; $P^*_{DC1} = 1.4004$;
$P^*_{DC5} = -0.3$;	$I^*_{DC4} = -0.0662$; $I^*_{DC5} = -0.0993$;
	NI = 3

AC-MTDC power flow with linear voltage droop and nonlinear voltage margin (VSCs connected to AC buses 1280, 1281, 1286, 1292 and 1313)

Specified Quantities		Power-Flow Solution	
Control Parameters		**DC Terminal Buses**	
$V_{DCmax} = 3.014$;		$V_{DC1} = 3.0280$; $V_{DC2} = 3.0271$; $V_{DC3} = 3.0269$; $V_{DC4} = 3.0272$;	
$V^*_{DChigh} = 3.013$;		$V_{DC5} = 3.0272$;	
$V^*_{DClow} = 3.012$;		DC power	Converter loss (%)
$V_{DCmin} = 3.009$;		$P_{DC1} = 1.2538$;	$P_{loss1} = 2.55$;
$R_1 = 20$; $R_2 = 15$;		$P_{DC2} = -0.6102$;	$P_{loss2} = 1.42$;
$R_3 = 1000$; $R_4 = 10$;		$P_{DC3} = -1.0$;	$P_{loss3} = 1.79$;
$R_5 = 10$;		$P_{DC4} = -0.4231$;	$P_{loss4} = 1.29$;
		$P_{DC5} = -0.5216$;	$P_{loss5} = 1.35$;
Converter	**Control Mode**	**VSCs**	
1,2	V-P droop	$\theta_{sh1} = -46.8584$; $\theta_{sh2} = -32.7337$;	
3	V-P droop with VM	$\theta_{sh3} = -5.9638$; $\theta_{sh4} = -32.0917$; $\theta_{sh5} = -33.3638$;	
4,5	V-I droop	$m_1 = 0.9940$; $m_2 = 0.9799$; $m_3 = 0.9619$;	
		$m_4 = 0.9746$; $m_5 = 0.9772$;	
$V_{1280} = 0.98$; $Q_{sh2} = 0.1$;		AC terminal buses	
$Q_{sh3} = 0.05$; $Q_{sh4} = 0.09$;		$\theta_{1280} = -39.7571$;	
$Q_{sh5} = 0.08$;		$V_{1281} = 1.0367 \angle -35.8685$; $V_{1286} = 1.0182 \angle -11.3285$;	
$P_{DCWF1} = 0.7$; $P_{DCWF2} = 0.6$;		$V_{1292} = 1.0326 \angle -34.2579$; $V_{1313} = 1.0362 \angle -36.0443$;	
		NI = 7	

characteristic, similar to the base case. The bus voltage profile for this study is shown in Figure 6.15. From Figure 6.15, it is observed that the bus voltage profile hardly changes except at the AC buses connected to the VSCs.

Case II: Study with Model B employing Linear V-P and V-I droop characteristics incorporating OWFs

TABLE 6.6

Study of European 1354-Bus System with 7-Terminal VSC-HVDC Network Employing Nonlinear DC Voltage Droop Control and Incorporating OWFs (Model B)

Base case power flow converged in seven iterations (NI=7)

$V_{1280} = 0.9554 \angle -35.2574$; $V_{1281} = 1.0354 \angle -37.3082$; $V_{1286} = 1.0156 \angle -12.3760$;

$V_{1292} = 1.0289 \angle -35.23066$; $V_{1313} = 1.0351 \angle -37.4755$;

(VSCs connected to AC buses 1280, 1281, 1286, 1292 and 1313)

Power flow of AC-MTDC system with OWFs

Given Quantities		AC Buses	VSCs
		Solution	
V_{DCav}	3.02	$\theta_{1280} = -36.8321$;	$\theta_{sh1} = -40.0721$; $\theta_{sh2} = -33.9561$;
V_{1280}	0.98	$\theta_{1281} = -36.0647$;	$\theta_{sh3} = -10.3547$; $\theta_{sh4} = -31.9230$; $\theta_{sh5} = -34.6543$;
P_{sh2}	0.4	$\theta_{1286} = -11.9962$;	$m_1 = 0.9686$;
Q_{sh2}	0.09	$\theta_{1292} = -34.0417$;	$m_2 = 0.9801$; $m_3 = 0.9612$; $m_4 = 0.9781$;
P_{sh3}	0.3	$\theta_{1313} = -36.2403$;	$m_5 = 0.9775$; $V_{DC1} = 3.0203$; $V_{DC2} = 3.0198$;
Q_{sh3}	0.08	$V_{1281} = 1.0366$;	$V_{DC3} = 3.0199$; $V_{DC4} = 3.0198$; $V_{DC5} = 3.0199$;
P_{sh4}	0.4	$V_{1286} = 1.0177$;	Wind farm terminals
Q_{sh4}	0.1	$V_{1292} = 1.0335$;	$V_{DC6} = 3.0202$;
P_{sh5}	0.3	$V_{1313} = 1.0362$;	$V_{DC7} = 3.0202$;
Q_{sh5}	0.07		
Rectifying power of offshore wind farms			DC power	Converter loss (%)
$P_{DCWF1} = 0.5$;			$P_{DC1} = 0.5510$;	$P_{loss1} = 1.60$;
$P_{DCWF2} = 0.4$;			$P_{DC2} = -0.4130$;	$P_{loss2} = 1.29$;
			$P_{DC3} = -0.3124$;	$P_{loss3} = 1.23$;
			$P_{DC4} = -0.4130$;	$P_{loss4} = 1.28$;
			$P_{DC5} = -0.3124$;	$P_{loss5} = 1.23$;
	NI = 7			

Computation of references V_{DC}^* from droop equations

$V_{DC1} = 3.0203$; $V_{DC2} = 3.0198$;

$V_{DC3} = 3.0199$; $V_{DC4} = 3.0198$; $V_{DC5} = 3.0199$;

$V_{DC6} = 3.0202$; $V_{DC7} = 3.0202$;

$V_{DCmax} = 3.014$; $V_{DChigh}^* = 3.013$;

$V_{DClow} = 3.012$; $V_{DCmin} = 3.009$;

$R_1 = 20$; $R_2 = 15$; $R_3 = 10$; $R_4 = 15$; $R_5 = 10$; $R_{max} = 30$;

$V_{DC1}^* = 3.0138$; $V_{DC2}^* = 3.0133$;

$V_{DC3}^* = 3.0134$; $V_{DC4}^* = 3.0134$;

$V_{DC5}^* = 3.0134$;

$P_{DC1}^* = 0.6810$; $P_{DC2}^* = -0.3153$;

$P_{DC3}^* = -0.1268$; $P_{DC4}^* = -0.1219$;

$P_{DC5}^* = -0.1168$;

NI = 4

This study is similar to the study of Table 6.2 except that the VSC terminal end line active powers P_{sh2}, P_{sh3}, P_{sh4} and P_{sh5} are specified in this study. At first, an AC-MTDC power-flow is carried out with two OWFs connected at the terminals '6' and '7' of the DC grid. The powers specified for these two OWFs at terminals

TABLE 6.7

Study of European 1354-Bus System with 7-Terminal VSC-HVDC Network Employing DC Slack-Bus Control and Incorporating OWFs

Base case power flow converged in seven iterations (NI=7)

$V_{1280} = 0.9554 \angle - 35.2574$; $V_{1281} = 1.0354 \angle - 37.3082$; $V_{1286} = 1.0156 \angle - 12.3760$;

$V_{1292} = 1.0289 \angle - 35.23066$; $V_{1313} = 1.0351 \angle - 37.4755$;

AC-MTDC Power Flow with OWFs (Slack-Bus Control)

Given Quantities	Power-Flow Solution	
Master converter	Master converter	
$V_{1280} = 0.98$;	$\theta_{sh1} = -42.0550$; $m_1 = 0.9824$;	
$V_{DC1} = 3.0$;	Slave converters	
	$V_{DC2} = 2.9994$; $V_{DC3} = 2.9994$; $V_{DC4} = 2.9995$; $V_{DC5} = 2.9994$;	
	$m_2 = 0.9844$; $m_3 = 0.9652$; $m_4 = 0.9770$; $m_5 = 0.9808$;	
	$\theta_{sh2} = -33.3004$; $\theta_{sh3} = -9.7072$; $\theta_{sh4} = -32.6959$; $\theta_{sh5} = -33.9965$;	
Slave converters	AC terminal buses	Wind farm terminals
$P_{sh2} = 0.5$;	$\theta_{1280} = -37.6895$;	$V_{DC6} = 2.9999$;
$Q_{sh2} = 0.06$;	$V_{1281} = 1.0365 \angle - 35.9458$;	$V_{DC7} = 2.9998$;
$P_{sh3} = 0.4$;	$V_{1286} = 1.0175 \angle - 11.9055$;	
$Q_{sh3} = 0.05$;	$V_{1292} = 1.0315 \angle - 34.3022$;	
$P_{sh4} = 0.3$;	$V_{1313} = 1.0361 \angle - 36.1221$;	
$Q_{sh4} = 0.04$;		
$P_{sh5} = 0.4$;		
$Q_{sh5} = 0.03$;		
Rectifying power of offshore wind farms	DC power	Converter loss (%)
$P_{DCWF1} = 0.5$;	$P_{DC1} = 0.7523$;	$P_{loss1} = 1.82$;
$P_{DCWF2} = 0.4$;	$P_{DC2} = -0.5137$;	$P_{loss2} = 1.35$;
	$P_{DC3} = -0.4130$;	$P_{loss3} = 1.29$;
	$P_{DC4} = -0.3123$;	$P_{loss4} = 1.23$;
	$P_{DC5} = -0.4130$;	$P_{loss5} = 1.28$;
	NI=8	

'6' and '7' are 0.6 and 0.5, respectively. The other specified quantities are given in rows 3–14 and column 1–2 of Table 6.3. The power-flow solution is shown in rows 3–14 and columns 3–4 of Table 6.3. Subsequent to the AC-MTDC power flow, the VSC droop equations are solved to calculate the reference values of droop lines.

The convergence characteristic corresponding to the study of rows 3–14 of Table 6.3 is shown in Figure 6.16. From Figure 6.16, it is observed that the proposed model demonstrates the quadratic convergence characteristics, similar to the base case.

The bus voltage profile of the study corresponding to rows 3–14 of Table 6.3 is shown in Figure 6.17. From Figure 6.32, it is observed that by incorporating OWFs in the DC grid of the integrated AC-DC system, the bus voltage profile of

TABLE 6.8

Study of European 1354-Bus System with 7-Terminal VSC-HVDC Network Employing Nonlinear DC Voltage Droop Characteristics and Incorporating OWFs (Model B)

Base case power flow converged in seven iterations (NI = 7)

$V_{1280} = 0.9554 \angle -35.2574$; $V_{1281} = 1.0354 \angle -37.3082$; $V_{1286} = 1.0156 \angle -12.3760$;

$V_{1292} = 1.0289 \angle -35.23066$; $V_{1313} = 1.0351 \angle -37.4755$;

(VSCs connected to AC buses 1280, 1281, 1286, 1292 and 1313)

Power flow of AC-MTDC system with OWFs

Given		Solution	
Quantities		**AC Buses**	**VSCs**
V_{DCav}	3.0	$\theta_{1280} = -35.9694$;	$\theta_{sh1} = -38.0795$; $\theta_{sh2} = -33.3124$;
V_{1280}	0.98	$V_{1281} = 1.0367 \angle -35.9452$;	$\theta_{sh3} = -9.6939$; $\theta_{sh4} = -32.5092$;
P_{sh2}	0.5	$V_{1286} = 1.0182 \angle -11.8789$;	$\theta_{sh5} = -35.0786$; $m_1 = 0.9687$; $m_2 = 0.9880$;
Q_{sh2}	0.1	$V_{1292} = 1.0326 \angle -34.1041$;	$m_3 = 0.9694$; $m_4 = 0.9816$; $m_5 = 0.9837$;
P_{sh3}	0.4	$V_{1313} = 1.0362 \angle -36.1349$;	$V_{DC1} = 3.0002$; $V_{DC2} = 2.9998$;
Q_{sh3}	0.09		$V_{DC3} = 2.9998$; $V_{DC4} = 2.9999$; $V_{DC5} = 2.9999$;
P_{sh4}	0.3		Wind farm terminals
Q_{sh4}	0.08		$V_{DC6} = 3.0003$; $V_{DC7} = 3.0002$;
P_{sh5}	0.2		Converter loss (%)
Q_{sh5}	0.07		$P_{loss1} = 1.44$; $P_{loss2} = 1.35$; $P_{loss3} = 1.29$;
P_{DC2}	−0.5138		$P_{loss4} = 1.23$; $P_{loss5} = 1.18$;
P_{DC3}	−0.4131		
P_{DC4}	−0.3124		
P_{DC5}	−0.2119		
$P_{DCWF1} = 0.6$;		DC power	
$P_{DCWF2} = 0.5$;		$P_{DCI} = 0.3514$;	
	NI = 7		

Computation of references V_{DC}^* from droop equations

$V_{DC1} = 3.0002$; $V_{DC2} = 2.9998$; $V_{DC1}^* = 2.9910$; $V_{DC2}^* = 2.9905$; $V_{DC3}^* = 2.9905$;

$V_{DC3} = 2.9998$; $V_{DC4} = 2.9999$; $V_{DC5} = 2.9999$; $V_{DC4}^* = 2.9907$; $V_{DC5}^* = 2.9907$;

$V_{DC6} = 3.0003$; $V_{DC7} = 3.0002$;   NI = 3

$R_1 = 20$; $R_2 = 15$; $R_3 = 10$; $R_4 = 15$; $R_5 = 10$;

the AC system does not change except at the AC buses connected to the VSCs (Figure 6.17).

Case III: Study with nonlinear DC voltage droop characteristics incorporating OWFs (Model A)

This study is similar to the study carried out in Table 6.3 except that VSC 3 now employs a nonlinear V-P droop with dead band. The droop control gains

TABLE 6.9
Study of European 1354-Bus System with 7-Terminal VSC-HVDC Network Employing Nonlinear DC Voltage Droop Control and Incorporating OWFs (Model B)

Base case power flow converged in seven iterations (NI = 7)

$V_{1280} = 0.9554 \angle -35.2574$; $V_{1281} = 1.0354 \angle -37.3082$;

$V_{1286} = 1.0156 \angle -12.3760$;

$V_{1292} = 1.0289 \angle -35.23066$; $V_{1313} = 1.0351 \angle -37.4755$;

(VSCs connected to AC buses 1280, 1281, 1286, 1292 and 1313)

Power flow of AC-MTDC system with OWFs

Given Quantities		AC Buses	VSCs
			Solution
V_{DCav}	3.02	$\theta_{1280} = -36.8319$;	$\theta_{sh1} = -40.0718$; $\theta_{sh2} = -33.9560$;
V_{268}	0.98	$\theta_{1281} = -36.0647$;	$\theta_{sh3} = -10.3547$; $\theta_{sh4} = -31.9230$;
P_{sh2}	0.4	$\theta_{1286} = -11.9962$;	$\theta_{sh5} = -34.6543$; $m_1 = 0.9686$;
Q_{sh2}	0.09	$\theta_{1292} = -34.0417$;	$m_2 = 0.9801$; $m_3 = 0.9612$;
P_{sh3}	0.3	$\theta_{1313} = -36.2402$;	$m_4 = 0.9781$; $m_5 = 0.9775$;
Q_{sh3}	0.08	$V_{1281} = 1.0366$;	$V_{DC1} = 3.0203$; $V_{DC2} = 3.0198$;
P_{sh4}	0.4	$V_{1286} = 1.0177$;	$V_{DC3} = 3.0199$; $V_{DC4} = 3.0198$; $V_{DC5} = 3.0199$;
Q_{sh4}	0.1	$V_{1292} = 1.0335$;	Wind farm terminals
P_{sh5}	0.3	$V_{1313} = 1.0362$;	$V_{DC6} = 3.0202$; $V_{DC7} = 3.0202$;
Q_{sh5}	0.07		Converter loss (%)
P_{DC2}	−0.4130		$P_{loss1} = 1.60$; $P_{loss2} = 1.29$;
P_{DC3}	−0.3124		$P_{loss3} = 1.23$; $P_{loss4} = 1.28$;
P_{DC4}	−0.4130		$P_{loss5} = 1.23$;
P_{DC5}	−0.3124		

Rectifying power of offshore wind farms

DC power

$P_{DCI} = 0.5510$;

$P_{DCWF1} = 0.5$;

NI = 7

$P_{DCWF2} = 0.4$;

Computation of references V_{DC}^* from droop eqns.

$V_{DC1} = 3.0203$; $V_{DC2} = 3.0198$;

$V_{DC3} = 3.0199$; $V_{DC4} = 3.0198$;

$V_{DC5} = 3.0199$; $V_{DC6} = 3.0202$;

$V_{DC7} = 3.0202$;

$V_{DCmax} = 3.014$; $V_{DChigh}^* = 3.013$;

$V_{DClow}^* = 3.012$; $V_{DCmin} = 3.009$;

$R_1 = 20$; $R_2 = 15$; $R_3 = 10$;

$R_4 = 15$; $R_5 = 10$; $R_{max} = 30$;

$V_{DC1}^* = 3.0138$; $V_{DC2}^* = 3.0133$;

$V_{DC3}^* = 3.0134$; $V_{DC4}^* = 3.0134$; $V_{DC5}^* = 3.0134$;

$P_{DC1}^* = 0.6809$; $P_{DC2}^* = -0.3153$;

$P_{DC3}^* = -0.1268$; $I_{DC4}^* = 0.0404$;

$I_{DC5}^* = 0.0388$;

NI = 4

TABLE 6.10

A Comparison of Convergence Features with Existing Models

Ref No.	Tolerance (p.u.)	No. of Buses AC	No. of Buses DC	NI
[92]	10^{-8}	29	5	Min:3
				Max:15
[88]	10^{-6}	9	4	6
		32	4	7
Proposed model	10^{-8}	1354	NIL	7
			7 (DC slack bus)	7 (unified)
				8 (sequential)
			7 (linear voltage droop)	7 (both unified and sequential)
			7 (nonlinear droop with dead band)	7 (both unified and sequential)
			7 (nonlinear DC voltage margin)	7 (both unified and sequential)

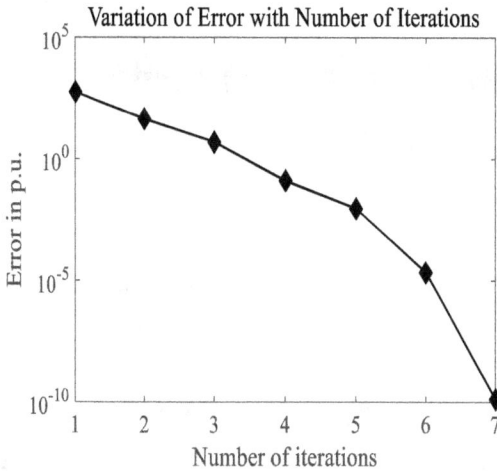

FIGURE 6.11 Convergence characteristic of the base case power flow in the European 1354-bus system.

are kept identical to the studies of Table 6.3. At first, the reference values for the droop lines are computed by solving the droop equations. The results are shown in rows 2–4 and column 2 of Table 6.4. Subsequently, an AC-MTDC power flow is conducted. The results are shown in rows 5–13 and column 3–4 of Table 6.4. From the power-flow results shown in Table 6.4, it is observed that the VSC connected at AC bus 1286 operates at point 'A' of the characteristic (Figure 6.3).

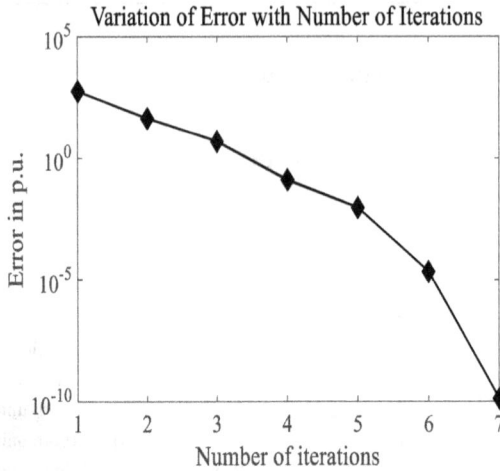

FIGURE 6.12   Convergence characteristic for the study of Table 6.1.

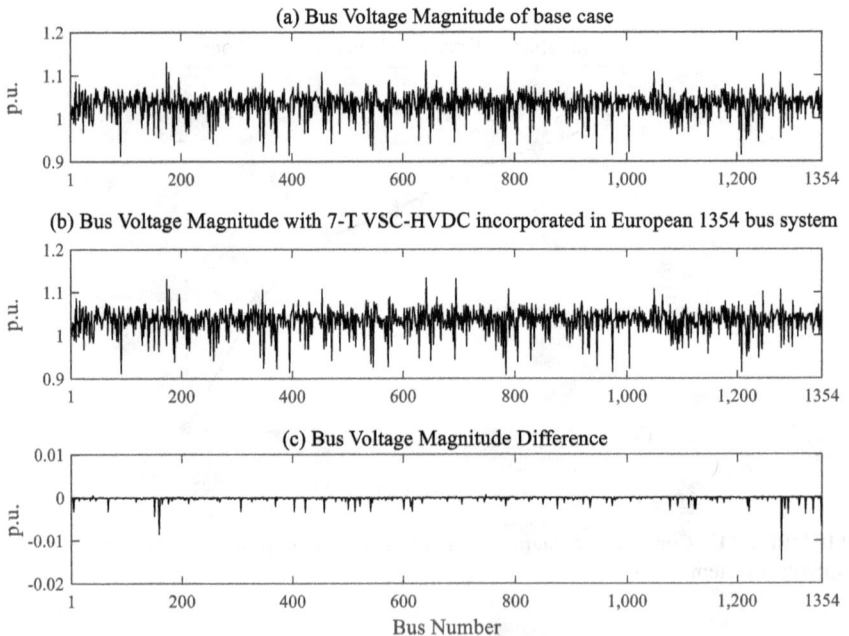

FIGURE 6.13   Bus voltage profile for the study of Table 6.1.

The convergence characteristic corresponding to the study of rows 5–13 of Table 6.4 is shown in Figure 6.18. From Figure 6.18, it is observed that the proposed model demonstrates excellent convergence characteristics, converging in seven iterations. The bus voltage profile for the studies of Table 6.4 is shown in Figure 6.19. From Figure 6.19, it is again observed that in the presence of MTDC

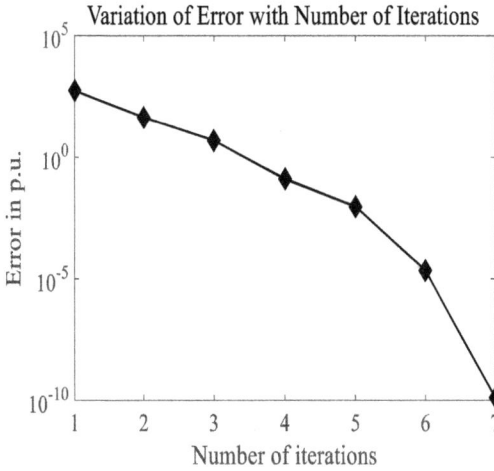

FIGURE 6.14   Convergence characteristic for the study of Table 6.2.

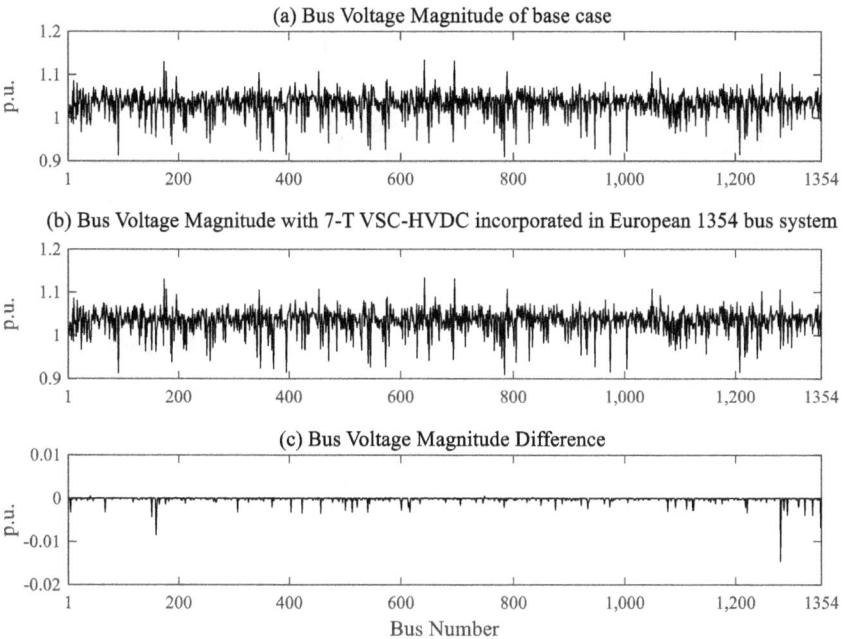

FIGURE 6.15   Bus voltage profile for the study of Table 6.2.

grid with OWFs, the bus voltage profile does not alter much except at the AC buses at which the VSCs are connected.

Subsequently, another study is carried out on the same AC-MTDC system but with the VSC connected at bus 1286 operated in the DC voltage margin control mode. The droop gain of the converter connected to AC bus 1286 is set to a

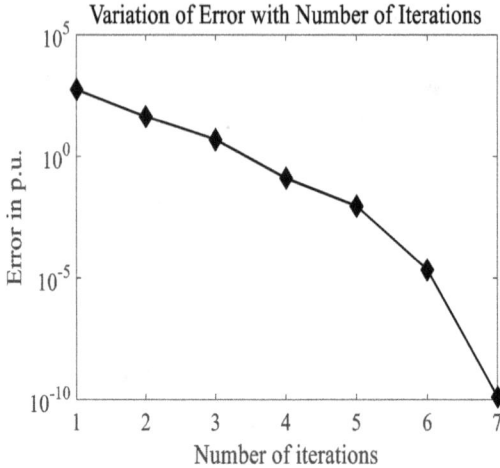

FIGURE 6.16   Convergence characteristic of rows 2–10 in Table 6.3.

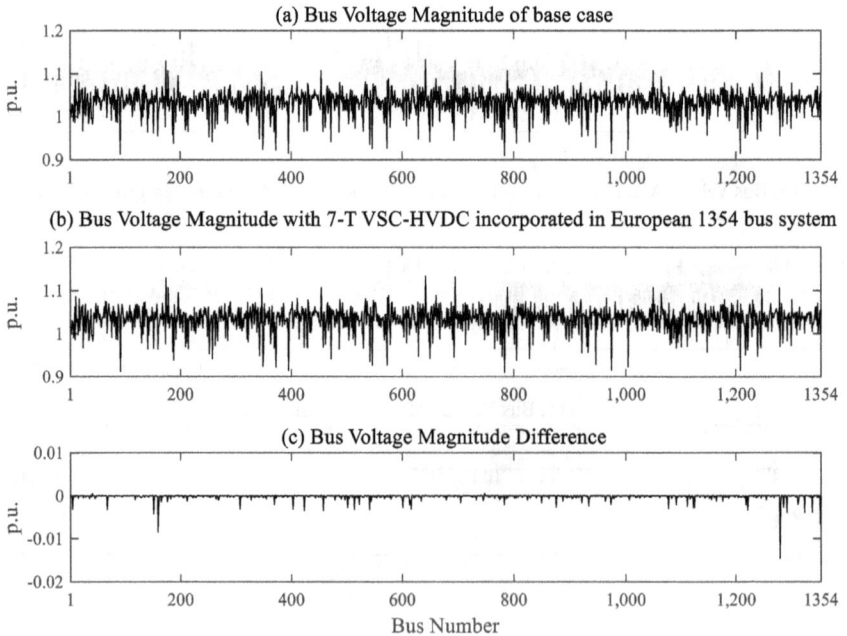

FIGURE 6.17   Bus voltage profile for the study of Table 6.3.

value of 1000 for voltage margin control. For voltage margin control, the maximum and minimum DC powers are set to values of 1.0 and −1.0 p.u., respectively. The power-flow solution is shown in Table 6.5. From Table 6.5, it is observed that the convergence characteristic of the proposed model is independent of the power injections from the OWFs and the location of the MTDC grid in the AC system.

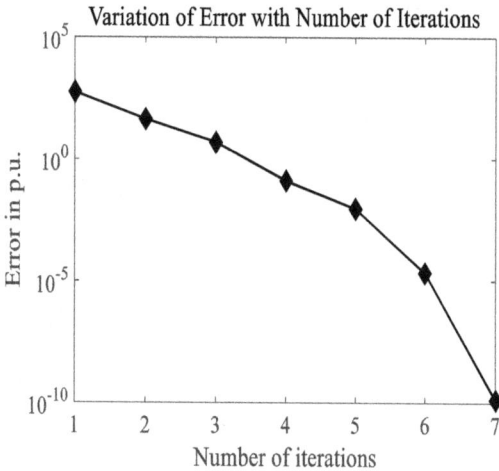

FIGURE 6.18   Convergence characteristic of rows 2–10 in Table 6.4.

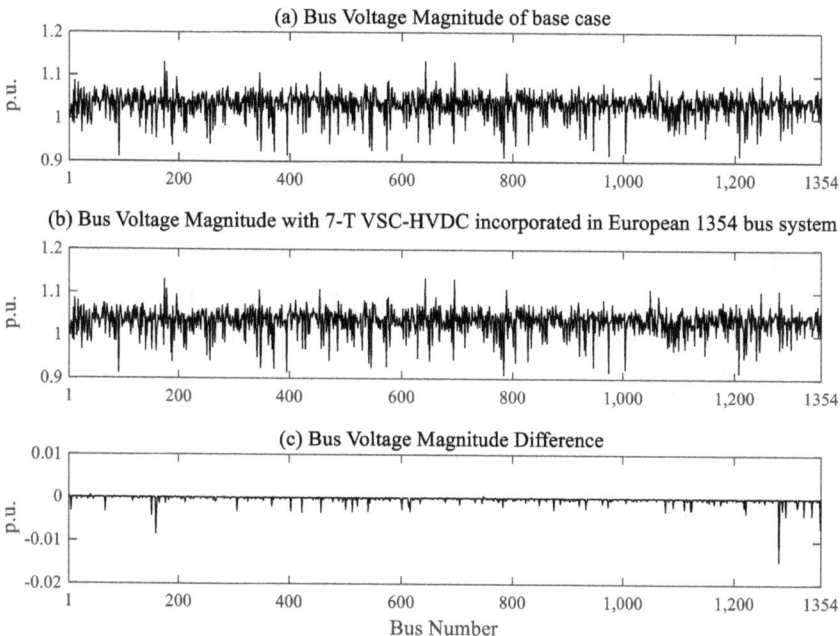

FIGURE 6.19   Bus voltage profile for the study of Table 6.4.

The convergence characteristics of rows 5–13 of Table 6.5 are shown in Figure 6.20. From Figure 6.20, it is observed that the proposed algorithm demonstrates excellent convergence characteristics, converging in seven iterations. The bus voltage profile for the study of Table 6.5 is shown in Figure 6.21. From Figure 6.21, it is again observed that in the presence of MTDC grid with OWFs,

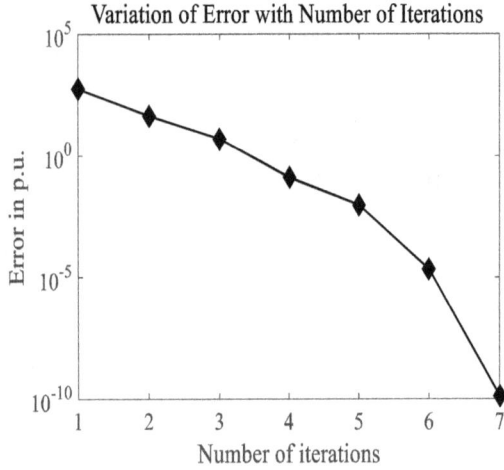

FIGURE 6.20 Convergence characteristic of rows 2–10 in Table 6.5.

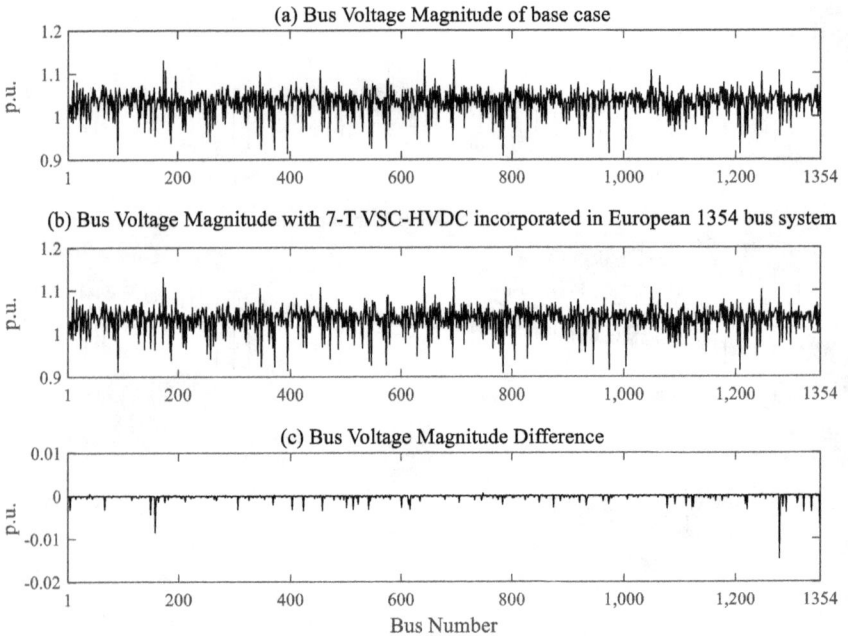

FIGURE 6.21 Bus voltage profile for the study of Table 6.5.

the bus voltage profile does not alter much except at the buses at which the VSCs are connected.

Case IV: Study with nonlinear DC voltage droop characteristics and incorporating OWFs (Model B)

Firstly, an AC-MTDC power flow is carried out with two OWFs connected at the terminals '6' and '7' of the DC grid. The specified quantities are given in

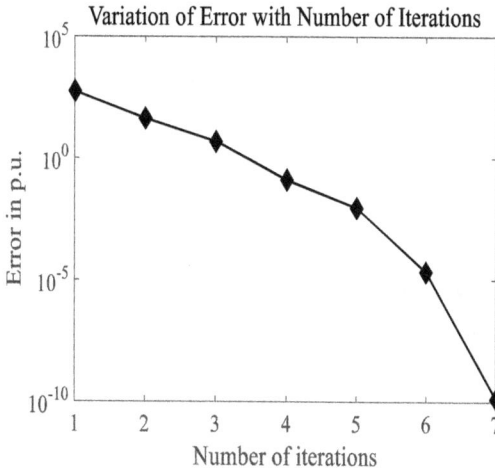

FIGURE 6.22   Convergence characteristic of rows 3–14 in Table 6.6.

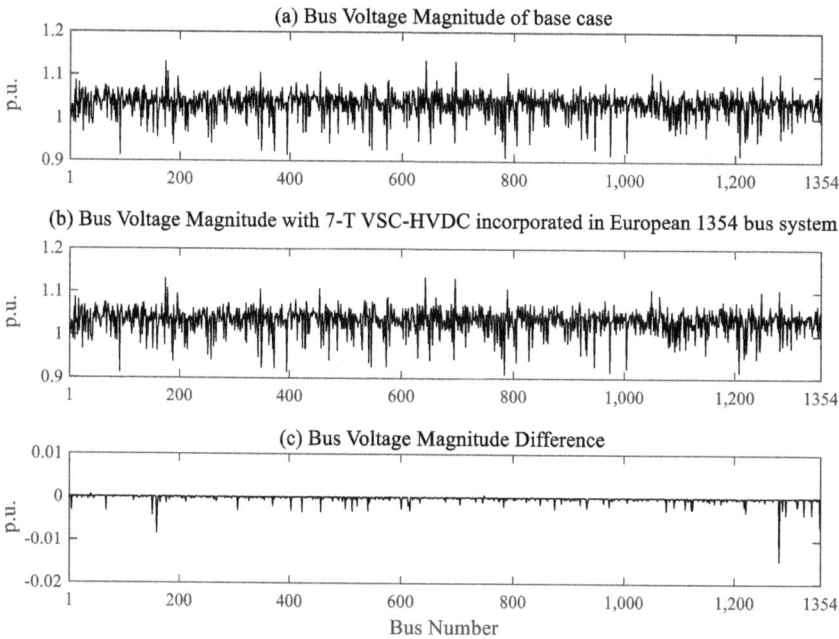

FIGURE 6.23   Bus voltage profile for the study of Table 6.6.

rows 3–14 and column 1–2 of Table 6.6. The power-flow solution is shown in rows 3–14 and columns 3–4 of Table 6.6. Subsequent to the AC-MTDC power flow, the VSC droop equations are solved to calculate the reference values of nonlinear droop lines.

The convergence characteristics of rows 3–14 of Table 6.6 are shown in Figure 6.22. From Figure 6.22, it is observed that the proposed algorithm demonstrates excellent convergence characteristics, converging in seven iterations. The bus voltage profile for the study of Table 6.6 is shown in Figure 6.23. From Figure 6.23, it is again observed that in the presence of MTDC grid with OWFs, the bus voltage profile does not alter much except at the buses at which the VSCs are connected.

6.6.3 STUDY WITH SEQUENTIAL POWER-FLOW MODEL OF EUROPEAN 1354 BUS TEST SYSTEM INTEGRATED WITH 7-TERMINAL MVDC NETWORK EMPLOYING DC SLACK-BUS CONTROL AND INTERFACED WITH OFFSHORE WIND FARMS (MODEL-B)

This study is similar to the study of Table 6.1 except the power-flow method. The sequential power-flow method is used to solve the power-flow equations in this study. The specified quantities are shown in rows 3–6 and column 1 of Table 6.7. The power-flow solution is shown in rows 3–6 and column 2–3 of Table 6.7.

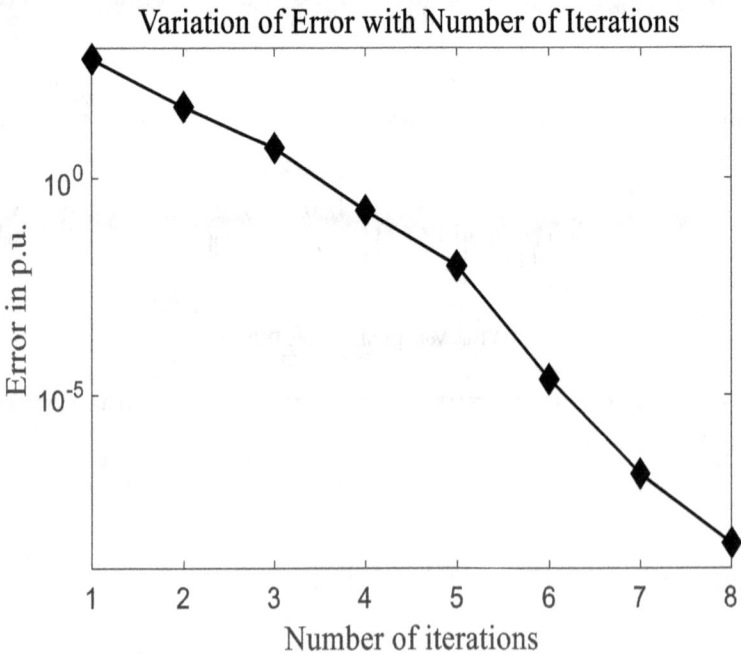

FIGURE 6.24   Convergence characteristic of rows 2–6 in Table 6.7.

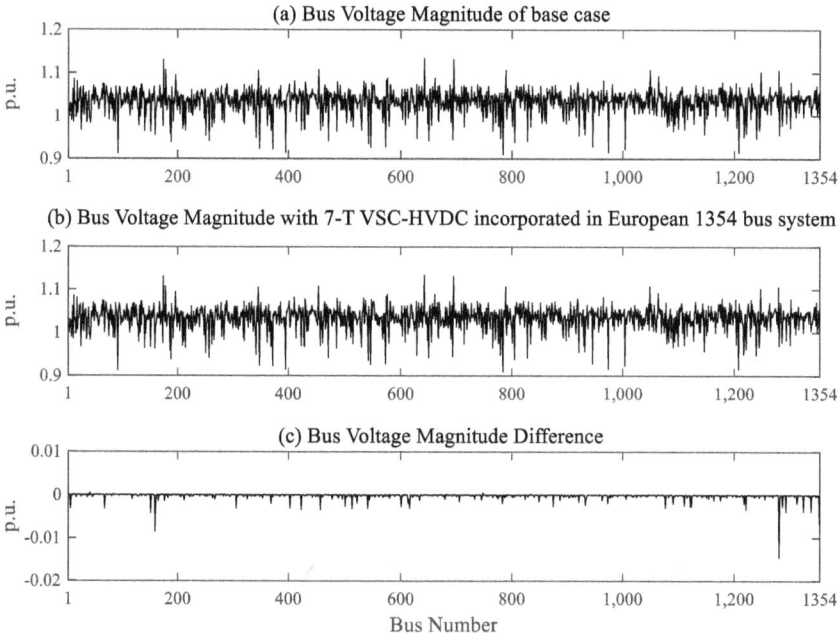

FIGURE 6.25   Bus voltage profile for the study of Table 6.7.

The convergence characteristics of rows 3–6 of Table 6.7 are shown in Figure 6.24. From Figure 6.24, it is observed that the proposed algorithm demonstrates excellent convergence characteristics, converging in eight iterations. The bus voltage profile for the study of Table 6.7 is shown in Figure 6.25. From Figure 6.25, it is again observed that in the presence of MTDC grid with OWFs, the bus voltage profile does not alter much except at the buses at which the VSCs are connected.

6.6.4 Study with Sequential Power-Flow Model of European 1354-Bus System Integrated with 7-Terminal MVDC Network Employing DC Voltage Droop Control and Interfaced with Offshore Wind Farms (Model B)

Based on the quantities specified, two AC-MTDC power-flow models have been developed. These models are named as 'Model A' and 'Model B' and have been described in Chapter 4 of this book.

Model B employing linear DC V-P and V-I voltage droop characteristics and interfaced with OWFs

The sequential method is used to solve the AC-DC power-flow equations. The powers specified for these two OWFs at terminals '6' and '7' are 0.6 and 0.5,

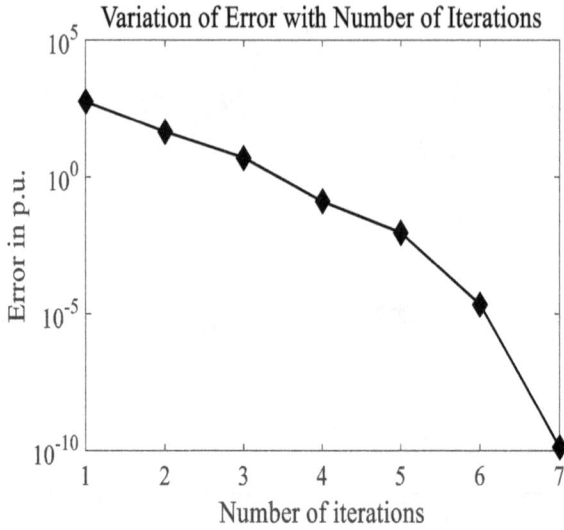

FIGURE 6.26 Convergence characteristic of rows 3–18 in Table 6.8.

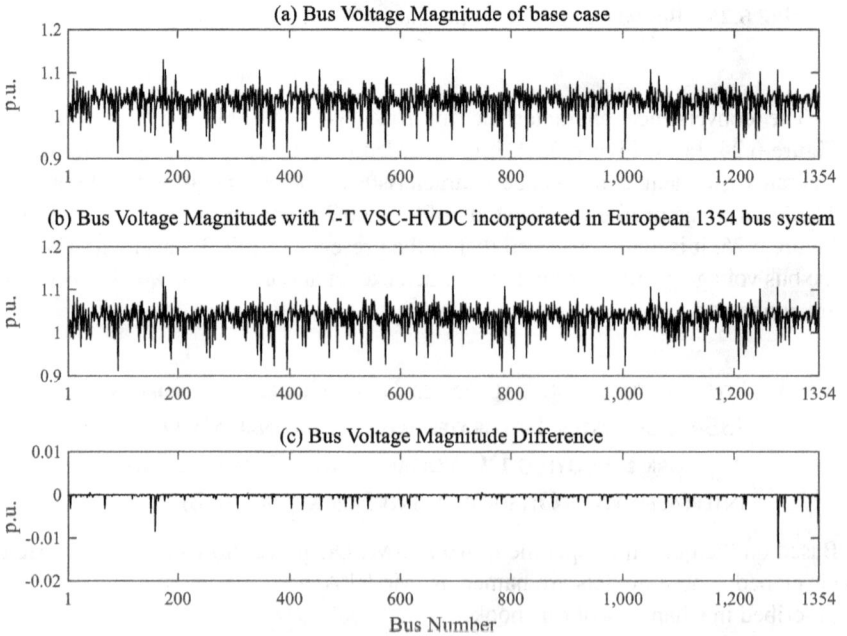

FIGURE 6.27 Bus voltage profile for the study of Table 6.8.

respectively. The other specified quantities are given in rows 3–18 and column 1–2 of Table 6.8. The power-flow solution is shown in rows 3–18 and columns 3–4 of Table 6.8. Subsequent to the AC-MTDC power flow, the VSC droop equations are solved to calculate the reference values of droop lines.

The convergence characteristics of rows 3–18 of Table 6.8 are shown in Figure 6.26. From Figure 6.26, it is observed that the proposed algorithm demonstrates excellent convergence characteristics, converging in eight iterations. The bus voltage profile for the study of Table 6.8 is shown in Figure 6.27. From Figure 6.27, it is again observed that in the presence of MTDC grid with OWFs, the bus voltage profile does not alter much except at the buses at which the VSCs are connected.

Study with nonlinear DC voltage droop characteristics and interfaced with offshore wind farms (Model B)

The sequential power-flow method is used to solve the AC-DC equations in this study. Firstly, an AC-MTDC power-flow is carried out with two OWFs connected at the terminals '6' and '7' of the DC grid. The specified quantities are given in rows 3–18 and column 1–2 of Table 6.9. The power-flow solution is shown in rows 3–18 and columns 3–4 of Table 6.9. Subsequent to the AC-MTDC power flow, the VSC droop equations are solved to calculate the reference values of nonlinear droop lines.

The convergence characteristics of rows 3–18 of Table 6.9 are shown in Figure 6.28. From Figure 6.28, it is observed that the proposed algorithm demonstrates excellent convergence characteristics, converging in eight iterations. The bus voltage profile for the study of Table 6.9 is shown in Figure 6.29. From Figure 6.29, it is again observed that in the presence of MTDC grid with OWFs,

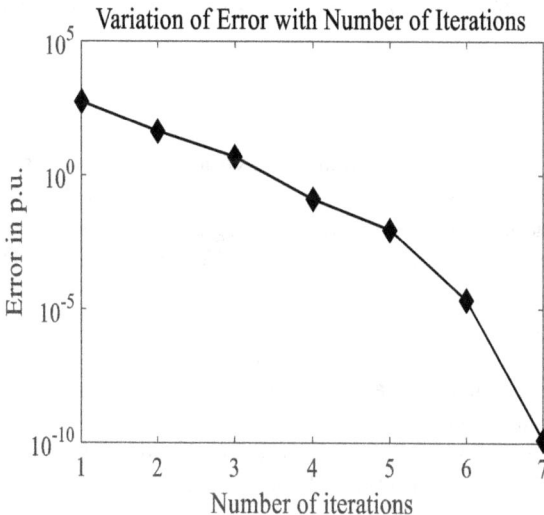

FIGURE 6.28   Convergence characteristic of rows 3–18 in Table 6.9.

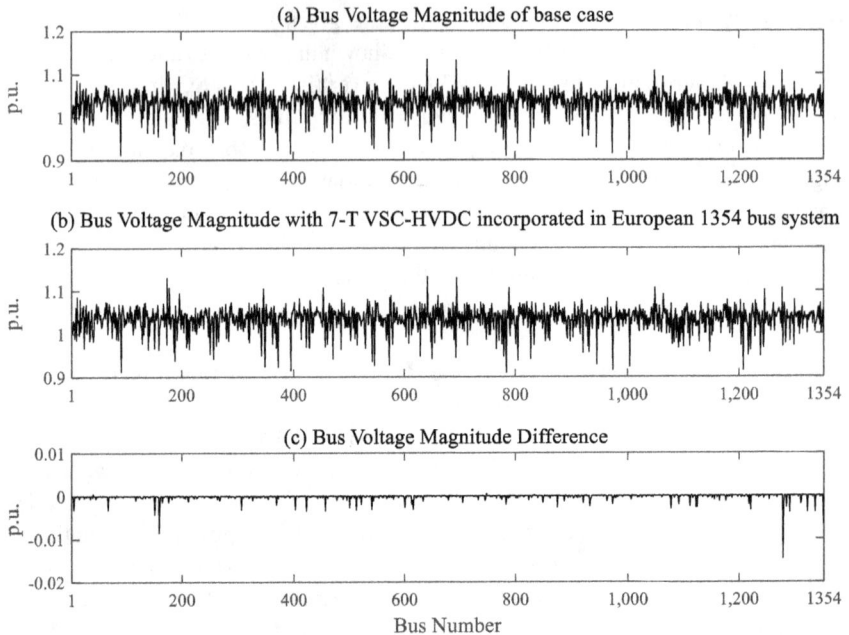

FIGURE 6.29 Bus voltage profile for the study of Table 6.9.

the bus voltage profile does not alter much except at the buses at which the VSCs are connected.

In addition, the proposed model has also been compared with some other existing models and a comparison of the convergence features is shown in Table 6.10.

6.7 SUMMARY

In this chapter, the power-flow model of hybrid AC-MTDC systems integrated with offshore wind farms has been developed. The proposed model was investigated by interfacing OWFs to different topologies of multi-terminal VSC-MTDC grids and integrating them with the European 1354-bus networks. Diverse MTDC grid control techniques including linear and nonlinear DC voltage droop control have employed. Both Models 'A' and 'B' were implemented for droop control. Droop Model 'B' facilitates the specification of both line end active and reactive power flows. Both unified and sequential power-flow results are incorporated. The model displays excellent convergence characteristics independent of the DC grid topology, the DC voltage droop control employed, the location of the MTDC grid and the OWF power injections. This validates the model.

Appendix
Derivations of Difficult Expressions

A.1 EXPRESSION FOR THE NET ACTIVE AND REACTIVE POWER INJECTIONS BY INCORPORATING THREE-TERMINAL LCC HVDC SYSTEM IN AN EXISTING AC SYSTEM USING UNIFIED METHOD

For AC buses 'i', 'j' and 'k', the rectifier is connected at bus 'i'. Two inverters '1' and '2' are connected at buses 'j' and 'k'. respectively.

For control strategy '1' and '5'

$$P_i = \sum_{p=1}^{n} V_i V_p Y_{ip} \cos\left(\theta_i - \theta_p - \phi_{ip}\right) + V_{DCR} \left(\frac{2\,V_{DCR} - V_{DCI1} - V_{DCI2}}{R_{DC}} \right) \quad \text{(A.1)}$$

$$Q_i = \sum_{p=1}^{n} V_i V_p Y_{ip} \sin\left(\theta_i - \theta_p - \phi_{ip}\right) + V_{DCR} \left(\frac{2\,V_{DCR} - V_{DCI1} - V_{DCI2}}{R_{DC}} \right) \tan\Phi_R$$

$$\text{(A.2)}$$

$$P_j = \sum_{p=1}^{n} V_j V_p Y_{jp} \cos\left(\theta_j - \theta_p - \phi_{jp}\right) - P_{DCI1} \quad \text{(A.3)}$$

$$Q_j = \sum_{p=1}^{n} V_j V_p Y_{jp} \sin\left(\theta_j - \theta_p - \phi_{jp}\right) + P_{DCI1}\tan\Phi_{I1} \quad \text{(A.4)}$$

$$P_k = \sum_{p=1}^{n} V_k V_p Y_{kp} \cos\left(\theta_k - \theta_p - \phi_{kp}\right) - P_{DCI2} \quad \text{(A.5)}$$

$$Q_k = \sum_{p=1}^{n} V_k V_p Y_{kp} \sin\left(\theta_k - \theta_p - \phi_{kp}\right) + P_{DCI2}\tan\Phi_{I2} \quad \text{(A.6)}$$

A.1.1 SOME TYPICAL ELEMENTS AND SUB-MATRICES OF JACOBIAN IN EQ. (2.21) FOR CONTROL STRATEGIES '1' AND '5' (UNIFIED METHOD)

$$\frac{\partial P_i}{\partial V_{DCI1}} = -\frac{V_{DCR}}{R_{DC}}; \quad \frac{\partial P_i}{\partial V_{DCI2}} = -\frac{V_{DCR}}{R_{DC}}; \quad \frac{\partial Q_i}{\partial V_{DCI1}} = -\frac{V_{DCR}}{R_{DC}}\tan\Phi_R;$$

$$\frac{\partial Q_i}{\partial V_{DCI2}} = -\frac{V_{DCR}}{R_{DC}}\tan\Phi_R;$$

$$\frac{\partial Q_i}{\partial \Phi_R} = V_{DCR}\left[\frac{2\,V_{DCR} - V_{DCI1} - V_{DCI2}}{R_{DC}}\right]\sec\Phi_R^2; \quad \frac{\partial Q_j}{\partial \Phi_{I1}} = P_{DCI1}\ \sec\Phi_{I1}^2;$$

$$\frac{\partial Q_k}{\partial \Phi_{I2}} = P_{DCI2}\ \sec\Phi_{I2}^2;$$

Sub-matrix of control strategy '1' for Model '1' is given below:

$$\frac{\partial f}{\partial X} = \begin{bmatrix} -kk_1 & -kk_1 & -V_i\cos\alpha_R & 0 & 0 & 0 & 0 & 0 \\ 1-kk_1 & 0 & 0 & -V_j\cos\gamma_{I1} & 0 & 0 & 0 & 0 \\ 0 & 1-kk_1 & 0 & 0 & -V_k\cos\gamma_{I2} & 0 & 0 & 0 \\ 0 & 0 & -V_i\cos\Phi_R & 0 & 0 & a_R\,V_i\,\sin\Phi_R & 0 & 0 \\ 1 & 0 & 0 & -V_j\cos\Phi_{I1} & 0 & 0 & a_{I1}\,V_j\,\sin\Phi_{I1} & 0 \\ 0 & 0 & 0 & 0 & -V_k\cos\Phi_{I2} & 0 & 0 & a_{I2}\,V_k\,\sin\Phi_{I2} \\ kk_2 & 0 & 0 & 0 & 0 & 0 & 0 & 0 \\ 0 & kk_2 & 0 & 0 & 0 & 0 & 0 & 0 \end{bmatrix}$$

$$kk_1 = \frac{X_c}{R_{DC}}; \quad kk_2 = -V_{DCR} + 2V_{DCI1};$$

Sub-matrix of control strategy '1' for Model '2' is given below:

$$\frac{\partial f}{\partial X} = \begin{bmatrix} -kk_1 & -kk_1 & -k_2V_i\cos\alpha_R & 0 & 0 & 0 & 0 & 0 \\ 1-kk_1 & 0 & 0 & -k_2V_j\cos\gamma_{I1} & 0 & 0 & 0 & 0 \\ 0 & 1-kk_1 & 0 & 0 & -k_2V_k\cos\gamma_{I2} & 0 & 0 & 0 \\ 0 & 0 & -k_2V_i\cos\Phi_R & 0 & 0 & k_2\,a_R\,V_i\,\sin\Phi_R & 0 & 0 \\ 1 & 0 & 0 & -k_2V_j\cos\Phi_{I1} & 0 & 0 & k_2\,a_{I1}V_j\,\sin\Phi_{I1} & 0 \\ 0 & 0 & 0 & 0 & -k_2\,V_k\cos\Phi_{I2} & 0 & 0 & k_2\,a_{I2}V_k\,\sin\Phi_{I2} \\ kk_2 & 0 & 0 & 0 & 0 & 0 & 0 & 0 \\ 0 & kk_2 & 0 & 0 & 0 & 0 & 0 & 0 \end{bmatrix}$$

where $k_1 = 3\dfrac{X_c\,n_b}{\pi\,R_{DC}}$; $k_2 = \dfrac{3\sqrt{2}\,n_b}{\pi}$; $kk_2 = -V_{DCR} + 2V_{DCI1}$;

Sub-matrix of control strategy '5' for Model '1'

$$\frac{\partial \mathbf{f}}{\partial \mathbf{X}} = \begin{bmatrix} -kk_1 & -kk_1 & -a_R V_i & 0 & 0 & 0 & 0 & 0 \\ 1-kk_1 & 0 & 0 & -a_{I1} V_j & 0 & 0 & 0 & 0 \\ 0 & 1-kk_1 & 0 & 0 & -a_{I2} V_k & 0 & 0 & 0 \\ 0 & 0 & 0 & 0 & 0 & a_R V_i \sin\Phi_R & 0 & 0 \\ 1 & 0 & 0 & 0 & 0 & 0 & a_{I1} V_j \sin\Phi_{I1} & 0 \\ 0 & 0 & 0 & 0 & 0 & 0 & 0 & a_{I2} V_k \sin\Phi_{I2} \\ kk_2 & 0 & 0 & 0 & 0 & 0 & 0 & 0 \\ 0 & kk_2 & 0 & 0 & 0 & 0 & 0 & 0 \end{bmatrix}$$

where $kk_1 = \dfrac{X_c}{R_{DC}}$; $kk_2 = -V_{DCR} + 2V_{DCI1}$;

Sub-matrix of control strategy '5' for Model '2'

$$\frac{\partial \mathbf{f}}{\partial \mathbf{X}} = \begin{bmatrix} -kk_1 & -kk_1 & -k_2 a_R V_i & 0 & 0 & 0 & 0 & 0 \\ 1-kk_1 & 0 & 0 & -k_2 a_{I1} V_j & 0 & 0 & 0 & 0 \\ 0 & 1-kk_1 & 0 & 0 & -k_2 a_{I2} V_k & 0 & 0 & 0 \\ 0 & 0 & 0 & 0 & 0 & -k_2 a_R V_i \sin\Phi_R & 0 & 0 \\ 1 & 0 & 0 & 0 & 0 & 0 & -k_2 a_{I1} V_j \sin\Phi_{I1} & 0 \\ 0 & 0 & 0 & 0 & 0 & 0 & 0 & -k_2 a_{I2} V_k \sin\Phi_{I2} \\ kk_2 & 0 & 0 & 0 & 0 & 0 & 0 & 0 \\ 0 & kk_2 & 0 & 0 & 0 & 0 & 0 & 0 \end{bmatrix}$$

where $k_1 = 3\dfrac{X_c \, n_b}{\pi \, R_{DC}}$; $k_2 = \dfrac{3\sqrt{2} \, n_b}{\pi}$; $kk_2 = -V_{DCR} + 2V_{DCI1}$;

A.1.2 INITIAL VALUES OF VARIABLES IN LCC-BASED HVDC SYSTEMS

TABLE A.1
Initial values of variables in LCC based HVDC systems

Variables in LCC HVDC System		Initial Values
DC bus voltage	V_{DC}^0	1 p.u. (model 1); 2.3 p.u. (model 2)
Converter transformer tap ratios	a_R^0, a_{I1}^0 and a_{I2}^0	1
Firing angle of rectifier	α_R^0	$\cos\alpha_R^0 = 1$
Extinction angles of inverters	γ_{I1}^0 and γ_{I2}^0	$\cos\gamma_{I1}^0 = 1; \cos\gamma_{I2}^0 = 1$
Power factors	$\cos\Phi_R^0, \cos\Phi_{I1}^0$ and $\cos\Phi_{I2}^0$	0.9

A.2 EXPRESSION FOR THE MAGNITUDE OF CONVERTER CURRENT (I_{sha}) FOR CONSIDERING CONVERTER LOSSES OF ATH VSC

From Eq. (3.1),

$$S_{sha} = V_{i+a-1} \, I_{sha}^* = V_{i+a-1} \left[\, y_{sha}^* \left(V_{sha}^* - V_{i+a-1}^* \right) \right] \tag{A.7}$$

$$= V_{i+a-1} \, m_a \, c \, V_{DCa} \, y_{sha} \, \angle \, \left(\theta_{i+a-1} - \theta_{sha} - \Phi_{sha} \right) - V_{i+a-1}^2 \, y_{sha} \, \angle - \Phi_{sha}$$

$$= V_{i+a-1} \, m_a \, c \, V_{DCa} \, y_{sha} \left\{ \cos\left(\theta_{i+a-1} - \theta_{sha} - \Phi_{sha} \right) + j \sin\left(\theta_{i+a-1} - \theta_{sha} - \Phi_{sha} \right) \right\}$$

$$- V_{i+a-1}^2 \, y_{sha} \left\{ \cos\Phi_{sha} - j \sin\Phi_{sha} \right\}$$

$$= V_{i+a-1} \, m_a \, c \, V_{DCa} \, y_{sha} \, \cos\left(\theta_{i+a-1} - \theta_{sha} - \Phi_{sha} \right) - V_{i+a-1}^2 \, y_{sha} \cos\Phi_{sha}$$

$$+ j \left\{ V_{i+a-1} m_a \, c \, V_{DCa} \, y_{sha} \, \sin\left(\theta_{i+a-1} - \theta_{sha} - \Phi_{sha} \right) + V_{i+a-1}^2 \, y_{sha} \sin\Phi_{sha} \right\}$$

$$= \alpha + j \, \beta$$

where $\alpha = V_{i+a-1} \, m_a \, c \, V_{DCa} \, y_{sha} \cos\left(\theta_{i+a-1} - \theta_{sha} - \Phi_{sha} \right) - V_{i+a-1}^2 \, y_{sha} \cos\Phi_{sha}$

and $\beta = V_{i+a-1} m_a \, c \, V_{DCa} \, y_{sha} \, \sin\left(\theta_{i+a-1} - \theta_{sha} - \Phi_{sha} \right) + V_{i+a-1}^2 \, y_{sha} \sin\Phi_{sha}$

$$S_{sha}^2 = \alpha^2 + \beta^2 = V_{i+a-1}^2 \, V_{sha}^2 \, y_{sha}^2 + V_{i+a-1}^4 y_{sha}^2 - 2 \, V_{i+a-1}^3 V_{sha} y_{sha}^2$$

$$\left\{ \cos\Phi_{sha} \cos\left(\theta_{i+a-1} - \theta_{sha} - \Phi_{sha} \right) - \sin\Phi_{sha} \sin\left(\theta_{i+a-1} - \theta_{sha} - \Phi_{sha} \right) \right\}$$

Now, using formula $\cos A \cos B - \sin A \sin B = \cos (A+B)$,

$$S_{sha}^2 = V_{i+a-1}^2 \, V_{sha}^2 \, y_{sha}^2 + V_{i+a-1}^4 y_{sha}^2 - 2 \, V_{i+a-1}^3 V_{sha} \, y_{sha}^2 \, \cos\left(\theta_{i+a-1} - \theta_{sha} \right)$$

$$= y_{sha}^2 \left[V_{i+a-1}^4 + V_{i+a-1}^2 \left(m_a c \, V_{DCa} \right)^2 - 2 \, V_{i+a-1}^3 \, m_a \, c \, V_{DCa} \cos\left(\theta_{i+a-1} - \theta_{sha} \right) \right]$$

$$S_{sha} = y_{sha} \left[V_{i+a-1}^4 + V_{i+a-1}^2 \left(m_a c \, V_{DCa} \right)^2 - 2 \, V_{i+a-1}^3 \, m_a \, c \, V_{DCa} \cos\left(\theta_{i+a-1} - \theta_{sha} \right) \right]^{1/2}$$

$$I_{sha} = \frac{S_{sha}}{V_{i+a-1}}$$

$$I_{sha} = y_{sha} \left[V_{i+a-1}^2 + \left(m_a c \, V_{DCa} \right)^2 - 2 \, V_{i+a-1} \, m_a \, c \, V_{DCa} \cos\left(\theta_{i+a-1} - \theta_{sha} \right) \right]^{1/2} \tag{A.8}$$

A.2.1 EXPRESSION OF POWER BALANCE EQUATION WITH CONVERTER LOSSES OF ATH VSC USING UNIFIED METHOD

From Eq. (3.14),

For master converter,

$$\left(m_a c\, V_{DCa}\right)^2 y_{sha}\cos\phi_{sha} - m_a c\, V_{DCa}V_{i+a-1}y_{sha}\cos\left(\theta_{sha} - \theta_{i+a-1} - \phi_{sha}\right)$$

$$+ \sum_{v=1}^{q} V_{DCa}V_{DCv}Y_{DCav} + a_1 + b_1 d_1 + c_1 d_1^{\,2} = 0 \qquad (A.9)$$

where $d_1 = y_{sha}\left[V_{i+a-1}^2 + m_a^2 c^2 V_{DCa}^2 - 2\,m_a\,c\,V_{DCa}V_{i+a-1}\cos\left(\theta_{i+a-1} - \theta_{sha}\right)\right]^{1/2}$

For slave converter,

$$\left(m_a c\, V_{DCa}\right)^2 y_{sha}\cos\phi_{sha} - m_a c\, V_{DCa}V_{i+a-1}y_{sha}\cos\left(\theta_{sha} - \theta_{i+a-1} - \phi_{sha}\right)$$

$$+ \sum_{v=1}^{q} V_{DCa}V_{DCv}Y_{DCav} + a_1 + b_1 \frac{\sqrt{P_{sha}^{sp\ 2} + Q_{sha}^{sp\ 2}}}{V_{i+a-1}} + c_1 \left(\frac{\sqrt{P_{sha}^{sp\ 2} + Q_{sha}^{sp\ 2}}}{V_{i+a-1}}\right)^2 = 0 \quad (A.10)$$

A.2.2 TYPICAL ELEMENTS IN JACOBIAN SUB-BLOCKS OF EQ. (3.23) USING UNIFIED METHOD

For master converter,

$$\frac{\partial f_{1a}}{\partial \theta_{sha}} = m_a c\, V_{DCa}V_{i+a-1}y_{sha}\sin\left(\theta_{sha} - \theta_{i+a-1} - \phi_{sha}\right) + b_1 y_{sha}\frac{d_3}{2d_2} + c_1 y_{sha}^2 d_3 \quad (A.11)$$

where $d_2 = \left[V_{i+a-1}^2 + m_a^2 c^2 V_{DCa}^2 - 2m_a c V_{DCa}V_{i+a-1}\cos\left(\theta_{i+a-1} - \theta_{sha}\right)\right]^{1/2}$

$$d_3 = -2m_a c V_{DCa}V_{i+a-1}\sin\left(\theta_{i+a-1} - \theta_{sha}\right)$$

$$\frac{\partial f_{1a}}{\partial V_{DCa}} = 2m_a^2 c^2 V_{DCa}y_{sha}\cos\phi_{sha} - m_a c V_{i+a-1}y_{sha}\cos\left(\theta_{sha} - \theta_{i+a-1} - \phi_{sha}\right)$$

$$+ \sum_{v=1}^{q} V_{DCv}Y_{DCav} + V_{DCa}Y_{DCaa} + b_1 y_{sha}\frac{d_4}{2\,d_2} + c_1 y_{sha}^2 d_4 \qquad (A.12)$$

where $d_4 = 2V_{DCa}m_a^2 c^2 - 2\,m_a c V_{i+a-1}\cos\left(\theta_{i+a-1} - \theta_{sha}\right)$

$$\frac{\partial f_{1a}}{\partial m_a} = 2m_a c^2 V_{DCa}^2 y_{sha}\cos\phi_{sha} - c V_{DCa}V_{i+a-1}y_{sha}\cos\left(\theta_{sha} - \theta_{i+a-1} - \phi_{sha}\right)$$

$$+ b_1\, y_{sha}\frac{d_5}{2\,d_2} + c_1 y_{sha}^2 d_5 \qquad (A.13)$$

$$d_5 = 2V_{DCa}^2 m_a c^2 - 2c V_{DCa} V_{i+a-1} \cos\left(\theta_{i+a-1} - \theta_{sha}\right)$$

For slave converter in PQ control operation,

$$\frac{\partial f_{1a}}{\partial \theta_{sha}} = m_a c V_{DCa} V_{i+a-1} y_{sha} \sin\left(\theta_{sha} - \theta_{i+a-1} - \phi_{sha}\right) \tag{A.14}$$

$$\frac{\partial f_{1a}}{\partial V_{DCa}} = 2m_a^2 c^2 V_{DCa} y_{sha} \cos\phi_{sha} - m_a c V_{i+a-1} y_{sha} \cos\left(\theta_{sha} - \theta_{i+a-1} - \phi_{sha}\right)$$

$$+ \sum_{v=1}^{q} V_{DCv} Y_{DCav} + V_{DCa}\ Y_{DCaa} \tag{A.15}$$

$$\frac{\partial f_{1a}}{\partial m_a} = 2m_a c^2 V_{DCa}^2 y_{sha} \cos\phi_{sha} - c\ V_{DCa} V_{i+a-1} y_{sha} \cos\left(\theta_{sha} - \theta_{i+a-1} - \phi_{sha}\right) \tag{A.16}$$

A.2.3 TYPICAL ELEMENTS IN JACOBIAN SUB-BLOCKS OF EQ. (3.30) USING SEQUENTIAL METHOD

From Eq. (3.27),

$$\frac{\partial f_{2a}}{\partial V_{i+a-1}} = -m_a c V_{DCa} y_{sha} \cos\left(\theta_{sha} - \theta_{i+a-1} - \phi_{sha}\right) - \left(P_{sha}^{sp\ 2} + Q_{sha}^{sp\ 2}\right)\left(\frac{-2\ R_{sha}}{V_{i+a-1}^3}\right) \tag{A.17}$$

From Eq. (3.28),

$$\frac{\partial f_{3a}}{\partial V_{i+a-1}} = -m_a c V_{DCa} y_{sha} \sin\left(\theta_{sha} - \theta_{i+a-1} - \phi_{sha}\right) - \left(P_{sha}^{sp\ 2} + Q_{sha}^{sp\ 2}\right)\left(\frac{-2\ X_{sha}}{V_{i+a-1}^3}\right) \tag{A.18}$$

A.2.4 INITIAL VALUES OF VARIABLES IN VSC-BASED HVDC SYSTEMS

TABLE A.2

Initial values of variables in VSC based HVDC systems

Variables in VSC HVDC System		Initial Values
DC bus voltage	V_{DCi}^0	3 p.u
DC voltage reference for droop line of ith VSC	$\left(V_{DCi}^*\right)^0$	3 p.u
Modulation index of ith VSC	m_i^0	1.0
Magnitude of output voltage phasor of ith VSC	V_{shi}^0	1.0 p.u
Phase angle of output voltage phasor of ith VSC	θ_{shi}^0	0 (degree)
Variable ith DC voltage source of IDCPFC	V_{DCsi}^0	0.001 p.u

A.3 EXPRESSION OF POWER BALANCE EQUATION WITH CONVERTER LOSSES OF ATH VSC CONSIDERING AS A MASTER CONVERTER BY INCORPORATING IDCPFC

From Eq. (5.13),

$$
\left(m_a c\, V_{DCa}\right)^2 y_{sha}\cos\phi_{sha} - m_a c\, V_{DCa} V_{i+a-1} y_{sha}\cos\left(\theta_{sha} - \theta_{i+a-1} - \phi_{sha}\right)
$$

$$
+ \sum_{v=1}^{q} V_{DCa} V_{DCv} Y_{DCav} + \sum_{v=1,\ v\neq1}^{z+1} V_{DCa} V_{DCs(v-1)} Y_{DCav} + a_1 + b_1 d_1 + c_1 d_1^2
$$

$$
= 0 \quad \text{if } a = 1,\ z = 2 \tag{A.19}
$$

$$
\frac{\partial f_{1a}}{\partial V_{DCa}} = 2m_a^2 c^2 V_{DCa} y_{sha}\cos\phi_{sha} - m_a c\, V_{i+a-1} y_{sha}\cos\left(\theta_{sha} - \theta_{i+a-1} - \phi_{sha}\right)
$$

$$
+ \sum_{v=1}^{q} V_{DCv} Y_{DCav} + V_{DCa}\, Y_{DCaa} + \sum_{v=1,\ v\neq1}^{z+1} V_{DCs(v-1)} Y_{DCav}
$$

$$
+ b_1 y_{sha}\frac{d_4}{2\, d_2} + c_1 y_{sha}^2 d_4 \tag{A.20}
$$

$$
\frac{\partial f_{1a}}{\partial V_{DCs(v-1)}} = \sum_{v=1,\ v\neq1}^{z+1} V_{DCa}\, Y_{DCav} \tag{A.21}
$$

$$
\frac{\partial P_{IDCPFC}}{\partial V_{DCsz}} = \left[V_{DC1} - 2\, V_{DCsz} - V_{DC(z+1)}\right] Y_{DC1(z+1)} \tag{A.22}
$$

$$
\frac{\partial P_{IDCPFC}}{\partial V_{DC(z+1)}} = -V_{DCsz} Y_{DC1(z+1)} \tag{A.23}
$$

Bibliography

1. J. Arrillaga and N.R. Watson, *Computer Modelling of Electrical Power Systems*, 2nd ed., John Wiley & Sons Ltd, Hoboken, New Jersey, United States, 2003.
2. P. Kundur, *Power System Stability and Control*, Tata McGraw-Hill Publishing Co. Ltd., New York, New York, United States, 2007.
3. K.R. Padiyar, *HVDC Power Transmission Systems*, 2nd ed., New Age International Publishers, Delhi, India, 2012.
4. V.K. Sood, *HVDC and FACTS Controllers – Applications of Static Converters in Power Systems*, Kluwer, Academic Publishers New York, Boston, Dordrecht, London, Moscow, 2004.
5. C.K. Kim, V.K. Sood, G.S. Jang, S.J. Lim and S.J. Lee, *HVDC Transmission: Power Conversion Applications in Power Systems*, John Wiley & Sons Ltd, Hoboken, New Jersey, United States, 2009.
6. X.F. Wang, Y.H. Song and M. Irving, *Modern Power Systems Analysis*, Springer, Berlin, Germany, 2008.
7. J. Arrillaga, Y.H. Liu and N.R Watson, *Flexible Power Transmission: The HVDC Options*, John Wiley & Sons Ltd, Hoboken, New Jersey, United States, 2007.
8. E. Acha, V.G. Agelidis, O. Anaya Lara and T.J.E. Miller, *Power Electronic Control in Electrical Systems*, Newnes, Butterworth, Oxford, 2002.
9. E. Acha, C.R Fuerte-Esquivel, H.A. Perez and C.A. Camacho, *FACTS Modelling and Simulation in Power Networks*, John Wiley, Hoboken, New Jersey, United States, 2004.
10. X.P. Zhang, C. Rehtanz and B. Pal, *Flexible AC Transmission Systems: Modelling and Control*, Springer, Berlin, Germany, 2006.
11. A. Yazdani and R. Iravani, *Voltage-Sourced Converters in Power Systems: Modeling, Control and Applications*, Wiley-IEEE Press, Hoboken, New Jersey, U.S, 2010, p. 171.
12. N.R Chaudhuri, B. Chaudhuri, R. Majumdar and A. Yazdani, *Multi-terminal Direct Current Grids: Modelling, Analysis and Control*, IEEE Press, Wiley, Hoboken, New Jersey, U.S, 2014.
13. N.G. Hingorani and L. Gyugyi, *Understanding FACTS: Concept and Technology of Flexible AC Transmission Systems*, IEEE Press, New York, 2000.
14. D.P. Kothari and I.J Nagrath, *Modern Power System Analysis*, 3rd ed., Tata McGrow Hill, New Delhi, 2003.
15. G. Asplund, "Application of HVDC light to power system enhancement", *IEEE Proceedings on PES Winter Meeting*, January 2000.
16. N. Flourentzou, V.G. Agelidis and G.D. Demetriades, "VSC-Based HVDC Power Transmission Systems: An Overview", *IEEE Transactions on Power Electronics*, vol. 24, no. 3, pp. 592–602, Mar. 2009.
17. C. Dierckxsens, "VSC MTDC systems with distributed slack bus", M.S. thesis, Katholieke Universiteit, Leuven, Belgium, 2010.
18. G. Daelemans, "VSC HVDC in meshed networks", Katholieke Universiteit Leuven, Belgium, Master's Thesis, p. 11, 2008.
19. J. Blau, "Europe Plans a North Sea grid", *IEEE Spectrum*, vol. 47, no. 3, pp. 12–13, Mar. 2010.

20. T.M. Haileselassie, K. Uhlen, "Power System Security in a Meshed North Sea HVDC Grid", *IEEE Proc., Invited Paper*, vol. 101, no. 4, pp. 978–990, Apr. 2013.

21. S.K. Chaudhary, R. Teodorescu and P. Rodriguez, "Wind farm grid integration using VSC based HVDC transmission: An overview", *IEEE Conference on Energy, 2030*, Nov. 2008, pp. 1–7.

22. C. Kim, V.K. Sood, G.S. Jang, S.J. Lim and S.J. Lee, *HVDC Transmission Power Conversion Applications in Power Systems*, John Wiley & Sons Ltd, Hoboken, New Jersey, United States, 2009.

23. R.T. Pinto, P. Bauer, S. Rodrigues, E. Wiggelinkhuizen, J. Pierik and B. Ferreira, "A Novel Distributed Direct Voltage Control Strategy for Grid Integration of Offshore Wind Energy Systems through MTDC Network", *IEEE Transactions on Industrial Electronics*, vol. 60, pp. 2429–2441, June 2013.

24. S. Cole, J. Beerten and R. Belmans, "Generalized Dynamic VSC MTDC Model for Power System Stability Studies", *IEEE Transactions on Power Systems*, vol. 25, no. 3, pp. 1655–1662, Aug. 2010.

25. E.P Arauzo, F.D. Bianchi, A. Ferre and O.G. Bellmunt, "Methodology for Droop Control Dynamic Analysis of Multiterminal VSC-HVDC Grids for Offshore Wind Farms", *IEEE Transactions on Power Delivery*, vol. 26, no. 4, pp. 2476–2485, Oct. 2011.

26. M.A. Penalba, A.E. Alvarez, O.G. Bellmunt and A. Sumper, "Optimum Voltage Control for Loss Minimization in HVDC Multi-Terminal Transmission Systems for Large Offshore Wind Farms", *Electrical Power Systems Research*, vol. 89, pp. 54–63, Aug. 2012.

27. O.G. Bellmunt, J. Liang, J. Ekanayake and N. Jenkins, "Voltage Current Characteristics of Multiterminal HVDC VSC for Offshore Wind Farms", *Electrical Power Systems Research*, vol 81, pp. 440–450, Feb. 2011.

28. K. Rouzbehi, A. Miranian, J.I Candela, A. Luna and P. Rodriguez, "A Generalised Voltage Droop Strategy for Control of Muti-terminal DC Grids", *IEEE Transactions on Industry Applications*, vol. 51, no. 1, pp. 607–618, Feb. 2015.

29. A.S.A. Khalik, A.M. Massoud, A.A. Elserougi and S. Ahmed, "Optimum Power Transmission Based Droop Control Design for Multi-terminal HVDC of Offshore Wind Farms", *IEEE Transactions on Power Systems*, vol. 28, no. 3, pp. 3401–3409, Aug. 2013.

30. X. Zhao and K. Li, "Droop Setting Design for Multi-terminal HVDC Grids Considering Voltage Deviation Impacts", *Electric Power Systems Research*, vol. 123, pp. 67–75, Feb. 2015.

31. L. Xu, B.W. Williams and L. Yao, "Multi-terminal DC transmission systems for connecting large offshore wind farms", *IEEE Proceedings on PES General Meeting-Conversion and Delivery of Electrical Energy, 21st Century*, pp. 1–7, July 2008.

32. R. da Silva, R. Teodorescu and Rodriguez, "Power delivery in MTDC transmission system for offshore wind power applications", *IEEE Proceedings on PES ISGT Europe*, pp. 1–8, Oct. 2010.

33. P. Mitra, L. Zhang and L. Harnefors, "Offshore Wind Integration to a Weak Grid by VSC-HVDC Links Using Power Synchronization Control: A Case Study", *IEEE Transactions on Power Delivery*, vol 29, no. 1, pp. 453–461, Feb. 2014.

34. L. Zhang, L. Harnefors and H.P. Nee, "Interconnection of Two Very Weak AC Systems by VSC HVDC Links Using Power Synchronization Control", *IEEE Transactions on Power Systems*, vol. 26, no. 1, pp. 344–355, Feb. 2011.

35. L. Zhang, L. Harnefors and H.P. Nee, "Modeling and Control of VSC HVDC Links Connected to Island Systems", *IEEE Transactions on Power Systems*, vol. 26, no. 2, pp. 783–793, May 2011.

36. A.E. Alvarez, J. Beerten, D.V. Hertem and O.G. Bellmunt, "Hierarchical Power Control of Multi-terminal HVDC Grids", *Electrical Power Systems Research*, vol. 21, pp. 207–215, Apr. 2015.

37. A. Moawwad, M.S. El Moursi, W. Xiao and J.L. Kirtley Jr., "Novel Configuration and Transient Management Control Strategy for VSC-HVDC", *IEEE Transactions on Power Systems*, vol. 29, no. 5, pp. 2478–2488, Sep. 2014.

38. J. Beerten and R. Belmans, "Analysis of Power Sharing and Voltage Deviations in Droop Controlled DC Grids", *IEEE Transactions on Power Systems*, vol. 28, no. 4, pp. 4588–4597, Nov. 2013.

39. T.M. Haileselassie and K. Uhlen, "Impact of DC Line Voltage Drops on Power Flow of MTDC Using Droop Control", *IEEE Transactions on Power Systems*, vol. 27, no. 3, pp. 1441–1449, 2012.

40. D. Jovcic and B.T. Ooi, "Developing DC Transmission Networks Using DC Transformers", *IEEE Transactions on Power Delivery*, vol. 25, no. 4, pp. 2535–2543, Oct. 2010.

41. Q. Mu, J. Liang, Y. Li and X. Zhou, "Power Flow Control Devices in DC Grids", *IEEE Power and Energy Society General Meeting*, 2012, pp. 1–7.

42. C. Barker and R. Whitehouse, "A current flow controller for use in HVDC grids", *Proc. IET Int. Conf. AC DC Power Transm.*, pp. 1–5, 2012.

43. E. Veilleux and B. Ooi, "Multiterminal HVDC with Thyristor Power-Flow Controller", *IEEE Transactions on Power Delivery*, vol. 27, no. 3, pp. 1205–1212, July 2012.

44. W. Chen, X. Zhu, L. Yao, X. Ruan, Z. Wang and Y. Cao, "An Interline DC Power-Flow Controller (IDCPFC) for Multiterminal HVDC System", *IEEE Transactions on Power Delivery*, vol. 30, no. 4, pp. 2027–2036, Aug. 2015.

45. W. Chen, X. Zhu, L. Yao, G. Ning, Y. Li, Z. Wang, W. Gu and X. Qu, "A Novel Interline DC Power Flow Controller (IDCPFC) for Meshed HVDC Grids", *IEEE Transactions on Power Delivery*, vol. 31, no. 4, pp. 2027–2036, 1719–1727, Aug. 2016.

46. E.B. Melendrez, L.E.U. Caballero and E.L.M. Goytia, "Sequential algorithm based model for the study of AC/DC power flow in VSC-MTDC systems", *IEEE International Autumn Meeting on Power Electronics and Computing, ROPEC*, pp. 1–5, 2015.

47. N. Narayanan and P. Mitra, "A Comparative study of a sequential and simultaneous AC-DC Power flow algorithms for a multi-terminal VSC-HVDC system", *IEEE ISGT*, pp. 1–6, 2013.

48. G. Zeng, T. Hennig and K. Rohrig, "Multi-terminal HVDC Modelling in Power Flow Analysis Considering Converter Station Topologies and Losses", *Energy Procedia*, Elsevier, vol. 80, pp. 123–130, 2015.

49. W. Feng, L.A. Tuan, L.B. Tjernberg, A. Mannikoff and A. Bergman, "A New Approach for Benefit Evaluation of Multiterminal VSC–HVDC Using a Proposed Mixed AC/DC Optimal Power Flow", *IEEE Transactions on Power Delivery*, vol. 29, no. 1, pp. 432–443, Feb. 2014.

50. A. Martinez, C. Esquivel, H. Perez and E. Acha, "Modeling of VSC-Based HVDC Systems for a Newton Raphson OPF Algorithm", *IEEE Transactions on Power Systems*, vol. 22, no. 4, pp. 1794–1799, Nov. 2007.

51. J. Cao, W. Du and H.F. Wang, "An Improved Corrective Security Constrained OPF for Meshed AC/DC Grids with Multi-Terminal VSC-HVDC", *IEEE Transactions on Power Systems*, Early Access Articles, pp. 1–11, 2016.

52. K. Rouzbehi, A. Miranian, A. Luna and P. Rodriguez, "DC Voltage Control and Power Sharing in Multi-terminal DC Grids Based on Optimal DC Power Flow and Voltage Droop Strategy", *IEEE Journal of Emerging and Selected Topics in Power Electronics*, vol. 02, no. 4, pp. 1171–1180, Dec. 2014.

53. L. Gyugyi, K.K. Sen and C.D. Schauder, "The Interline Power Flow Controller Concept: A New Approach to Power flow Management in Transmission Systems", *IEEE Transactions on Power Delivery*, vol. 14, no. 3, pp. 1115–1123, July 1999.

54. Y. Zhang, Y. Zhang and C. Chen, "A Novel Power Injection Model of IPFC for Power flow Analysis Inclusive of Practical Constraints", *IEEE Transactions on Power Systems*, vol. 21, no. 4, pp. 1550–1556, Nov. 2006.

55. S. Bhowmick, B. Das and N. Kumar, "An Advanced IPFC Model to Reuse Newton Power Flow Codes", *IEEE Transactions on Power Systems*, vol. 24, no. 2, pp. 525–532, May 2009.

56. S. Bhowmick, *Flexible AC Transmission Systems (FACTS): Newton Power-Flow Modeling of Voltage-Sourced Converter-Based Controllers*, CRC Press, 2016, Chapter 2.

57. A.R. Bergen and V. Vittal, *Power System Analysis*, Pearson Education, 2004.

58. J.J. Grainger and W.D. Stevenson, *Power System Analysis*, Tata McGrow-Hill, India, 1994.

59. J.D. Glover, M.S. Sarma and T.J. Overbye, *Power System: Analysis and Design*, 4th ed., Thomson Learning, India, 2007.

60. J. Arrillaga, B.J. Harker and K.S Turner, "Clarifying an ambiguity in recent AC-DC load flow formulations", *IEEE Proc. C*. 127, pp. 324–325, 1980.

61. J. Arrillaga and P. Bodger, "Integration of HVDC Links with Fast Decoupled Load Flow Solutions", *IEEE Proceedings*, vol. 124, no. 5, pp. 463–468, May 1977.

62. Y.K. Fan, D. Niebur, C.O. Nwankpa, H. Kwanty and R. Fischl, "Multiple Power Flow Solutions of Small Integrated AC-DC Power Systems", *IEEE Proc. Int. Symp. Circuits and Systems*, Geneva, Switzerland, pp. 224–227, May 2000.

63. H.A. Perez, E. Acha and C.R.F. Esquivel, "High Voltage Direct Current Modelling in Optimal Power Flows", *Electrical Power and Energy Systems*, pp. 157–168, Mar. 2008.

64. S. Messalti, S. Belkhiat, S. Saadate and D. Flieller, "A New Approach for Load Flow Analysis Integrated AC-DC Power Systems Using Sequential Modified Gauss Seidel Methods", *European Transactions on Electrical Power*, vol. 22, no 4, pp. 421–432, May 2012.

65. J. Yu, W. Yan, W.Y. Li, C.Y. Chung and K.P. Wong, "An Unfixed Piecewise Optimal Reactive Power Flow Model and its Algorithms for AC-DC Systems", *IEEE Transactions on Power Systems*, vol. 23, no. 1, pp. 170–176, Feb. 2008.

66. U. Kilic and K. Ayan, "Optimal Power Flow Solution of Two Terminal HVDC Systems Using Genetic Algorithm", *Electrical Engineering*, pp. 65–77, 2014.

67. U. Kilic and K. Ayan, "Optimizing Power Flow of AC-DC Power Systems Using Artificial Bee Colony Algorithms", *Electrical Power and Energy Systems*, vol. 53, pp. 592–602, Dec. 2013.

68. U. Kilic, K. Ayan and U. Arifoglu, "Optimizing Reactive Power Flow of HVDC Systems Using Genetic Algorithms", *Electrical Power and Energy Systems*, vol. 55, pp. 1–12, Feb. 2014.

69. M.M. Djesus, D.S. Martin, S. Arnaltes and E.D. Castronuovo, "Optimal Reactive Power Allocation in an Offshore Wind Farm with LCC HVDC Link Connection", *Renewable Energy*, vol. 40, pp. 157–166, Apr. 2012.

70. H.N.P. Srivastava, A.K. Sinha and L. Roy, "Fast Second order AC-DC Load Flow", *Electrical Machines and Power Systems*, pp. 233–247, 1995.

71. J. Arrillaga and P.S. Bodger, "AC–DC Load Flows with Realistic Representation of the Converter Plant", *Proceedings of the Institution of Electrical Engineers*, vol. 125, pp. 41–46, Jan. 1978.

72. D.A. Braunagel, L.A. Kraft, J.L. Whysong, "Inclusion of DC Converter and Transmission Equations Directly in A Newton Power Flow", *IEEE Transactions on Power Apparatus and Systems*, vol. PAS-95, pp. 76–87, Feb. 1976.

73. M.M. El-Marsafawy and R.M. Mathur, "A New, Fast Technique for Load Flow Solution of Integrated Multi Terminal DC/AC Systems", *IEEE Transactions on Power Apparatus and Systems*, vol. PAS-99, pp. 246–255, Feb. 1980.

74. T. Smed, G. Andersson, G.B. Sheble and L.L. Grigsby, "A New Approach to AC/DC Power Flow", *IEEE Transactions on Power Systems*, vol. 6, no. 3, pp. 1238–1244, Aug. 1991.

75. C.M. Ong, A.H. Nejad, "A General Purpose Multi-terminal DC Load Flow", *IEEE Transactions on Power Apparatus and Systems*, vol. PAS 100, no. 7, pp. 3166–3174, July 1981.

76. J. Reeve, G. Fahmy and B. Stott, "Versatile Load Flow Method for Multi Terminal HVDC Systems", *IEEE Transactions on Power Apparatus and Systems*, vol. PAS-96, pp. 925–933, June 1977.

77. J. Mahseredjian, S. Lefebvre and D. Mukhedkar, "A Multiterminal HVDC Load Flow with Flexible Control Specifications", *IEEE Transactions on Power Systems*, vol. PWRD-1, no. 2, pp. 272–282, Apr. 1986.

78. Q. Ding, T.S. Chung and B. Zhang, "An Improved Sequential Method for AC/MTDC Power System State Estimation", *IEEE Transactions on Power Systems*, vol. 16, no. 3, pp. 506–512, Aug. 2001.

79. C. Liu, B. Zhang, Y. Hou, F.F. Wu and Y. Liu, "An Improved Approach for AC-DC Power flow Calculation with Multi Infeed DC Systems", *IEEE Transactions on Power Systems*, vol. 26, no. 2, pp. 862–869, May 2011.

80. K.R. Padiyar and V.K. Raman, "A general method for power flow analysis in MTDC systems", *ACE 90 Proceedings of XVI Annual Convention and Exhibition of the IEEE in India*, pp. 146–150, 1990.

81. C. Liu, A. Bose and Y. Hou, "Discussion of the Solvability of HVDC Power Flow with a Sequential Method", *Electric Power Systems Research*, vol. 92, pp. 155–161, Nov. 2012.

82. Q.F. Ding and B.M. Zhang, "A New Approach to AC/MTDC Power Flow", *APSCOM IEEE*, vol. 2, 1997, pp. 689–694.

83. Padiyar, K.R. and Kalyanaraman, "Power Flow Analysis in MTDC-AC Systems – A New Approach", *Electric Machines and Power Systems*, vol. 23, pp. 37–54, 1995.

84. H. Fudeh and C.M. Ong, "A Simple and Efficient AC-DC Load Flow Method for Multiterminal DC Systems", *IEEE Transactions on Power Apparatus and Systems*, vol. PAS-100, no. 11, pp. 4389–4396, Nov. 1981.

85. M.E. El-Hawary and S.T. Ibrahim, "A New Approach to AC–DC Load Flow Analysis", *Electric Power Systems Research*, vol. 33, pp. 193–200, 1995.

86. C. Angeles-Camacho, O.L. Tortelli, E. Acha and C.R. Fuerte-Esquivel, "Inclusion of a High Voltage DC-Voltage Source Converter Model in a Newton Raphson Power Flow Algorithm", *IET Proceedings – Generation Transmission and Distribution*, vol 150. no. 6, pp. 691–696, Nov. 2003.

87. X.P. Zhang, "Multiterminal Voltage-Sourced Converter-Based HVDC Models for Power Flow Analysis", *IEEE Transactions on Power Systems*, vol. 19, no. 4, pp. 1877–1884, Nov. 2004.

88. M. Baradar and M. Ghandhari, "A Multi-Option Unified Power Flow Approach for Hybrid AC/DC Grids Incorporating Multi-Terminal VSC-HVDC", *IEEE Transactions on Power Systems*, vol. 28, no. 3, pp. 2376–2383, Aug. 2013.
89. R.Z. Chai, B.H. Zhang, Z.Q. Bo and J.M. Dou, "A Generalized Unified Power Flow Algorithm for AC/DC Networks Containing VSC Based Multi-terminal DC Grid", *Powercon, CP2771*, pp. 2361–2366, 2014.
90. L. Gengyin, Z. Ming, H. Jie, L. Guangkai and L. Haifeng, "Power Flow Calculation of Power Systems Incorporating VSC-HVDC", *Proceedings on International Conference on Power System Technology (PowerCon)*, vol. 2, pp. 1562–1566, Nov. 2004.
91. J. Beerten, D.V. Hertem and R. Belmans, "VSC MTDC Systems with a Distributed DC Voltage Control – A Power Flow Approach", *IEEE Trondheim Power Tech*, 2011, pp. 1–6.
92. W. Wang, M. Barnes, "Power Flow Algorithms for Multi-Terminal VSC-HVDC With Droop Control", *IEEE Transactions on Power Systems*, vol. 29, no. 4, pp. 1721–1730, July 2014.
93. J. Cao, W. Du and H.F. Wang, "Minimization of Transmission Loss in Meshed AC/DC Grids with VSC-MTDC Networks", *IEEE Transactions on Power Systems*, vol. 28, no. 3, pp. 3047–3055, Aug. 2013.
94. L. Xiao, Z. Xu, T. An and Z. Bian, "Improved Analytical Model for the Study of Steady State Performance of Droop-controlled VSC-MTDC Systems", *IEEE Transactions on Power Systems, Early Access*, 2016.
95. E. Acha, B. Kazemtabrizi and L.M. Castro, "A New VSC-HVDC Model for Power Flows Using the Newton-Raphson Method", *IEEE Transactions on Power Systems*, vol. 28, no. 3, pp. 2602–2612, Aug. 2013.
96. A.P. Martinez, C. Esquivel and C. Camacho, "Voltage Source Converter Based High Voltage DC System Modelling for Optimal Power Flow Studies," *Electric Power Components and Systems*, vol. 40, no 3, pp. 312–320, Jan. 2012.
97. R. Wiget and G. Anderson, "DC Optimal Power Flow Including HVDC Grids", *IEEE Electrical Power and Energy Conference*, pp. 1–6, 2013.
98. M. Baradar, M. Ghandhari and D.V. Hertem, "The modeling of multi-terminal VSC-HVDC in power flow calculation using unified methodology", *Proc. IEEE PES Int. Conf. Exhib. Innovative Smart Grid Technology. (ISGT Europe)*, pp. 1–6, Dec. 2011.
99. J. Lei, T. An, Z. Du and Z. Yuan, "A General Unified AC/DC Power Flow Algorithm with MTDC", *IEEE Transactions on Power Systems, Early Access*, 2016.
100. E. Acha and L.M. Castro, "A Generalised Frame of Reference for the Incorporation of, Multi-terminal VSC HVDC Systems in Power Flow Solutions", *Electrical Power Systems Research*, vol. 136, pp. 415–424, July 2016,
101. T.M. Haileselassie and K. Uhlen, "Power flow analysis of multi-terminal HVDC networks", in *Proc. Power Tech*, Trondheim, Norway, pp.1–6, 2011.
102. J. Beerten, S. Cole and R. Belmans, "Generalized Steady-State VSC MTDC Model for Sequential AC/DC Power Flow Algorithms", *IEEE Transactions on Power Systems*, vol. 27, no. 2, pp. 821–829, May 2012.
103. J. Beerten, S. Cole and R. Belmans, "Implementation aspects of a sequential AC/DC power flow computation algorithm for multi-terminal VSC HVDC systems", *IET International Conference*, pp. 1–6, 2010.
104. J. Beerten, S. Cole and R. Belmans, "A sequential AC/DC power flow algorithm for networks containing multi-terminal VSC-HVDC systems", *Proc. IEEE PES General Meeting*, pp. 1–7, July 2010.

105. G. Daelemans, K. Srivastava, M. Reza, S. Cole and Belmans, "Minimization of Steady State Losses in Meshed Networks using VSC HVDC", *IEEE Power and Energy Society General Meeting*, pp. 1–5, 2009.

106. M.J. Carrizosa, F.D. Navas and G. Damm, "Optimal Power Flow in Multi-terminal HVDC Grids with Offshore Wind Farms and Storage Devices", *Electrical Power and Energy Systems*, vol. 65, pp. 291–298, 2015.

107. S. Khan and S. Bhowmick, "Impact of Selection of DC Base Values and DC Link Control Strategies on Sequential AC-DC Power Flow Convergence", *Frontier in Energy*, Springer, vol. 9, no. 4, pp. 399–412, 2015.

108. S. Khan and S. Bhowmick, "Impact of DC Link Control Strategies on the Power Flow Convergence of Integrated AC DC Systems", *AIN Shams Engineering Journal*, Elsevier, vol. 7, no.1, pp. 249–264, 2016.

109. S. Khan and S. Bhowmick, "Effect of DC Link Control Strategies on Multi-terminal AC-DC Power Flow", *Advances in Electrical Engineering, Hindawi*, vol. 2015, pp. 1–15, 2015.

110. S. Khan and S. Bhowmick, "A Novel Power-Flow Model of Multi-terminal VSC-HVDC Systems", *Electrical Power System Research*, Elsevier, vol. 133, pp. 219–227, 2016.

111. S. Khan and S. Bhowmick, "Generalised Power Flow Models for VSC Based Multi-terminal HVDC Systems", *International Journal of Electrical Power and Energy Systems*, Elsevier, vol. 82, pp. 67–75, 2016.

112. S. Khan and S. Bhowmick, "A Comprehensive Power-Flow Model of Multi-Terminal PWM based VSC-HVDC Systems with DC Voltage Droop Control", *International Journal of Electrical Power and Energy Systems*, Elsevier, vol. 102, pp. 71–83, 2018.

113. S. Khan and S. Bhowmick, "A Generalized Power-Flow Model of VSC Based Hybrid AC-DC Systems Integrated with Offshore Wind Farms", *IEEE Transactions on Sustainable Energy, IEEE*, vol. 10, no. 4, pp. 1775–1783, 2019.

114. S. Khan and S. Bhowmick, "A Unified AC-MTDC Power Flow Algorithm with IDCPFC", *Arabian Journal for Science and Engineering*, Springer, vol. 44, no. 8, pp. 6795–6804, 2019.

115. http://www.ee.washington.edu/research/pstca/

116. http://amfarid.scripts.mit.edu/Datasets/SPG-Data/index.php

Index

For Product Safety Concerns and Information please contact our EU
representative GPSR@taylorandfrancis.com
Taylor & Francis Verlag GmbH, Kaufingerstraße 24, 80331 München, Germany